Communications
for Tomorrow

Published with the Aspen Institute for Humanistic Studies

Communications for Tomorrow

Policy Perspectives for the 1980s

Edited by   Glen O. Robinson

Foreword by   Joseph E. Slater

PRAEGER PUBLISHERS
Praeger Special Studies

New York   •   London   •   Sydney   •   Toronto

Library of Congress Cataloging in Publication Data
Main Entry under title:
Communications for tomorrow.

Bibliography: p.
Includes index.
1. Telecommunication—United States—Addresses, essays, lectures.
2. Telecommunications—Law and legislation—United States—Addresses, essays, lectures. 3. Telecommunication—Social aspects—United States—Addresses, essays, lectures. I. Robinson, Glen O. II Aspen Institute for Humanistic Studies.

HE7781.C64   1978   384'.0973   78-13474
ISBN 0-03-046546-X
ISBN 0-03-046541-9 pbk.

PRAEGER PUBLISHERS, PRAEGER SPECIAL STUDIES
383 Madison Avenue, New York, N.Y., 10017, U.S.A.

Published in the United States of America in 1978
by Praeger Publishers
A Division of Holt, Rinehart and Winston, CBS Inc.

9 038 98765432

©1978 by Aspen Institute Program
on Communications and Society

All rights reserved

Printed in the United States of America

Acknowledgement

This volume is the first report of the Project on Communications Policy of the Program on Communications and Society of the Aspen Institute for Humanistic Studies.

Communications for Tomorrow is the result of the very generous assistance of the Ford Foundation and the John and Mary R. Markle Foundation, which helped define the project and its key issues, identify many of the participants and contributors who were involved, and provided the support that has made the entire project possible. The volume was also aided significantly by the Rockefeller Brothers Fund, the Charles F. Kettering Foundation and the Edna McConnell Clark Foundation.

A special note of gratitude is owed to McGeorge Bundy, Fred Friendly and David Davis of the Ford Foundation and to Lloyd Morrisett and Jean Firstenberg of the Markle Foundation for their personal interest and contributions to this project. William Dietel and Robert Bates of the Rockefeller Brothers Fund gave substantial time and support, as did Robert Chollar and Kent Collins of the Kettering Foundation and Merrell Clark of the Clark Foundation.

Foreword

In a complex, fast-acting world, communication has become the glue that holds society together. Contemporary communications systems and styles also create problems and tensions. The importance of global communications among governments, businesses, organizations and individuals is unique to our age. Never before in history have mankind's well-being and survival hung so precariously on accurate and swift communications.

This phenomenon is both the cause and the result of massive changes in our communications technology. Television, satellites, laser-transported messages are only better-known examples of technology that did not exist a generation ago. Today, the world is dependent on them.

In communications, emphasis on technology always seems to be put ahead of concern about quality of substance and content. Technology has outpaced the formation of attitudes, understanding and public policy. We need policies that will promise to use the new communications technology most effectively and fairly for the benefit of the world's society.

This book is concerned with the creation of such policies. The need is clear; the solutions, still to be determined. The communications needs of today's world are varied. People need news of the world because now what happens in all the world affects their lives. Corporations must have access to rapid and far-reaching information. Governments need to communicate with their constituencies and with each other.

The way societies tend to think about communication policy reflects their very natures. In the United States, the emphasis is on speed and efficiency, and on open and fair access to information and the basic principles of pluralism and diversity. There is a concern for privacy and confidentiality and most people and institutions try to restrain malicious behavior and misuse of information. There is a desire for freedom of speech and freedom of choice.

How do we translate these needs and wishes into public policy? How do we agree on objectives, limitations and methods?

The Aspen Institute Project on Communications Policy, from which this book emerged, began in the spring of 1976 under the direction of Glen O. Robinson, Special Adviser to the Aspen Institute and Professor of Law at the University of Virginia, and of Roland Homet, then director of the Institute's Program on Communications and Society. Institute Trustees and Fellows have contributed to the process out of which the book has taken shape. In recent months, Marc U. Porat, Executive Director of the Program, has played a key role.

The project was conceived as a major effort to examine broadly (a) the social, economic and legal implications of the new communications systems; (b) the policy options they offer; and (c) the capability of existing policymaking institutions to address them.

The project is the latest evolution of the Aspen Institute's long-term commitment to discuss and seek consensus on the most urgent issues relating to Communications and Society. Under this charge, the Institute's Communications and Society Program has examined the possibilities of upgrading media content and standards; the relationships between government and media; the dangers to creativity and freedom in communications; the social impact of the media, especially television; the relationship between communications and learning, and other issues.

In its recent phase, the Program has concentrated a major share if its attention on the clearly emerging problem of communications policy. It has been concerned especially with the imaginative regulation of new systems of communication; the development of new and responsible communication policies; and the promotion of public-service applications of new technologies. It has aimed to recommend public policy choices and recommendations for institutional change.

This work has been undertaken not only at the Institute's center in Aspen, Colorado, but in Washington D.C., Berlin, Japan, London, and Toronto. On the basis of many international consultations and workshops, a special task force was formed at the end of 1976 to prepare a series of essays exploring major communications policy issues. Out of that effort, this book was born.

The volume helps to move the Institute into the third phase of its Program on Communications and Society, in which re-examination of basic philosophical issues in communications as well as definitions of emerging policy choices will be linked to our overall objectives, including that of enhancing the status of the individual in society. I commend it to anyone who, aware of the great technological revolution which delineates our age, wishes to understand how members of tomorrow's society will communicate with each other.

Joseph E. Slater
President
Aspen Institute for Humanistic Studies

Preface

The Aspen Project on Communications Policy was conceived in the Spring of 1976 as an effort to explore broadly, and in a systematic way, the social, economic and legal implications of new communications systems, the policy options which they present and the capability of existing policymaking institutions to address them. The detailed design of the Project was an evolutionary product of numerous consultative conferences with representatives of the communications industries, government leaders, spokesmen for public interest groups, and others. On the basis of these consultations and extensive deliberations among the Aspen Project group, a special Task Force was established at the end of 1976, comprised of the following members with varied professional backgrounds in law, economics, physics, engineering, mathematics, and political science. (Positions first listed are those held during the period of the Task Force's existence. The positions shown in parenthesis were assumed after the work of the Task Force was completed).

Walter Baer, Physicist, The Rand Corporation. Author of monographs and studies on cable communications and telecommunications subjects.

Stanley Besen, Professor of Economics, Rice University (presently Co-Director, Network Study, Federal Communications Commission). Author of numerous articles on various aspects of broadcasting policy.

Raymond Bowers, Physicist, Director of Program on Science, Technology, and Society, Cornell University. Author of books on the video-telephone and mobile radio communications and other monographs on communications and science.

Anne Branscomb, Attorney, Vice President, Kalba-Bowen Associates. Author of numerous articles on broadcasting and communications policy.

Douglass Cater, President, Observer International, London, and Trustee Aspen Institute. Author of books on politics and communications. Former Director, Aspen Institute Program on Communications and Society.

Forrest Chisman, Political Scientist. Associate Director, Aspen Institute Program on Communications and Society (presently Director of Plans and Policy Coordination, National Telecommunications Information Administration, U.S. Department of Commerce).

Kan Chen, Professor of Electrical Engineering, University of Michigan.

Henry Geller, Attorney. Fellow, Aspen Institute Program on Communications and Society (presently Assistant Secretary of Commerce for Communication and Information, National Telecommunications Information Administration, U.S. Department of Commerce). Former General Counsel, FCC. Author of numerous articles on broadcast regulation.

Ernest Gellhorn, Dean, Arizona State University, College of Law (presently Dean, University of Washington School of Law). Author of books on administrative and antitrust law, and numerous articles in those fields.

Henry Goldberg, Attorney, Verner, Liipfert, Bernhardt, and McPherson. Former General Counsel of the Office of Telecommunications Policy, author of articles on broadcast regulation, specialist in communications law.

Roland Homet, Attorney, Fellow, Aspen Institute Program on Communications and Society. Former Director, Aspen Institute Program on Communications and Society.

Leland Johnson, Economist, The Rand Corporation (presently Associate Administrator for Policy Analysis and Development, National Telecommunications Information Administration, U.S. Department of Commerce). Author of articles and monographs on regulatory economics and communications policy.

W. Kenneth Jones, Professor of Law, Columbia University School of Law. Former Commissioner New York Public Service Commission. Author of books and articles on antitrust and regulatory policy.

William Lucas, Social Scientist, The Rand Corporation (presently Associate Administrator for Telecommunication Applications, National Telecommunications Information Administration, U.S. Department of Commerce). Author of monographs on broadcasting, cable, and social service applications of communications media.

Roger Noll, Professor of Economics, California Institute of Technology. Author of books on economics of television and economic regulation, and articles on regulatory economics.

Anthony Oettinger, Professor of Applied Mathematics, Director, Harvard University Program on Information Technologies, Member, Massachusetts Cable Commission. Author of books and monographs on information technology and policy.

Bruce Owen, Professor of Economics, Stanford University (presently Associate Professor in the Graduate School of Business Administration, Adjunct Associate Professor in the School of Law, and Director of the Center for the Study of Regulation of Private Enterprise). Fellow, Aspen Institute Program on Communications and Society. Former Chief Economist, Office of Telecommunications Policy. Author of books on television economics, free speech, and numerous monographs and articles on communications regulation.

Marc Porat, Economist. Executive Director of the Aspen Institute Program on Communications and Society. Author of monographs and articles on economics of information and communications.

Glen Robinson, Special Adviser, Aspen Institute and Professor of Law, University of Virginia School of Law (presently, also, U.S. Ambassador to the 1979 World Administrative Radio Conference). Chairman of Aspen Institute Project on Communications Policy. Former Commissioner, FCC. Author of books on administrative law, public land management, articles on communications and on administrative law.

Also contributing to this volume are the following:

Stuart Brotman (law student University of California Berkeley);

xi

Kim Degnan (graduate student University of California Berkeley); and Daniel Polsby (Professor of Law, Northwestern).

The work of the Task Force was to prepare and critique a series of essays exploring major communications policy issues which were seen as emerging over the next two decades and the institutional capacities for dealing with them. This book is a product of that work.

The book is organized into five parts. Part one explores some of the general economic, social, and technological trends in communications services and facilities in our "information society." Part two deals with some of the key issues of industry structure and regulatory boundaries that are implicated by emerging trends in communications. Part three evaluates social service applications of the "new electronic media" and the impact of those applications on present communications media. Part four examines major government roles and institutional capabilities for addressing communications policy issues. Finally, a review of, and commentary on, the issues exposed in the first four parts is provided in part five.

Since it was not the primary concern of the Aspen Project to reach final judgments as to how the policy issues should be resolved, or what institutional changes should be made, we offer no single set of conclusions or recommendations. We have not suppressed the individual opinions or recommendations of the authors and, explicitly or implicitly, such personal views appear in all of the essays in this book. But we have not sought any institutional consensus on any of the matters discussed.

Although this book represents the completion of an initial phase of the Aspen Institute Policy Project, we are continuing to explore both domestic and international policy issues; separate reports of those studies will be forthcoming. In present and future endeavors, our aims and our expectations are, somewhat contrarily, both modest and ambitious—ambitious insofar as we have taken a large and diversified realm of public policy as our purview, modest because we seek no results other than to promote and provoke critical discussion of these important matters.

This book reflects the contributions of many people—not only members of our Task Force and our sponsors, who have been mentioned, but others too numerous to identify. We are grateful to all of them. Special thanks are due C. J. Cross and Fran Sills, without whom this book would have remained just an idea. Finally, I am personally indebted to Dean Emerson Spies and the University of Virginia for allowing me the time to devote to this enterprise.

Glen O. Robinson

Contents

ACKNOWLEDGEMENT v
FOREWORD vii
PREFACE ix

Part One COMMUNICATION POLICY IN AN INFORMATION AGE

CHAPTER ONE Communication Policy in an Information Society
by Marc U. Porat 3

CHAPTER TWO Telecommunications Technology in the 1980s
by Walter Baer 61

Part Two COMMUNICATIONS INDUSTRY STRUCTURES AND REGULATORY BOUNDARIES

CHAPTER THREE Boundaries to Monopoly and Regulation in Modern Telecommunications
by Leland Johnson 127

CHAPTER FOUR International Telecommunication Regulation
by Henry Goldberg 157

Part Three APPLICATIONS OF THE NEW ELECTRONIC MEDIA

CHAPTER FIVE Pluralistic Programming and Regulatory Policy
by Benno Schmidt 191

CHAPTER SIX The Role of Print in an Electronic Society
by Bruce Owen 229

CHAPTER SEVEN Telecommunications Technologies and Services
by William Lucas 245

CHAPTER EIGHT Communications for a Mobile Society
by Ray Bowers 275

CHAPTER NINE Electronic Alternatives to Postal Service
by Henry Geller and Stuart Brotman 307

Part Four GOVERNMENT INSTITUTIONS AND POLICYMAKING PROCESSES IN COMMUNICATIONS

CHAPTER TEN	The Federal Communications Commission by Glen O. Robinson	353
CHAPTER ELEVEN	The Executive Branch by Forrest Chisman	401
CHAPTER TWELVE	The Judicial Role by Glen O. Robinson	415
CHAPTER THIRTEEN	The Role of Congress by Ernest Gellhorn	445

Part Five COMMUNICATIONS ISSUES, INSTITUTIONS, AND PROCESSES: AN OVERVIEW

CHAPTER FOURTEEN	Communications for the Future: An Overview of the Policy Agenda by Glen O. Robinson	467
CHAPTER FIFTEEN	Institutions for Communications Policymaking: A Review by Daniel Polsby and Kim Degnan	501
INDEX		517

Part 1.
Communication Policy in an Information Age

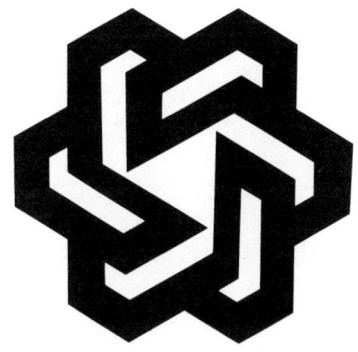

1.
Communication Policy in an Information Society

Marc U. Porat

Communication in an Information Society

Communication policy encompasses broad economic, legal, technical, social, and institutional issues. The purpose of this essay is to position the field in the context of an emerging information-based economy.

At the outset, I present the economic evidence showing that the U.S. is changing from an industrial society to an "information society," and suggest that problems relating to communication and information will come to occupy a central place on the policy agenda. I then discuss two historically useful views for understanding the process of change—technological determinism and ideological doctrine. The first perspective focuses on scientific and technological revolutions. In the modern context, the invention of telecommunications and electronic computing qualify as epochal technological leaps, precipitating a host of changes in the economic, institutional, and political conditions. The second perspective vests ideology as the main engine of social change, where a system of ideas guides society's economic and political destiny. Using both perspectives, I offer a selection rule for structuring the communication and information policy agenda: suggesting how to assign priorities to the many issues that arise. Finally, using the selection rule, I identify those communication and information issues that will capture the policy agenda in the 1980s. Some of the issues are treated in depth in this book. Others remain for future scholarship.

The Emergence of an Information Economy: Economic Evidence

In 1967, 25.1 percent of the U.S. Gross National Product (GNP) originated with the production, processing, and distribution of information goods and services sold on markets. In addition, the purely informational requirements of planning, coordinating, and managing the rest of the economy generated 21.1 percent of GNP. These informational activities engaged more than 46 percent of the work force, which earned over 53 percent of all labor income. On the strength of these findings, we call ours an "information economy." Let us examine how these findings were developed.

The Information Occupations

The structure of the work force is one basic indicator of a nation's stage of development. Agricultural activities, which engaged nearly 50 percent of U.S. labor in the 1860s, now occupy less than 4 percent. Industrial activities, which engaged nearly 40 percent of the work force in the early 1940s, now occupy around 20 percent. And information occupations, which engaged only 10 percent of the work force at the turn of the century, now account for 46 percent of all jobs.

Bell summarizes how a transition to a post-industrial economy affects the work force:[1]

> In preindustrial societies—still the condition of most of the world today—the labor force is engaged overwhelmingly in the extractive industries: mining, fishing, forestry, agriculture. Life is primarily a game against nature....Industrial societies—principally those around the North Atlantic littoral plus the Soviet Union and Japan—are goods-producing societies. Life is a game against fabricated nature. The work has become technical and rationalized. The machine predominates, and the rhythms of life are mechanically paced....A post-industrial society is based on services.

Policy in an Information Society

What counts is not raw muscle power, or energy, but information (emphasis added).

Stating precisely who is an information worker and who is not is a risky proposition. Obviously, every human endeavor involves some measure of information processing and cognition; intellectual content is present in every task no matter how mundane. It is, after all, the critical difference between humans and animals that the former can process symbolic information quite readily while the latter cannot.

We are trying to get at a different question: Which occupations are *primarily* engaged in the production, processing, or distribution of information as the output, and which perform tasks where information handling is only ancillary? I have developed a conceptual scheme for classifying information workers, presented as an overview in *Table 1-1*.[2]

Table 1-1
Typology of Information Workers and 1967 Compensation[a]

	Employee Compensation ($ Millions)
Markets for information	
Knowledge producers	46,964
Scientific and technical workers	18,777
Private information services	28,187
Knowledge distributors	28,265
Educators	23,680
Public information disseminators	1,264
Communication workers	3,321
Information in markets	
Market search and coordination specialists	93,370
Information gatherers	6,132
Search and coordination specialists	28,252
Planning and control workers	58,986
Information processors	61,340
Nonelectronic based	34,317
Electronic based	27,023
Information infrastructure	
Information machine workers	13,167
Nonelectronic machine operators	4,219
Electronic machine operators	3,660
Telecommunication workers	5,288
Total information	243,106
Total employee compensation	**454,259**
Information as percentage of total	*53.52%*

Note: (a) Based on 440 occupational types in 201 industries. Employee compensation includes wages and salaries and supplements.

Source: Computed using BLS Occupation by Industry matrix and Census of Population average wages. See Porat (2), vol. 1.

Communications for Tomorrow

The first category includes those workers whose output or primary activity is producing and selling knowledge. Included here are scientists, inventors, teachers, librarians, journalists, and authors. The second major class of workers covers those who gather and disseminate information. These workers move information within firms and within markets; they search, coordinate, plan, and process market information. Included here are managers, secretaries, clerks, lawyers, brokers, and typists. The last class includes workers who operate the information machines and technologies that support the previous two activities. Included here are computer operators, telephone installers, and television repairers. In 1967, the nation's information workers accounted for about 46 percent of the work force and about 53 percent of total employee compensation.

The trend toward a predominance of information workers has been persistent since the 1940s (see *Figure 1-1*). The information work force in 1860 comprised less than 10 percent of the total. By 1975, the information workers surpassed the noninformation

**Figure 1-1
Time Series of U.S. Labor Force (1860-1980),
Two Sector Aggregate by Percent**

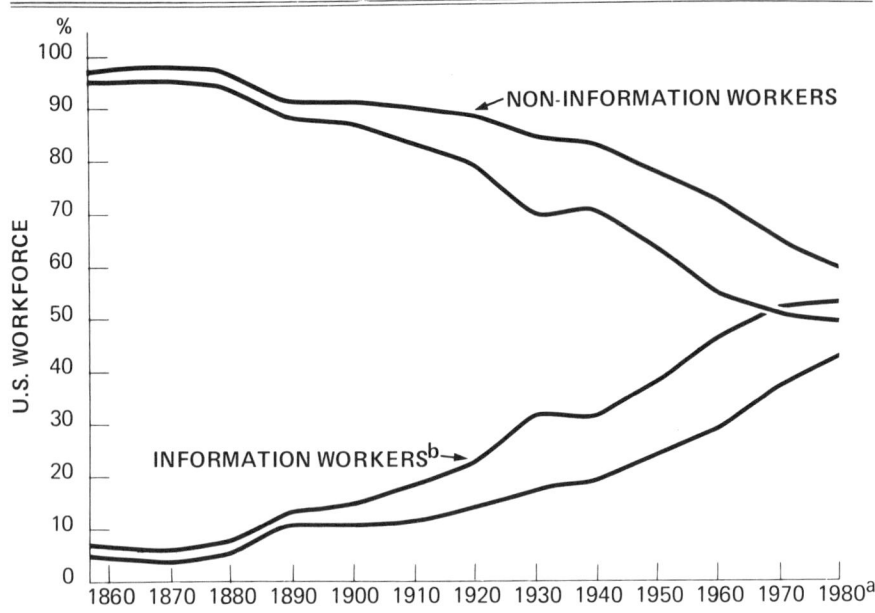

Notes: (a) 1980 projections supplied by the Bureau of Labor Statistics (unpublished).
(b) Bandwidth reflects restrictive vs. inclusive definitions of information workers.

Source: Porat (2), vol. 1.

group. The crossover in employee compensation occurred much sooner, as information occupations tend to earn a higher average income. *Figure 1-2* shows the passage of the U.S. economy through three distinct stages.

Figure 1-2
Four Sector Aggregation of the U.S. Work Force by Percent, 1860-1980

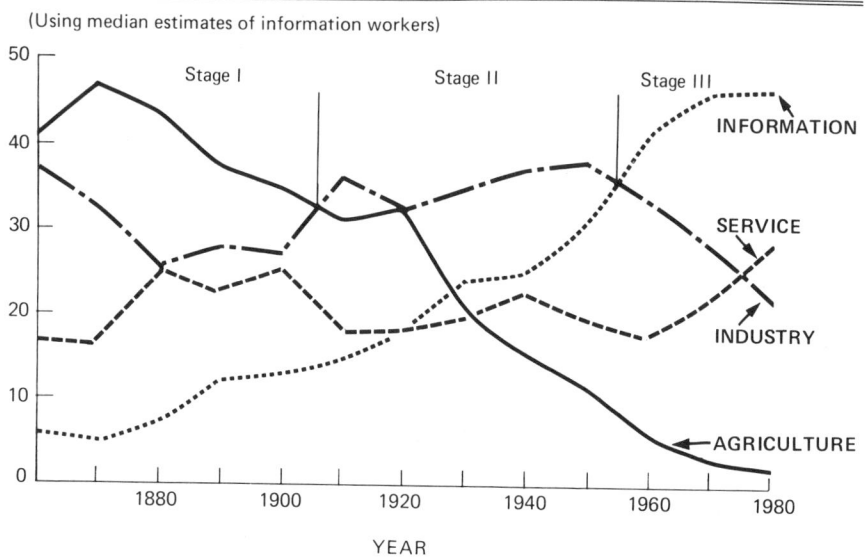

Source: Porat (2), vol. 1.

In Stage I (1860-1906), the largest single group in the labor force was agricultural. By the turn of the century, industrial occupations began to grow rapidly, and became predominant during Stage II (1906-1954). In the current period, Stage III, information occupations comprise the largest group.

During Stage II, the industrial share of the work force reached a peak of around 40 percent and has been declining precipitously since then. It plunged from a dominant position in 1950, when it was twice the size of the information work force, to the present, when it is only half the size of the information work force. This reversal was extremely rapid, and the trend has only recently been abated.

In Stage III, the current period, the number of information workers grew at almost the same rate as the overall work force (net rates of 1.01 percent for 1960-1970, and 0.04 percent projected for 1970-1980). New entrants to the labor force who could not be absorbed into information occupations moved into the service sector, including a large contingent into the medical service sector.

It is unlikely that the information work force will increase at the pace of the 1940s and 1950s again. Resurgence in the information occupations is likely only if new types of information industries are launched by entrepreneurs—information utilities, search services, storage and retrieval services, computer-based diagnostic services of everything from cars to hearts, facsimile and electronic mail transmission services, specialized microcomputer programming services, and so on. As of the 1970s, private and public bureaucracies are glutted with information workers. No more can be easily absorbed.

The Information Activity and GNP

We now turn to measuring the share of GNP that originates with the production and distribution of information in the U.S. economy, dividing all industries into "primary," "secondary," and "noninformation" sectors.[3]

The primary information sector includes all industries that produce information machines or sell information services on established markets. This sector provides the technical infrastructure for a variety of information processing and transmission activities. It also offers information for sale as a commodity. Included here are such diverse industries as computer manufacturing, telecommunications, printing, mass media, advertising, accounting, and education. This is a productive locus of an information-based economy.

The secondary information sector includes the informational activities of the public bureaucracy and private bureaucracies. The private bureaucracy is that portion of every noninformation firm that engages in purely informational activities, such as research and development, planning, control, marketing, and recordkeeping. It includes the costs of organizing firms, maintaining markets, developing and transmitting prices, monitoring the firm's behavior, and making and enforcing rules. The public bureaucracy includes all the informational functions of the federal, state, and local governments. Governments perform myriad planning, coordination, deciding, monitoring, regulating, and evaluating activities. These are the information overhead costs to the private economy. The rest of the economy, net of the information activities, is measured as the "noninformation" sector.

Defining a primary information sector. The end product of all information service markets is knowledge. An information market enables the consumer to know something that was not known beforehand; to exchange a symbolic experience; to learn or relearn something; to change perception or cognition; to reduce uncertainty; to expand one's range of options; to exercise rational choice; to evaluate decisions; to control a process; to communicate an idea, a fact, or an opinion. An information market may trade topical knowledge with a very short useful life, or it may produce long-lasting knowledge. It may involve specialized or unique knowledge, useful only to one person in one situation, or it may be a public good available to all simultaneously and useful in many contexts. It could be extremely costly to produce, or it might involve only simple processing and transmission. Information production could be a lengthy process spanning a whole lifetime (such as invention), or it could be a burst of data occurring in a millionth of a second.

The primary information sector includes hundreds of industries that in some way produce, process, disseminate, or transmit knowledge or messages. The unifying definition is that the goods and services that make up the primary sector must be fundamentally valued for their information producing, processing, or distributing characteristics. If the informational aspect is ancillary to the noninformational aspect, the good or service is eliminated from my calculation.

Defining a secondary information sector. Not all information services produced in the economy are sold in primary information markets, but are produced by noninformation firms and governments for internal consumption. Every noninformation firm produces and consumes a variety of informational services internally: planning capability, financial control and analysis, a communications network, computer processing, typing, filing, and duplication services. The private bureaucracies consume a tremendous amount of both capital and human resources in producing these overhead information services. Their inputs are computers, facsimile machines, laboratory equipment, office buildings, office machines, telephones, and file cabinets. The bureaucracies hire managers, research scientists, programmers, accountants, typists, and librarians. These information capital and labor resources are organized into production units that play a purely informational role. Large corporations often create a planning group, R&D group, electronic data processing group, and so on. Each unit has a well defined technology with recognizable inputs and outputs and can be conceptualized as a "quasi-firm" embedded within a noninformation enterprise. Such quasi-firms' information producing, processing, and distributing activities are ancillary to or in support of the main productive activity.

For example, when an automobile manufacturer installs an in-house data processing facility, hires programmers and analysts, leases equipment, and constructs a private data network, a "quasi electronic data processing firm" has been created within the firm. In may respects, the economics of the quasi-firm are indistinguishable from those of an established independent data processing vendor and the two are seen as interchangeable.

Consolidated accounts of the two information sectors. The share of GNP that originates in the primary and secondary sectors can be precisely measured by integrating several data bases in the economic censuses.[4] Once the definitions are established, the measurement effort is one of reaggregating the national economic statistics.

The task of measuring the primary information sector share of GNP is quite straightforward, and can be accomplished at the seven-digit Standard Industrial Classification (SIC) level. Over 6,000 goods and services are included (see *Appendix*). In 1967, over 25 percent of GNP originated in the primary information sector, mostly from business services ($43 billion), components of finance ($41 billion) and portions of general government ($40 billion).

The secondary information sector is measured by tearing firms apart (in an accounting sense) into an information activity and a noninformation activity. The informational side of the firm sells its services on a fictitious account to the noninformation side. The value added by the secondary quasi-firms is defined as the employee compensation paid to information workers in noninformation industries plus the capital services of information machines (see *Appendix*). In 1967, over 21 percent of GNP originated in the secondary information services produced within firms and governments for internal comsumption.

Communications in an Information-Based Society

A historical perspective might help us place communication policy in context. Communication has clearly been a part of every civilization in every stage of development. But the importance of communication, relative to other aspects of life, has taken on a new meaning: communication is central to an information-based economy.

During the millenia that spanned the agricultural age, the major arena of economic activity was the farm. The central fact on the farm was agricultural technology; political organization followed from the economic surplus created by the farm; and the overwhelming majority of the work force was engaged in the production of food. The outstanding "policy" issues of the day revolved around agriculture—land tax, farm price and income, land inheritance, agricultural improvement of the infrastructure (e.g., irrigation ditches, dams, canals).

The industrial age began with the mechanization of farm machines. The systematic application of engine power (whether water, horse, or steam driven) signalled a leap in the amount of goods produced for each day of human labor. Fixed capital spelled the difference, usually embodied in what was to become the symbol of the era—an industrial factory. Massive amounts of industrial surplus created wealth at an unprecedented rate.

The transformation from an agricultural to an industrial society was sudden, taking no more than one hundred years in the United States (1800-1900) and arguably much less (1840-1890). The shock was felt throughout the society and economy—cities swelled in size and squalor, the farm population collapsed, and a new set of social values (e.g., mass culture) emerged. Note our description of an industrial society: The major arena of economic activity is the factory; the central technological facts are machines for manipulating matter and energy; the owner of capital is the new lord and captain; and the overwhelming majority of the work force is engaged in industrial activities.

The outstanding domestic policies of the day revolved around industrial organization and practices—antitrust, labor law, natural resources, and improvement of the industrial infrastructure (e.g., railroads, highways, energy grids, and the financial markets).

The United States is now an information-based economy. The emergence of an information economy at its heart implies three changes in our social and economic organization. First, most people are divorced from the physical aspects of agricultural, industrial, or crafts activities; instead, the intellectual or information-processing abilities of people are at a premium. Second, an information economy is largely bureaucratic in nature; decisions are made and power is wielded not by many independent firms competing for a market, but by relatively insulated private and public bureaucracies. Third, the information technologies (computers and communications) are key components of our technical infrastructure.

Note our description of an information society. The major arena of economic activity are the information goods and service producers, and the public and private (secondary information sector) bureaucracies. The leaders of an information society are not large landholders or industrialists, but a highly educated and mobile "information elite"—scientists, technicians, managers, and professionals. A majority of the work force is engaged in the production, processing, transmission, and distribution of information goods and services.

The major policies of the day address two kinds of issues: (1) those relating to the arrangements of the information infrastructure itself, e.g., common carrier policy, competition in the related telecommunication markets, and investment policies in new technologies, and (2) those relating to the application of information technologies across all sectors of the economy, including the postal service, publishing, finance, media, education, manufacturing, transportation, medicine, energy, and governance. The first collection of problems we know as conventional "communication policy." The second collection, about which little has been written, is called "information policy."

Two Perspectives on the Policy Process

A communication issue can emerge on the policy agenda in various places: in Congress, through hearings and the legislative process; in the Executive Branch, through presidential or agency attention; in the independent regulatory commissions, through initiatives of the regulated companies, public appellants, or commission-initiated inquiries; and in the courts. How do issues arise? How are they organized into an agenda? Before we consider the communication or information policy issues of the day, we will try to develop a more general framework for understanding how issues are born and how we might establish priorities for our beleaguered public servants.

An issue, where economic or social interests are under contention, rises to the level of policy when private means of conflict mediation are insufficient. Most normal issues are resolved through the established practices of the market, through binding legal precedent, through common sense (consensus), or through many types of voluntary contracts and agreements. However, in those few cases where private means of conflict resolution are inadequate, issues surface in the public arena.

The two classical philosophies about the political process are technological determinism and ideological control. Each has its strengths and inadequacies, and, at the end, a synthesis of the two seems most relevant for understanding how to organize the communication and information policy agenda.

Technological Determinism:
Big Wheels and Little Wheels

Technological determinism is a Hegelian concept that cyclically rises and falls from intellectual favor. I find it useful for understanding communication policy in a long-run context. The "model" is mechanistic, and presumes an "engine" of change that drives a complicated ensemble of large and small wheels. The fundamental engine of economic and social change is technological change. In an economic context, the large wheels of conflict are radically technological, the small wheels are the firms that rise and fall, or even industries that are born or die. These large wheels, often enmeshed with the wheels of the natural and physical sciences (and mathematics), drive the smaller wheels of the marketplace, institutions, and ideas.

The model has six aspects: (1) technology, (2) prices, (3) institutions, (4) policy, (5) the law, and (6) public opinion. The causal progression starts with a change in technology (the big wheels) whose effects then filter through to the smaller wheels.

Technology. The material facts are society's endowments of labor, capital, and natural resources. The unique combinations of those resources we call "technology." Technological change is defined as recombinations of labor, capital, and natural resources in ways that augment the productivity of the resources, or expand the frontiers of social output. For example, we combine the silica from sand, the skills of a mathematician, and a modest amount of capital to produce a new economic fact—computer technology. That unique mix (at the material level) sets a large wheel in motion, which eventually causes broad realignments of prices, institutions, laws and policies.

Prices. The immediate consequence of a productive innovation is realized as a change in prices. New ways of performing an economic activity replace the old as prices are compared, and as rational decisions are taken. Most of the management science is concerned with this second set of wheels in our causal model. By example, an economist and operations researcher specializing in production theory are concerned with tuning a firm's technology (and inevitably the prices of its goods and services) in response to existing technical conditions. A firm which is unresponsive to technical and price substitution cannot effectively maintain its market share. The first wheel set in motion by technology—technical substitution—works through the price system.

Institutions. Once the process of substitution has begun, institutional forces (smaller wheels yet) come into play. The theme is market realignment. Entrants, hampered by the lack of financial and regulatory cushions, wage uphill battles. Established firms control the strategic hills, and enjoy the fruits of long-standing market domination. Past rewards for work well done (both financially and politically) are expended while the established firm decides how to cope with the new technical realities. Two options emerge: either adapt institutionally, or suffer a traumatic demise as the wave of technology washes away the previously impregnable industrial bulwark. Corporations, like living organisms, rarely choose suicide. Adaptation is the single solution, and the institutional issue becomes one of process: How quickly does the institution change, and how is the change to be effected with a minimum loss of market share? The name of the game is entrenchment. The manager's responsibility to the stockholder is to stall the progress of the entrant until the firm can set its own direction and exploit its own resources.

Policy. The instruments of institutional entrenchment (and its counterpart, institutional assault) are well known and varied. We have classified them for convenience into three types: through market forces; through unlawful actions; and by working both sides of the fence. We are now ready to uncover the policy wheel in this machination.

Most devices for gaining and defending market shares are strictly legal, falling under the rubric of free market competition. In this first mode, firms compete openly, and the most clever or aggressive wins. Even though the initial bargaining power held before the contest often determines the outcome, the role of policy is minimal.

Other devices, such as predatory pricing, are clearly unlawful. Although illegality is often circumstantial (dependent on the likelihood of apprehension and the cost of punishment), and certain illegalities are conducted with virtual impunity, public policy is unambiguous in these instances. Where the behavior is illegal and the normal burden of proof satisfied, relief is eventually granted by the courts or regulatory agencies.

The most interesting mode of institutional assault and defense, one that poses the greatest challenge to public policy, is the middle ground between strictly legal and strictly illegal: the region where public policy is ambiguous, inadequate, or obsolete. In such cases, rules and regulations are stretched to the limit, and always interpreted to the benefit of the interpreter. When all parties to a conflict self-righteously point to a federal law or regulation to justify their behavior, public policymaking is implicated. The "superstructure" mediates between competing claims, legitimizes institutional change, and accommodates, *post hoc,* to changes that are driven by the infrastructure.

It seems clear that alibis for quasi-legal behavior should be quickly eliminated by reforming public policy. That, however, is difficult to accomplish. Those who benefit from the current rule will argue that the rule is a pristine example of clear logic. Indeed, an entrenched firm might argue that a protective rule

Policy in an Information Society

needs to be fortified. Those who lose from the current rules will publicly proclaim their anguish and accuse the rule of destroying the spirit of free enterprise. Both sides use hyperbole because the very ambiguity of the contested rule's worthiness creates the room for contention. In the midst of such a storm where no fact is undisputed and every opinion is portrayed as fact, the policymaker is truly a troubled soul.

The really interesting areas of conflict have technical, legal, economic, and social dimensions that cannot be easily separated. Moreover, the underlying technical conditions are likely to be volatile, leaving the policymaker in the hapless position of building a castle on a sand dune. The winds of change keep altering the landscape.

Law. The instruments of public policy are judicial and administrative. When a conflict has risen to the level of policy, and when an appropriate deal is struck between the contestants, the instruments of the superstructure come into play. The legal system is the membrane through which technological facts are filtered into economic realities. That which begins as a scientific innovation eventually emerges as new social rules. Where this transmutation causes molting pains, the legal and political system is activated as a salve: laws are enacted, administrative rules implemented.

Public opinion. The last element of the model rides atop the superstructure and is called "public opinion." After all the technical, economic, institutional, legal, and political accommodations are played out, public judgment and ratification are called into play.

This view differs from the conventional, where public opinion is an essential, if not the leading, element of the political process. In our mechanistic conception, opinions and attitudes are formed in reaction to deeper realities. At the shallowest level, least connected to the infrastructure, we have fads. Fads live and die without affecting life for very long. Their charm is precisely defined by their evanescence. Fads should be of least relevance to the policy process, although their more sinister cousin, mob rule, emerges as a potent policy instrument from time to time.

The Role of Ideology

This completes the mechanistic view of the policy process, a kind of technological determinism that is a useful but incomplete model. We have neglected the causal force of ideology, and thus have developed a rather arid framework for understanding policy formulation.

The ideational system normally captures most of the public's attention. It is alluring to throw ideas and images into a colliseum and watch the struggle. Ideology enters critically in every stage of policymaking, and vitally affects the direction of technology, economics, and the law.

First, at the technological level, it is the ideational system that partly determines which research is conducted, and consequently which types of knowledge are produced. Science policy encourages certain forms of R&D and ignores others. Once a direction for basic research has been set, applied research follows, as does development, production, and marketing. By example, if a decision is taken to implement high-power communication satellites and low-cost ground stations, the pace of third world rural economic development might quicken. All the derivative economic and institutional impacts would thence occur. The decision to pursue the policy or not is ideological.

Second, at the economic level, a technical innovation that is introduced to the productive system may be accepted or rejected on ideational rather than economic grounds. By example, computer-assisted instruction was rejected by users for a variety of noneconomic reasons. That rejection precluded institutional conflicts (e.g., traditional public schooling versus curriculum development by private firms), and no conflicts were raised to the level of policy.

Third, at the institutional level, ideas may determine the basic direction of corporate and governmental action, uncoupled from the dictates of technological or economic realities. By example, the implementation of the Freedom of Information Act was a purely ideological policy which undoubtedly affects the workings of both the government and the economy. The impulse was rooted in a social value (openness and accountability), not in technology or economics. The very notion of institutional organization is heavily laced with ideological issues such as ownership, control and internal policy. Therefore, as material realities filter through institutions, the ideological system serves as an influential screen.

Fourth, at the policy level, we witness the replacement of technological determinism by politics. An ideological notion here carries more weight than a historical trend, at least in the short run. The policy process can stave off tidal waves of technological change, simply by resorting to ideological arguments or rhetoric. The political process killed a plan to create a government-wide computer network (FEDNET) on the legitimate grounds that it made a mockery of the right to privacy. FEDNET was technologically feasible and economical, but a potential civil liberties nightmare.

Fifth, at the legal level, once again the system of ideas (as constructed by long vertical lines of precedent) impose their own *raison d'etre.* A technical innovation can be retarded or accelerated depending on how the legal system responds. This is especially true in communications, where the courts are a major force in shaping national policy.

Sixth, at the public opinion level, the attitudes of consumers and voters clearly affect the outcomes of market and political forces. To the extent that consumer sovereignty and political independence of mind determine decisions, ideas can dominate deeper underlying forces. But to the extent that ideas themselves are born of responses to changing economic and social conditions, (e.g., the demise of traditional morality and ethics as a function of twentieth century urbanization), it is not clear which dominates—the ideas or the historical context.

The Two Flows of Causation

Causation clearly works in two directions simultaneously: from technology to institutions to legal and political decisions; and from political and legal ideas back to technological choice and institutional structure. The policy process acts as an intermediary in both flows, mediating real or perceived conflicts and defining the boundaries of acceptable behavior. In a long-run sense, technology is cast as the big wheel of social change, and ideological tenets are the little wheels. If little wheels spin hard enough, with sufficient force and consistency, they surely affect the big wheels. But, following the mechanical analogy, if a big wheel moves only slightly, its effects on the little wheels can be devastating. Witness the development of nuclear technology on the "little" wheels of defense ideology. Witness the development of electrical power on the "little" wheels of lifestyle. Witness the development of the printing press on the "little" wheels of universal literacy. Even, as argued by Harold Innis, witness the development of papyrus on the Egyptian concept of administration.

The fruitful parts of ideology, from a policy viewpoint, are those which affect the creation and direction of technological and economic facts and the derivative policies relating to industry structure. Other ideological debates, although they generate generous amounts of heat, are less useful. Their resolution is based on the force of rhetoric, not on the force of analysis. As such, there is less that the analyst can contribute.

This is not a doctrine of technological determinism. After all, ours is a period where the impact of technology is being assessed with sharper judgment, and where social planning and control join in partnership with market forces to encourage or deter technology. Some outstanding recent examples are the meritorious battles around the SST, breeder reactors, food additives, automobile safety, home and office construction, and genetic research. My principle states simply that technological facts alter economic and social reality, and that it is the responsibility of the political process to guide technology. Certain politicians, social activists, and enlightened corporate executives fully embrace this concept of "social planning"—not in the gray and unappealing Soviet model, but in a historically unique manner, owing allegiance to the principle that society is the keeper of its own destiny. There are many possible futures, not one, and what we end up with is critically dependent on the technological decisions made in the present.

The first philosophical issue facing the policymaker is to develop a causal model that relates technology to social change (through the linkage of economics and law). This is a Herculean task, and is best relegated to the private world of mental rather than formal models. The second problem is to develop positions on issues of value or ideology. Generally, two people might agree on the model but disagree on the parameters. By raising the level of discussion to one of values, we can disagree on those things that are ideological, while agreeing on those that are free of ideology.

In cases where ideological issues have material or technological components, policy decisions can make the greatest political impact. For example, "teletext" services such as the British ViewData system, when introduced in the United States, will undoubtedly be laden with First Amendment issues. Does the First Amendment freedom of the press extend to electronic forms of information-giving? (See Benno Schmidt's discussion in Chapter 5.) Or is the "electronic newspaper" going to fall under the regulatory hand of the FCC (as Bruce Owen fears in Chapter 6)?

An issue that seems purely ideological on its face may actually hinge on variables that are technical or economic. By focusing on the ideological component, policymaking is grounded in a system of ideas divorced from the imperative of science and technology. Conversely, an issue that seems purely technical or economic may imply a critical choice of values. A technical decision not mindful of its ideological implications is unacceptable.

Policymakers hence must tread a fine line between social design and technological determinism. To do nothing in the face of changing technology is to admit of determinism. To inject ideology into an emerging technological system is to dabble in social design. Especially in the fields of communication and information, where the technological big wheels are turning at a fast clip and where the implications are largely shielded from public awareness, this choice is disheartening. The policymaker has to understand the ideological implications of technical decisions. The policymaker should also understand that an ideology can be expressed as a willful rearrangement of technologies and markets.

In the next section, we review the process by which an ideal policy agenda is structured, and offer a method for organizing the communication and information policy agenda.

Structuring a Policy Agenda

Structuring an agenda is essentially a problem of scarcity: time, energy, and political capital are all in short supply and must be expended with care. Assuming that issues emerge at random in the policy arena, how are we to establish priorities for the policymaking process?

Conventionally, a politician senses the various pressures and noises that are brought to bear, and responds according to their intensity. This rule may, by chance, work well—the politician responds to the important problems and ignores the unimportant. But it can only work perfectly where there is a perfect relationship between noise and importance. Those that yell the loudest may not be yelling about the important issues.

The cynical view is that the importance of an issue bears little relation to the clamor that attends its introduction. The populist view is that important problems (pluralistically defined) are those forced onto the public servants by their constituents. Noise and pressure present a limited, although enduring, strategy for structuring the policy agenda.

A Selection Rule

An alternate strategy, presented in the following passage, is divided into five parts: (1) the test for technology push, (2) the test of ideological or value relevance, (3) the test of maximum economic (interindustry) impact, (4) the test of the market forces, and (5) the test of political feasibility. In an ideal world, the rule would shape the agenda selection process. In the real world, where the timing of legislative initiatives is only weakly correlated with the need for legislation, the following is at best construed as a policy guideline. It might better serve as a philosophical touchstone for those officers in the Executive Branch who sport a badge saying "communication and information policymaker." As with every rule, the exceptions are as interesting and as definitive of the process as the rule itself.

Technology as the main engine of social change. Communication and information technologies are among the fastest changing technologies today because of intense research and development activity (as discussed at length in Walter Baer's chapter). There is no shortage of articles showing the dramatic drop in costs of information transmission, processing, storage, retrieval, and equipment. The starting point of communication and information policy should be in those areas where technical substitution is likely to occur and where the "push" is compelling and cannot be easily ignored by any sector of the economy.

For example, we might agree that direct broadcast satellites (DBS) are an efficient new form of information distribution; we might further agree that this phenomenon will increase the volume of commerce and cultural exchange; and we might agree that the introduction of DBS will tend to "open" cultures and reduce the importance of geographic boundaries as the cost of communication drops. But depending on our ideological precepts or values, we might take very different positions on the desirability of DBS. An ultranationalist position might see DBS as a threat to national integrity, and a crack in the armament against neocolonialist incursions. A country actively engaged in multinational commerce and exportation might see DBS as a great boon in making the planet a smaller, more familiar place. Same model, different parameters, different outcomes. The first part of our rule for policy planning is:

> Structure a policy agenda by focusing first on those policy issues that are driven by a technology or by a convergence of several technologies.

Ideological or value relevance. Technological decisions are rarely void of ideological or value questions. But some decisions, such as establishing a PBS "fourth network" or relaxing the regulatory boundary between communication and computing, are vested with greater ideological moment than others.

Some technological pushes in the field of information and communication could have no serious ideological or value implications, occurring at a purely technical level and with few visible social impacts. At best, the innovation might benefit one firm at the expense of another. Looked at from afar, we might see a brief struggle between competing capitalists. But no major intersectoral flows of either labor or capital emerge; no major institutional issues arise; and certainly no social or behavioral issues come into conflict. I hasten to add that examples of purely technical changes void of consequences at the level of superstructure are hard to find. The second part of our policy planning role is:

> Focus on those technological issues that trigger the most far-reaching or deepest ideological choices. Put aside those issues where the ideological or value choices are remote.

Interindustry impact. Wassily Leontief, Nobel laureate in economics, has reduced to numbers the intuition that an event which touches one industry ripples through the entire economy. Small pebbles can sometimes generate tidal waves of change in related industries; and seemingly large boulders can vanish without a trace. The determining factor is structural depth, the degree of linkage or "interconnectedness" between one industry and the rest of the economy. Shallow industries cannot vitally affect the economy; deep ones can. From a policy perspective, we should worry more about the structurally deep industries (e.g., energy, telephone, computing), than the structurally shallow (e.g., clothing, recreation). The third step is:

> After selecting those issues that satisfy the first two criteria—technology push and vested with ideological ramifications—discard the issues that are relatively void of interindustry effects and focus on those with broad impacts. Select those issues whose impacts are likely to generate the greatest ripple effects.

Market focus. The proper role of governmental intervention in private markets has developed into a philosophical *cause celebre*. Some see it as a basic value issue, central to modern political economy. The line between the liberal and authoritarian social democracies, say the conservatives, runs roughly parallel to the line between market domination and political domination of the economic system. More governmental intervention means less personal freedom. This old idea has its share of new converts. In part, the conversion has been a baptismal of fire, generated by the sparks of a social collision with the government's bad judgments. In part, the conversion is simply an admission of laziness—it is easier to blame Washington for our ills than to face up to some very tough problems of market neglect, excess, or failure.

The liberal side has traditionally held that the government is the trustee of the public welfare. The basic conditions of wealth or poverty and employment or unemployment are tied directly to economic activity. And as welfare derives from commerce, government should be the holder of commercial license. Since the turn of the century, intervention by government into affairs of commerce has cured an embarrassing litany of corporate abuse. It is not a question of intervention versus laissez-faire; it is a question of the degree, timeliness, precision, fairness, and consistency of the intervention.

Note two curious shifts in the old dichotomies. We now see a new liberal commitment to the tenets of competition. It is now the public-interest groups that urge greater competition in telecommunications, greater diversity in communications, lesser governmental regulation of the media. We now see a rush by some established corporations to encourage and extend governmental regulation. Competition, agree the conservatives and the liberals, increases output, decreases prices and profits, and spurs innovation. Depending on one's point of view, that is either desirable or abhorrent.

The relevant axis is no longer conservative-liberal; not when chairmen of corporate boards plead for greater regulation, and when disaffected populists rail against government intervention. The question now is: How well is the system functioning? Traditional liberals could as easily warm to private solutions as to public solutions, as long as the basic liberal values are met: That opportunities for a satisfactory "quality of life" are fair (within broad bounds) and insensitive to race or sex. By symmetry, traditional conservatives welcome or reject governmental influence, not *per se,* but as such intervention furthers or retards the basic conservative values.

The outstanding lesson of the last 20 years is that market forces are not as bad at achieving social equity as liberals once thought; nor are they as good at achieving economic efficiency as conservatives once thought. There is room for coalition.

The presumptions about regulation have changed. Although we are all more cautious about rushing to governmental solutions, we are also less charitable in granting that the government that governs least governs best. Problems do not disappear if firms are left to maximize profits and individuals to maximize their welfare: If we have learned anything of late, it is that most of us operate slightly irrationally, slightly in ignorance, and very much in a world of gross, preexisting imperfections. Doing good in a parochial way does not aggregate, across all firms and individuals, to doing good for society. This is especially true when so much of what we decide is conditioned by what we cannot easily control: by pressures of population, by natural resource shortages, by nationalist impulses, or by foreign decisions.

The fourth step is therefore indeed a difficult one. The policymaker must decide how, when, and where to intervene, and with what form and intensity of intervention. Governmental intervention that occurs too soon or too late can damage more than benign neglect. An intervention that is misplaced, that is improperly or imprecisely targeted, can backfire and wreak more havoc than inaction. An intervention that fails to strike a balance of fairness will be destroyed by the injured party's noncooperation. And, of course, an intervention that is rigid cannot survive into the future, because if any principle is immutable, it is that everything changes.

After narrowing the candidates for policy intervention by applying the first three tests, the fourth step is:

> Begin with the presumption that market forces yield an approximation of fair and efficient solutions, and look only at those issues in which the market solution is inadequate in a clearly definable way. The governmental intervention can then occur if a policy can be devised that is timely, targeted, and at the proper level of intensity.

Communications policy history has shown that the fourth test is very difficult to pass.

Political feasibility. After great deliberation about the shape, purpose, and contents of the cart, we finally arrive at the horse. To those in the Washington environment, the bottom line to any question is political feasibility: Can it be done? What is the payoff? What are the liabilities? I hope that by putting this question last, and not first, the discourse has been broadened to include the distant horizon of social values. Those impatient with the reflective license implied in the first four steps can now relish their just reward: Politics is the bottom line.

A successful administration is always mindful of which issues are politically viable and which are too hot to handle. How well this time-honored sieve sorts through the issues in communication and information policy is not an idle question. Fortunately, these fields are so technical and abstruse that the press of profane politics is lighter here than in other areas. In fact, the problem heretofore has been the opposite: Congress and the White House separately decided that communications is not a terribly rewarding business and generates little political capital.

The politics of the feasible imposes itself on the policymaking process in two senses. In the Type I error, the political process may generate issues that fail to meet the four previous criteria. In the Type II error, politics may cause ears to deafen to those issues that do deserve political attention.

A recent instance of the Type I error occurred during the political process surrounding the birth of the National Telecommunication and Information Administration.[5] The "information age," say some, is upon us, yet the Executive Branch until 1978 had no clear machinery to formulate and execute policy in these fields. Some observers therefore were heartened by the timeliness of the President's decision to organize communication and information policy under one roof. Yet the President's wishes, communicated through an executive order, evoked a concert of shrill and largely meaningless yelps of agony from the various executive agencies and congressional committees who feared a loss of turf. Through brilliant applications of bureaucratic experience, the substantive issues in communication and information policy were drowned in a tidal wave of interagency and congressional committee bickering. The turf battles involved two cabinet members, six line agencies, and scores of interagency meetings. Finally, after almost a year, the executive order creating the NTIA was signed.

The Type II error is exemplified by the massive political indifference to the dynamics in the telecommunications common carrier industry. There is no "crisis" in telephony, other than a massive assault by young high-technology companies on the competitive edge of the marketplace. The Bell System and the independents incontrovertibly provide the best telephone service in the world: The backlog for telephone hookup is nil; dial tone happens without fail; the other phone rings; and telephone service consumes a paltry fraction of average household income (1.7 percent). Therefore, by waggish definition, there is no political interest. Yet the telephone industry is a vitally important component of the nation's information infrastructure. The shape and economic characteristics of the industry affect all sectors of society—firms, governments and individuals. The problem, until lately, has been that politics failed to register much interest in telephony. That is understandable. Even the most enthusiastic communications expert reluctantly admits that energy, food, defense, welfare, and health issues are more deserving of public debate than internal cross-subsidy between telephone long-lines and local plant. Considerable technical and economic sophistication is required to realize that decisions made today about our communication and information resources play an important role in solving society's larger problems.

With some trepidation, we summarize the fifth selection rule:

> After selecting the issues that (1) are driven by technology, (2) are vested with ideological or value issues, (3) implicate the greatest

number of industries and economic sectors, and (4) are not likely to be solved by market forces, allow the political feasibility test to structure the policy agenda.

It is a unique policymaker that can adhere to the analytic discipline suggested above before applying the "bottom line" test. But if step five preempts the first four, will we ever ask the right questions?

Communications and Information Policy: An Agenda for the 1980s

In this section, we discuss four classes of issues. If the dramatic developments in communication and information are seen as a technological "bomb," dropped into a industrial society, then the following is descriptive of how one might analyze the event: (1) those issues which are directly concerned with the shape and performance of the information infrastructure; (2) the first wave of impact, on those industries which are information intensive; (3) the second wave, on the noninformation industries and on the process of governance; and finally (4) the third wave, on broad social and personal values.

Communications Policy: The Technical Infrastructure

The heart of the communication policy issues is a scheme of social control of the structure and performance of communication industries: common carriers, specialized common carriers, value-added networks, satellite facilities and services, telecommunication equipment, television and radio broadcasting, cable TV, pay TV, citizens band, mobile radio, etc.

The economic stakes in the communications game are gigantic. In 1977, AT&T's gross revenues were over $36 billion. As a "nation state," AT&T surpassed the Gross National Product of 118 of the 145 nations holding memberships in the U.N.[6] More than 10 percent of all investment (gross capital formation) in the United States is accountable to the telephone industry.

When we add those figures the revenues of some other major information firms, we begin to appreciate the enormous stakes involved:

Communications for Tomorrow

Table 1-2
Revenues and Profits of Selected Information Firms

	($ Millions) Revenues 1977	Profits 1977
American Telephone & Telegraph	36,112.1	4,455.8
American Broadcasting Company	1,616.9	109.8
ITT	13,194.2	562.7
CBS	2,776.3	182.0
Control Data	1,493.4	62.4
Digital Equipment	1,262.3	128.0
Gannett	537.2	69.4
Harris	770.5	46.2
Honneywell	2,911.1	134.3
IBM	18,133.2	2,719.4
Knight-Ridder Newspapers	751.7	60.2
McGraw-Hill	659.0	51.4
NCR	2,521.6	143.6
National Semiconductor	425.8	14.4
New York Times	511.2	26.1
RCA	5,921.8	247.0
Raytheon	2,818.3	113.2
Sperry Rand	3,475.4	164.8
Texas Instruments	2,046.5	116.6
Time	1,249.8	90.5
Times Mirror	1,143.7	96.1
Warner Communications	1,143.8	66.9
Xerox	5,076.9	406.6

Source: *Business Week*, March 20, 1978.

It is instructive to recall that most of the firms listed above were either infants or yet unborn at the turn of the century. Yet by 1977, two of the three most profitable U.S. corporations were in the information business: AT&T with profits of $4.5 billion, and IBM, with profits of $2.7 billion. (The other firm was General Motors, with profits of $3.3 billion.) Even Exxon, the fourth most profitable corporation in 1978, entered the information industry through a subsidiary that manufacturers computer terminals.

However, despite the stakes, the economic turf wars ranging around the communication and information industries hardly fire the citizens' and politicians' blood to the boiling point. Not when issues of controlling inflation, reducing unemployment, passing national health insurance, reforming the tax and welfare systems, curing our foreign policy blues, and saving porpoises so consistently preempt the policy agenda. Simultaneously a blessing and a curse, communication policy is low in the nation's consciousness.

Communication policy in an information society is akin to transportation policy in an industrial society. The design, structure, and performance of the technical infrastructure is at stake, and that is a crucially important economic fact. The decisions being made about the information infrastructure not only affect the communication industry giants themselves, but also all the industries that rely on the infrastructure for their lifeblood. By analogy, policy decisions made in the 1870s about railroad pricing policy affected the viability of midwestern farms relative to their eastern competitors. In setting transportation policy, one simultaneously set agriculture policy. Similarly, decisions made about telecommunications now affect a host of other industries, such as banking, airlines, publishing, retail manufacturing, and energy. Here we arrive at the first wave of "information policy."

Information Policy and the Information Industries

The communications problems in the 1980s and beyond will come from the field of information policy, where applications of communication and computer technologies upset conventional industry lines. Old industrial practices will be rendered technically obsolete, and the oft-discussed phenomenon of "convergence" will result in a multitude of economic and social problems.

The first impact of information technology is especially felt by information-intensive industries, following the inescapable logic that technical substitution follows function. The primary information sector industries, which are in the vanguard of technological upheaval, will experience the greatest and most severe policy problems. These include the postal service, banking and finance, securities exchanges, education, newspaper and book publishing, libraries, and legal services.

Scenario: The year is 2028. The information infrastructure is a composite of many entities offering three classes of goods and services: information processing, information transmission, and information terminals. From the user's point of view, the system is completely integrated, in that all the entities interconnect at will, and distinctions between the vendors are technically invisible. A highly competitive market develops in all three arenas.

Communications for Tomorrow

In this world, these industries essentially use the same infrastructure, and offer competing services: telephone, message services, and teletext; electronic mail, electronic banking, and retail catalog shopping; broadcast television, cable television, and pay television; electronic newspapers and information utilities; personal mobile radio telephone. The turf problems will, as one might imagine, be severe. It is likely that the normal workings of the market will not properly adjust to the technological push. Even though these functions are purely informational, their traditional industry lines are separate and rigid. A banker, a newspaper publisher and the Postmaster General do not fancy themselves as being in the same business. Yet they are all information "brokers," specializing in the retail packaging and distribution of (unlike) information services. Function and form are converging, driven by a convergence of technologies. Their enterprises are on a collision course: The infrastructures of electronic funds transfer, electronic publishing, and electronic mail services are essentially identical. Each uses a combination of information processing and transmission resources, arranged in a network, accessible by similar machines called "terminals." In time, a single terminal in the office, shop, or home may accommodate many services—and therein lies both the promise and the threat.

Should banks be allowed to enter the electronic retail "catalog shopping" market? Should the telephone company be allowed to establish a message service or a teletext service? Should electronic newspapers and information utilities be regulated? Should the postal service enter the electronic mail industry as a monopolist? And if not, what should society do to mitigate the catastrophic effects of the USPS's obsolescence? Should the Communications Workers of America (800,000 telephone employees) and the Letter Carrier's Union (600,000 postal employees) merge? Should broadcasting companies be allowed to own cable television companies and operate pay television? Should society invest in libraries as the "information utility" of the future, or should that function be left to the private sector? Should personal mobile telephones (wristwatch size) be given the spectrum now used by television and radio? Should the government require all common carriers to interconnect universally? Should copyright over information products be extended to computer software and to home recording of television programs? Who should appropriate the float when transactions are conducted instantaneously by EFT—the bank, the retailer, or the consumer?

Policy in an Information Society

The policy challenges are enormous, considered either from the narrow viewpoint of profits and employment, or from the larger social judgment of what is desirable. At present, information policy issues are largely understood as turf or interindustry conflicts, with a bottom line of profits and jobs. The less practical criteria—social welfare and the legacy we leave to future generations—have hardly been addressed.

Information Policy and the Noninformation Industries

The second wave of change is substitution by information labor and capital in the noninformation industries, especially in those aspects that are particularly information intensive. By example, computer technology is now at the heart of many processes in all manufacturing sectors: numerical control, sensing and process control, and switching and fine-tuning of electrical equipment. As microprocessor technology improves, the most ordinary objects (such as air conditioners and automobiles) will adopt "smart centers" for control. The small picture is a composite of the evolutions of mechanical objects; the large picture is the further encroachment on the industrial work place and work force by informational machines and workers.

The most immediate impacts will be felt in the "reservations" industries, such as airlines, automobiles, and hotels; the health care and medical industries; retail industries, especially nationwide department stores and grocery chains; and the large manufacturing industries, such as automobiles and chemicals. These industries, whose product is essentially not informational, rely critically on rapid, low-cost communication and computing.

The substitution will continue to erode the role of noninformation workers, and elevate the centrality of the information workers. The information bureaucracies, both private and public, will increasingly rely on information technologies and management science techniques. The secretary-cum-typewriter will be converted into a "word processor" where the essential distinction between typing, filing, document preparation and even mailing will be lost; and the office will increasingly be an information center that depends on information technology.

Information Policy and the Broader Social and Economic Impacts

The first two classes of information policy derive from applications of communication and information technology in both the information and noninformation industries. The third class of information policy is somewhat remote from communications

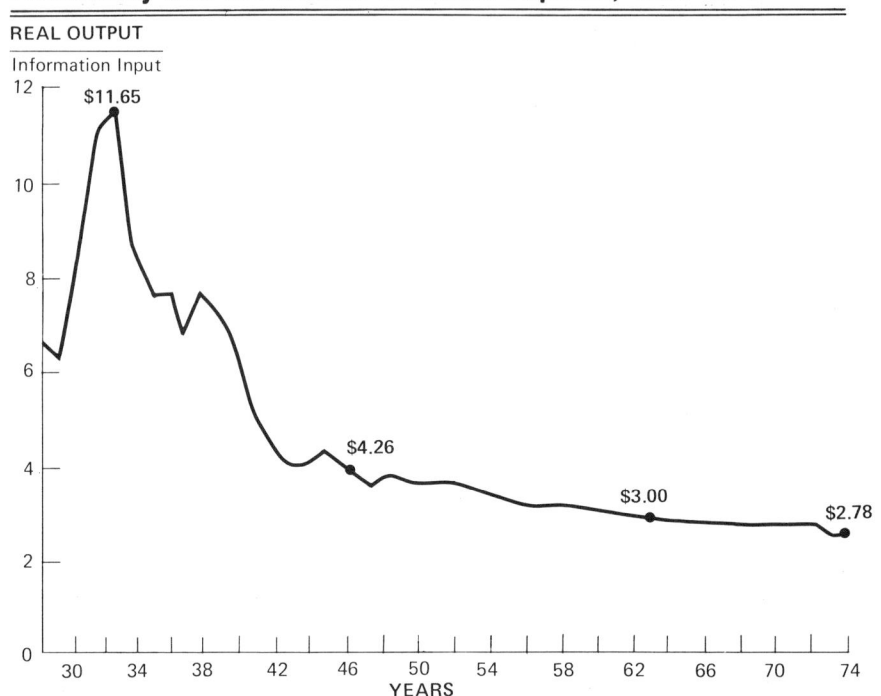

**Figure 1-3
Productivity and Information Overhead Expense, 1928-1974**

REAL OUTPUT
Information Input

- $11.65
- $4.26
- $3.00
- $2.78

YEARS

per se. The issues here relate more to the nature of an "information society," and to rules regarding the control and dissemination of information.

In this section, we discuss: (1) the productivity of information resources, (2) property rights to information, and (3) the question of social equity in the distribution of information.

Productivity and information resources. Both the public and private bureaucracies display an insatiable appetite for information, imposed internally and by other bureaucracies. The secondary information sector discussed previously represents society's allocations of information labor and capital to overhead activities. In 1967, more than $149 billion was expended by private firms for all the information activities ancillary to the delivery of a noninformation good or service. In addition, the information overhead of federal, state, and local governments amounted to nearly $19 billion.

Using data estimated from the National Income and Product Accounts, I constructed a ratio showing the amount of information overhead "input" used in the provision of goods and services. The ratio contains in the denominator the total information overhead bill for managing the noninformation industries, the

Policy in an Information Society

information industries, and federal, state, and local governments.[7] The numerator contains the "real value" of all goods and services produced in the economy, net of the information overhead necessary to bring them to market.

The resulting index, measured for the period 1929-1974, reveals an intriguing phenomenon.

In 1930, about $6.43 worth of goods and services were produced by $1.00 worth of information overhead. When the Depression struck, firms began slashing costs in response to a collapse in aggregate demand. The first victims of a general cutback were the technically nonessential information occupations. Machine operators cannot be laid off unless management decides to shut down the machine itself, because the capital to labor ratio is technically fixed. But clerks, secretaries, and managers are not tied to a specific production process. They are dispensable. Between 1930 and 1933, these information workers were the first economic luxury to hit the street (sometimes literally, in the more celebrated cases on Wall Street). The economy stumbled along with very little information overhead, squeezing $8.05, then $11.38, and finally $11.65 (in 1933) in real goods and services output from each $1.00 of secondary information input.

As the Depression rumbled away, firms rehired their information workers and the productivity index began to drop. During WW II, every bureaucrat, manager, and secretary was generating on average about $5.00 of goods and services per dollar of wages, about half as much as in the Spartan days of the Depression. At the close of the War, the boys came home and became "corporation men," and women in the industrial work force curiously returned to the traditional housewife role. The information work force exploded, driving down the productivity index. In the 1950s, each dollar in the overhead sector resulted in $3.50 of output. The index has dropped steadily ever since. By 1974, the secondary information (overhead) sector supported $2.78 of goods and services per dollar input—one quarter the amount in 1933.

A National Commission on Paperwork Reduction recently produced some bothersome statistics. Small businesses in the United States spent $20 billion in 1970 meeting the federal government's information requirements. The chairman of the board of Eli Lilly Company complained, "We spend more man-hours filling out government forms or reports than we do on research for cancer and heart disease combined.[8]

The total paperwork bill involved in filling out government forms responding to regulatory rule-makings, litigating antitrust suits, and lobbying for industry bills is enormous. AT&T claims that its cost of defending the Justice Department's antitrust case has exceeded $1 billion.[9]

Every organization, whether private or public, is saddled with overhead. Management is the brain, eyes, and ears of the central nervous system of the organism. It is crucial that accurate and timely information flows through the organism, lest irrational, aimless, or "epileptic" behavior results. No organization could long survive long without information. The question we raise here, as a matter of both public and corporate policy, is deciding how much information is enough—establishing a mechanism to sort out which informational activities are productive and which are useless or even damaging.

Secondary information can be divided roughly into three functional categories: (1) inventive activities, (2) coordinating activities, and (3) satisfying consumers' demand for complexity. In the next sections, we shall offer three explanations for the apparent decline in productivity.

Inventive activities. In a Schumpeterian world,[10] knowledge differentials explain the difference between success and failure. Firms that are unwilling to respond quickly to scientific and technological innovations are destined to decay and vanish. Economic growth—whether at the level of the firm or at the national level—is sensitive to the rate of investment in research and development. However, not all R&D is socially productive, although it may be privately beneficial. The untold sums of money spent in General Motors, Lever Brothers, and General Foods on product differentiation all count as R&D, and are so represented in the economic accounts and in front of congressional committees investigating the steady decline in U.S. productivity. Although product differentiation R&D (including its twin sister, advertising) helps generate monopoly rents, it does very little for true economic growth: it adjusts market shares, but does not increase the total pie.

The federal government, as a monopsonist of R&D, can influence the direction of R&D by using the procurement instrument with some delicacy. The government could, for example, devise a scheme of subsidizing R&D that encourages economic growth and market competition and taxing R&D that clearly does not. Such a policy would, in time, render the "overhead sector" involved with inventive activities more productive. This topic, however, extends beyond this paper.

Internal control of information. By far the greatest amount of information is generated internally by corporate bureaucracies. Information in this context is a nonmarket good, since it does not command an explicit market price. Determining the "optimal" amount of information is hence an impossible task. A production manager in a factory can know exactly how much a certain manufacturing task costs; an operations research analyst may specify with decimal-point accuracy the mix between machine A and machine B. But it is quite unlikely the chief executive officer knows how much the information resources of the firm cost, or how productive they are.

The folklore of bureaucratic life is rich with tales of empire-building, turf-skirmishing, and "the battle for slots." Once a bureaucracy is in place, it tries to fend off encroachment for budget by other bureaucracies, and to establish itself as a central player in the life of the organization. The best way to capture centrality is by being better informed than others. This is the first incentive for amassing information resources—people (managers, secretaries), machines (computers), and services (telecommunication, duplication). The best way to defend centrality, once captured, is by ensuring that everyone outside the bureaucracy is slightly mystified and overwhelmed by the complexity within. As Marx observed,[11]

> Bureaucracy is a circle no one can leave. Its hierarchy is a *hierarchy of information.* The top entrusts the lower circles with an insight into details, while the lower circle entrusts the top with an insight into what is universal, and thus they mutually deceive each other...

This deception can be accomplished by generating ominous piles of computer printout, shelves of manuals on administrative procedures, a constant stream of memoranda and reports, and other manifestations of activity in search of meaning.

All of this secondary information activity is costly, and one is tempted to ask if the secondary sector can be compressed. From a private point of view, it is desirable to squeeze out as much productivity from the information workers as possible. But from a social point of view, if rational management prevails and "office of the future" concepts succeed, what will society do with the newly unemployed legion of middle managers and secretaries? Our entire economic system has become accustomed to an information labor-intensive style of management. Even if we can reverse the trend, would that be desirable?

The nation has amassed an army of sophisticated and educated information workers equipped with an impressive array of information machines—computers, telecommunications satellites, pocket programmable calculators, word processors, and terminals. Half of what this army does is useless—akin to digging holes and filling them. But the other half is vital, spelling the difference between economic survival and destruction of the organization. An outstanding information policy problem—both in the private and public sector—is to learn how to distinguish the useful from the useless.

This task is extremely difficult, because the very notion of "productivity" is alien when applied to information. The productivity of a bricklayer or a machinist (or anyone working with physical goods or processes) can be measured. Ten bricks per minute is twice as fast as five, and therefore the wages can easily reflect the marginal product. However, information is not so straightforward. One good idea may take a month, a year, or a lifetime to develop. One cannot meaningfully measure productivity by counting the number of decisions per day, or memoranda per month. But without a metric, how does one measure the value of information? That remains an unsolved problem.

It is difficult to distinguish useful information from useless information inside an organization. As a hedge against the unexpected, every manager has an incentive to ask for another report or another set of figures.[12] Once the information is developed, it is stored for posterity. One is loathe to throw out the baby with the bathwater; so one computerizes both baby and bathwater, and hopes that the problem will thus vanish.

Demand for complexity. Possibly the most telling explanation for the secular drop in the productivity index is sociological, not economic. The consumer at the turn of the century had simple tastes, happily matching a relatively simple lifestyle. Henry Ford proudly offered America a wide choice in modern, rapid transportation, as long as it was a black two-door Model T. Today's consumer would hardly rejoice at that offer. Choice has ascended to the status of a basic freedom, synonymous with the good life itself. The auto manufacturers today offer countless permutations at their produce. In addition, the automobile is competing intermodally with airlines, trains, buses, motorcycles, bicycles, and skateboards. What are the productivity implications of this demand?

First, all of this complexity and variety imposes a staggering information burden on the economy. Each widget in inventory implies a paper trail of research, development, design, manufacturing, distribution, marketing, and accounting. Without the computer, such complexity could not be economically supported. But even with capital intensification, the costs of handling diversity is enormous.

Second, the modern consumer has come to demand a "cushion" of protection from life's events. In the simpler days, a fall from a horse might result in a visit by the town doctor, or it might not. Today, a car accident triggers a police report, an insurance report, an investigation of the claim, ambulance and hospital services, diagnostic tests, medical insurance claims, and so on *ad nauseum*.

Third, the voter has come to demand from government a generous level of protection and social control of industry, possibly without appreciating the accompanying price tag. We all want a clean environment. However, that "cushion" against industrial waste implies mountains of environmental impact statements, testing, registration, monitoring, and evaluation programs. The information burden is shouldered by industry, which passes it on to the consumer. This example can be replicated in every sector of the economy.

Unfortunately, the information cost of diversity and of "cushions" cannot be unbundled from the price. If they could be, one wonders whether we would vote (with our dollars) for as much complexity, or as much protection.

Property rights to information. Our economic and legal systems, indeed the Constitution itself, are all predicated on explicit notions of property rights. However, information does not easily lend itself to ascription of rights. In this section, we consider the policy implications of that growing realization.

Information is unlike most other ecomonic goods. The same piece of information can be simulataneously owned by two people without denying either the benefits of ownership. Certain types of information can be infinitely reproduced with very low resource costs. Information does not depreciate with use; to the contary, certain types of information (theoretical knowledge) increase in value the more they are used. Informational services, unlike personal services, do not vanish when the service ceases. Also, unlike personal services, informational services can be stored in inventory.

The most serious characteristic of information is that it lends itself so poorly to the classical economic and legal concepts of property rights. One cannot easily own information, because the act of theft is difficult to detect and even more difficult to prove. As simultaneous ownership is possible, there is no clear way of claiming or proving sole ownership.

Information that is unused is useless, but by using information, the "owner" reveals its content, either directly or inferentially. Information leaks in a most vengeful and uncontrollable way. Exclusion—a principle that separates private from public goods—is achieved only at some cost. In the case of computerized information, attempts to clamp the label "private property—keep out" on trade secrets, state secrets, and personal secrets are sometimes as effective as hiding honey in a colander.

The issue of property rights to information deserves serious study from the legal and economic communities. It has already cropped up in three policy contexts: the right to privacy; disclosure of public information; and the production and dissemination of governmental information.

The right to privacy. Privacy is a traditional social value, cherished by the frontiersman as well as the New England Puritan. Had Louis Brandeis' sensibility not been offended in 1890, he and Samuel Warren may not have written the first distinguished legal text on the subject:[13]

> Recent inventions and business methods call attention to the next step which must be taken for the protection of the person, and for securing to the individual what Judge Cooley calls the right 'to be let alone.' Instantaneous photographs and newspaper enterprise have invaded the sacred precincts of private and domestic life; and numerous mechanical devices threaten to make good the prediction that 'what is whispered in the closet shall be proclaimed from the housetops.'

The question of privacy revolves around property rights to information; or, more accurately, facing up to the almost impossible task of guaranteeing privacy in a highly technological environment.

The first problem with treating private information as property is that most personal information is released during a voluntary economic transaction. From the first moment of life to the last, we leave behind a massive trail of information. The *sine qua non* of taking a loan, writing a check, issuing an insurance policy, applying for a job, visiting the doctor's office, getting a driver's license, and registering for college is that the individual freely divulges personal data. Firms do not pay individuals for their personal information, even though it has both direct value (no information means no sale and no profit) and indirect value. In fact, some firms, such as banks, impose explicit transactions charges for handling the paperwork involved in assuming possession of personal information. Firms use personal data for a variety of analytic tasks, such as marketing studies, and sell personal information to users of mailing lists. Yet individuals produce the information at the cost to the eventual owner, and receive almost no protection against abuse of privacy.

Once the information is voluntarily created and exchanged, it becomes the property of the institution. If the institution is the federal government, the Privacy Act of 1974 protects the individual from certain kinds of release. The government "extends" the individual's property right over data which have physically left the person's possession. The law has loopholes in areas involving law enforcement and national security, and it imposes a costly administrative burden on the Executive Branch. But on the whole, the theory and practice of privacy protection over federally held records is established. If however, the information is held by a private firm, the individual has very little recourse in extending property rights over its use. For example, in *United States v. Miller*, the Supreme Court held that any federal agent has the right of access to private banking records. Furthermore, neither the agent nor the bank is under legal compulsion to notify the individual that his or her record is being inspected—before or after the event. Simple prenotification would allow the individual an opportunity to verify the accuracy of the record (or, argue the Departments of Justice and Treasury, to destroy incriminating evidence). Notification after the event would inform an individual that an investigation is under way, and allow an opportunity to correct the record should it contain misinformation (or, argue the law enforcement advocates, to place obstructions in the path of the investigator). The problem of governmental access to records held by the private sector, especially medical and financial records, is currently under legislative and executive review. The laws and administrative rules promulgated will serve as models for other arenas involving governmental access, such as personnel and educational files.

The last frontier involves the trade of personal information between private firms. In these cases, any claim of property rights by individuals is pure pretense. Once private information is given to a firm, the individual essentially loses control over its use. Even such sanctioned institutions as hospitals freely turn over (or sell) personal data to insurance and pharmaceutical companies and private detectives.

The value conflict in all these cases, aside from the ascription of property rights (which is an instrument, not an issue), is the tradeoff between managerial efficiency and a civil liberty. The problem of privacy became urgent with the advent of information technology. Overnight, where once no record existed, the imperative and ability to create a paper trail arose. A cash transaction leaves no trace; a check used to leave no trace—until the computerized clearinghouse operation and microfiche technologies came along. Now, more than 50 percent of all personal financial transactions are engraved in a permanent record.

The technologists who claim that the problem can be solved by technology are wrong. Freedom and efficiency (one skirts carefully around Luddite themes) are in a zero-sum game when privacy is at stake. Witness a hundred cases, including the one in which an ultra-secure Department of Defense computer system was cracked by a Mitre Corporation team in less than a day.[14]

This is a clear example of technology heightening the salience of a social value. A social and political decision about privacy will soon be made, and it will not involve dismantling computers and teleprocessing centers.

Disclosure of public information. The Freedom of Information Act (FOIA) was born out of an elemental democratic value—the public's right to know about the workings of its government. That value, coupled with a growing social demand for accountability, paved the way for the passage of FOIA in 1967. The theory behind the act is that information created or held by officials of the Executive Branch is the property of the public; and that aside from several specific exemptions, the public has standing to sue for any information it wants.

The problem has arisen in the interpretation of the exemptions, which is in fact an issue of property rights. For example, if a firm turns over a trade secret to a government agency (e.g., in conjunction with a procurement bid), the law clearly enjoins the release of that information to others. Competitors cannot use the FOIA as a weapon by unfairly gaining access to information and destroying the damaged firm's competitive edge. Without such protection, high technology firms would refuse to do business with the government, and the public would lose.

However, the definition of "trade secret" is ambiguous, as is the government's credibility in safeguarding the information. If firms impose a wall around all dealings with government, then the spirit of the FOIA is totally undermined. The problem is one of balance, judgment, and negotiation; that process is still continuing ten years after the FOIA was enacted.

The problem is made worse because the government's internal leak-plugging machinery is questionable, and occasional bureaucratic lapses of rigor and competence have occurred. An official may apologize for leaking a piece of information or for giving away trade secret data by mistake, but that is little solace to the injured party. Information, unlike horses, cannot be recaptured and tethered in the barn.

The ability of the government to govern properly is at stake. For example, the Federal Trade Commission is vested with the responsibility to promote competition and squelch anticompetitive practices. One such practice, predatory pricing, is accomplished in a multiproduct firm by internal cross-subsidy. Profits are siphoned from one product to support the losses incurred by the purposely underpriced product. The only way to prove that a firm is engaging in predatory pricing is by gaining access to "line-of-business" data. Information about revenues, prices, costs, and profits—product by line—reveals the pattern of cross-subsidy. The FTC has met with fierce resistance from the private sector, which argues that the line of business requirements impose an unconscionable economic burden on the firm, and that improper or accidental release of those data would cause grievous harm. The industry arguments are compelling. Firms have produced burden estimates in the tens of millions of dollars,[15] and using the cry "get the government off our back" have swayed public opinion. Also, wrongful release has happened, and has caused harm.

Interestingly, both defenses have technological implications. First, most large firms invest in some type of a management information system (MIS). Once data are entered into the MIS, the can be reconfigured in any way desired. As the basic economic data required by the FTC for its line-of-business reports already exist in the MIS, the cost of generating a report is not excessive. The second defense relating to wrongful release is almost unassailable. The government has a hard time guarding secrets. The modern credo should be, "The assumption is that any information written or stored will eventually find the light of day; secrecy is the exception." Firms have every right to exclaim concern over their proprietary data held by a government agency. However, no unfair advantage or disadvantage would occur if all firms were required to report the same kinds of data, and the FTC were to publish such data on an industry-by-industry basis. The benefits of release would not be captured by one firm; rather, the industry as a whole would become more competitive.

The proposal is radical, and will never come to pass. It points out, however, the information policy dilemma posed by the introduction of information technology. Before management information systems were developed, it would have been unthinkable for the government to demand such detailed information—firms did not have it, and the government could not have used it. But with computers, our private and public appetite for information has created a policy issue.

The politics of information. Another difficult area is with the release of data produced by the government to those who want it for private gain. One of the outstanding economic spin-offs from the space program was the development of earth resource satellites, EROS and ERTS. These advanced information systems, using sensors, cameras, communication, and computers, yield economically valuable data on agriculture, fishery, forestry, and energy resources.

This type of information is produced by the government because there is a clear and understandable market failure at work. The private cost of producing the information far outweighs the private gain, as no firm has the capacity to launch a private ERTS satellite nor has a monopoly such that it could capture all the benefits of the information in the relevant markets. However, the social burden of launching a satellite, given that it represents a marginal cost to the space program, is overshadowed by the public benefit.[16] This rationale for the government's role as a producer of information was first argued by Kenneth Arrow[17] in a seminal article on the economics of information. The rationale, first used to justify government support of inventive activities, applies here also.

The government's attitude regarding public production of information is unclear. No information policy exists that requires the government to produce information when private markets fail to do so. However important as that issue is, it has been eclipsed by another: information policy regarding the release or disclosure of information.

The government produces and retains vast amounts of economically useful information. It is therefore besieged by two types of complaints: those who demand access to the information because it has private economic value, and those who would enjoin the government from releasing the information because it would constitute competition to the private sector.

In the first case, we may find an oil company suing under FOIA for data on energy resources, such as geological surveys produced by ERTS, because those data would help the company bid for drilling leases. In the second case, we may find a publisher who sells computer database information trying to stop a government agency that freely disseminates similar information. Or we may find an information vendor making a healthy profit by obtaining information from the public domain, packaging it in some fashion, and reselling it for private gain.

All of these questions revolve around the property rights to information. At this point, the society is working out these information policies on a case-by-case basis.

Policy in an Information Society

Social Equity: The Information Poor and the Information Rich

Each historical stage of economic development has its unique class structure. In the earlies agricultural societies, we find the chieftan ruling over a relatively communal society: There is not enough surplus generated to enable a more elaborate class division. Aside from the lord or priest, knowledge, skill levels (e.g., literacy and technical information), and wealth are evenly distributed.

In feudal agriculture, we find that ownership of land demarks the rich from the poor. The "landed gentry" absolutely control the existing political and economic systems. We also find a sharp division in the literacy rates between the rich (who are largely literate) and the poor (who are not). But the dominant variable is ownership of land. One's access to the written word could not compensate for family background.

In the early industrial society, the ownership of land is democratized, and the feudal structure is completely dismantled. The new locus of economic wealth and power is the factory. In this context, the ownership of capital dominates all other variables, including that of basic educational levels. The acquisition of superior trade skills, especially those with industrial machines, is the road to upward mobility. But without capital, few craftsmen and small businessmen ascend to the heights of industry.

In an information society, the ownership of capital recedes to the background, as ownership of land receded during the last great transition from an agricultural to an industrial society. The professional technocrat-scientist-manager assumes an ascendant role.[18] Success and failure in an information society are functions of one's ability to acquire, organize, and use information. Ownership of land or capital are, of course, determinants of success, but they share the stage with education and the facility to learn and regenerate. The technocrat as the new leader enjoys vertical mobility; generally has the good life; is extremely well educated, tested, and trained for both verbal and quantitative skills; is rewarded for prowess with abstract concepts and symbolic behavior; and is keenly attuned to new information about the world or his or her profession. Above all, the technocrat knows how to acquire information. An invisible college, an "old-boys' (girls') network," a consultant, a telephone call, a library request—somehow the information is available.

In stark contrast, 40 percent of high school graduates in Florida failed the math portion of the functional literacy test—they could not answer basic questions or grasp basic concepts.[19] In New York City even some high school teachers and principals are functionally illiterate. Some incoming freshmen at Ivy League colleges perform at the ninth grade reading level or below, and are immediately placed in remedial reading and math courses.

The nation is in the midst of an epidemic of functional illiteracy. It is not clear whether the problem has been unnoticed and unmeasured for decades, or whether the problem is newly aggravated. In either case, the interpretation of the data is alarming.

This issue suggests two areas of concern: the ability of functionally illiterate workers and consumers to function in an information society, and the role of the media (including advanced telecommunications systems) in mitigating the problem.

The effects of functional illiteracy are being felt in all sectors. We select the banking industry as one example. The rush of new low-cost computing and communication devices makes inevitable the coming of electronic funds transfer systems. The only question is when EFTS will arrive and how the marketplace for financial transactions will look after EFTS is in place.

The banking industry needs a steady supply (or oversupply) of clerks to keep pace with its rapidly intensifying use of information services and information machines (autotellers, automatic deposit and payment). The industry needs people who can be quickly trained to perform numerical and procedural tasks, who are comfortable with information machines, and who can learn new information tasks easily and with minimum error. That does not describe the average incoming class of high school graduates. The relevant work force is less able to cope with information, and is hopelessly unequipped to accommodate to the requirements of the banking industry. Forecast: higher turnover, higher training costs, higher error rates, and a higher labor bill at a time when the capital-to-labor ratio is supposed to shift toward capital.

The banking industry loses also from the inability of consumers to interact with the modern bank. Just as the family grocery store yielded to the department store and the street-corner franchise, so is the small bank yielding to the large department bank and the street-corner teller. In both cases, the consumer is more on his own. The equivalent of pushing the shopping cart—hence shifting labor costs onto the consumer—is the automated teller. The greater the involvement of the consumer in controlling financial transactions without the help of a clerk, the more efficient the operation from the bank's viewpoint. However, if the average consumer is unable to cope with complex instructions and equipment, the game is lost. The banking indus-

Policy in an Information Society

try cannot swamp the consumer's capacity to participate in the bank's own procedures.

The illustration of the banking industry's problems can be repeated for all institutions, both public and private. As the enterprise of management becomes more information intensive, the demand for skilled labor increases. But with an apparent failure of the educational system to quash functional illiteracy, managers should expect to find a sizable contingent of only marginally productive workers. Furthermore, institutions should be mindful of a massive inability on the part of consumers (or constituents) to cope with complicated instructions.

A more conventional issue in communication and educational circles is the function of the media in imparting knowledge and distributing information. We know that the less educated watch more television than the educated, and that children stare at television for more hours than they look at a blackboard. These findings are inconclusive as to the effects of television on the process of acculturation and maturation.

The problem at hand—functional illiteracy—is both aggravated and mitigated by television. On the one hand, a whole generation has seized on television as the primary information source. The command of the language has decayed, as has reasoning ability, because people who do not read well also do not write well. On the other hand, the average 14 year-old knows more about life (e.g., sex, intrigue, violence, and relationships) than his counterpart 100 years ago. The television addict is sophisticated at least at the level of fantasy, in a way that a television nonviewer is not. Television both robs the viewer of time and compresses time. Television dulls the mind, but also fills it full of rich images. The challenge, in the context of functional illiteracy, is to use the television media in a way that causes learning to take place.

The most immediate and pressing challenge is professional and vocational retraining. We inherited "the degree" as a curious institutional artifact from the public school system. Once the degree is achieved, the formal learning process ostensibly stops, and the rest is accomplished informally, on the job. For those who have learned how to learn, the informal system is quite pleasant. But for those who lack the facility, new information is absorbed with great difficulty.

An Information Underclass?

We end this section on a difficult note. It would be tragic for the United States to enter its third major era of social evolution saddled with a new underclass. But by all signs, we are in danger of doing so.

In the agricultural and industrial ages, women and blacks were relegated to the low-paying jobs. Vertical mobility was extremely limited. The work—in the fields or the factory—was physically and mentally brutal. Now, the cruel conditions of the factory have been largely banished. But has the underclass been eliminated? Within the information work force, there are creative, rewarding professions, and there are the routine jobs offering limited opportunities for personal growth or income.

The endless rows of women on the weaving looms have been replaced by endless rows of women in the modern "information factories"—check clearinghouses, insurance claims, billing, and accounting departments—where the only difference is the addition of air conditioning and computer terminals. Even though physical prowess is of no value in an information society, the discrimination of the past is cascading forward.

We noted previously that our information bureaucracies are glutted. This observation poses some serious problems for society. Any time that economic expansion is frozen, one must look at the existing distribution of income, wealth, and opportunity. These distributions in the information sector reflect the society as a whole. Although women make up a larger proportion of the information work force than they do of the industrial and service workforce (50 percent vs. about 24 percent), their tasks are generally less prestigious, more routine, less career oriented and less lucrative. The story for blacks and other racial minorities is even more grim. Their participation in the information sector is quite small, and their share of high-paying jobs is almost nil. Those who do not develop information skills are forced into those aspects of labor that are not as lucrative.

It would be a great waste if we enter the information society by repeating the injustices of the past. Our present trajectory will produce, by the turn of the century, a new underclass—the information poor.

The Global Communications and Information Picture

If small is necessarily beautiful, our planet has become a dazzling gem. We regularly see, hear, and feel events as they occur anywhere in the world. Live color television broadcasting by satellite has collapsed space; we can telephone our European friends almost as easily as our next-door neighbors. Our computers chat internationally without pause. A stream of information accompanies each commercial transaction. As the trade web increases in complexity, so does the communications web. International arbitrage is accomplished in seconds; contracts are sent across the world by facsimile in minutes; a steady stream of business letters, financial data, and invoices cross national boundaries. The trade winds are now electromagnetic.

The United States is the diva in this operatic Babel, upstaging and dominating an admiring but slightly jealous supporting cast. The U.S. media—television, radio, journalism—dominate the cultural communications field, while U.S. telecommunications and computers dominate the technological field, and U.S. knowledge and technique dominate the management field. But ultimately, we too are dependent on the goodwill, friendship, trade, and natural resources of the rest of the world. Interdependence, for better and worse, perfectly describes the communications and information arena. As Harlan Cleveland reminds, the opposite of interdependence is not independence, but dependence.

International communication and information problems are born of this uneasy tension between interdependence and dependence. To understand the international communication and information issues in their proper political contexts, it is useful to organize them into three conventional frameworks:

(1) *Problems between the first world and the third world.* Under the North-South rubric, we consider the issues in "the new world information order," and those relating to communication and economic development.
(2) *Problems between the first world and the second world.* Under the East-West rubric, we consider issues of national security and human rights.
(3) *Problems within the first world.* Under the Western alliance rubric, we consider the problems with our trading partners.

The larger framework of international relations controls the language and prejudgments in all arenas, be they defense or human rights policies. As such, it is not surprising that the complaints and suspicions that attend foreign relations at large also cascade into the communication field.

Before highlighting some of the policy issues, we note with some regret the logic that dictated this conventional method of organizing the field. The promise of communication is that it breaks geographical and political boundaries. The images of Hungary in 1956 or Dallas in 1963, or the Sea of Tranquility in

1969, or My Lai in 1970, or Munich in 1972 were global events. The world suffered or rejoiced as one community. But idealism is an incorrect starting point for an analysis of political reality. At the singular moment in history when political barriers are rendered meaningless by information technology, barriers to information flow are being fortified. The same tired dogmas that attend our relations with the third world, the second world, and our trade partners in the first world are replicated with fidelity in the communication and information arena.

Sixteen issues are highlighted in the following passage; many more could have been presented. My purpose is limited to extracting some of the essential conflicts, and demonstrating that the problems of our information society have global implications.

Problems Between the First and Third Worlds

The North-South dialog is essentially between the haves and the have-nots: the developed industrial economies of Western Europe and North America on the one side, and the aligned and nonaligned developing nations on the other side. The less developed countries have called for a New International Economic Order, described in a U.N. Declaration of Principles. The guiding principle behind the new order is a commitment by the developed nations to close the economic gap between North and South.

Mirroring the North-South dialog, this basket contains two clusters of issues: (1) the perennially durable "free flow of information" problems; and (2) the role of communication and information in promoting economic development.

The New World Information order. The United States clearly dominates the global information flows of television, film, radio, and wire services. The third world, in a series of international conferences, has complained that the flow of information is one way, and that the flood of U.S. media constitutes a modern form of "cultural imperialism." The complaint is that information about the third world is distorted and misleading because foreign correspondents and bureaus impose alien standards and faulty perceptions in their reportage. Also, information imported into a third world nation is usually insensitive to the needs and conditions of that country. Consequently, some nations have started to exercise control over information flows, both in and out of their borders. The United States, by contrast, has steadfastly defended the principle of "free flow," and this political clash has given birth to a number of issues.

Policy in an Information Society

The problems of free flow of information have arisen in a particularly sharp manner during U.N. discussions on direct broadcast satellites (DBS). A communications satellite, placed in geostationary orbit 22,300 miles above the equator, is theoretically capable of cost-effective transmission of video signals to remote regions. The fear expressed by some third world nations is that an uncontrolled DBS would encourage indiscriminate broadcasting of television and film programs that might insult the culture or injure the security of a receiving nation. Our diplomatically unsatisfying response was that U.S. media would not knowingly behave in that fashion. A more compelling answer is technical: The DBS "footprint"[20] can be designed to carefully circumvent broadcast spillovers into nations concerned about such incursions.
tions concerned about such incursions.

During the course of the debate, the position of the third world hardened on the issue of prior consent—insisting that sovereign nations have a right and an obligation to control the incoming and outgoing flows of cultural communication. The U.S. has argued in defense of "internationalizing" the concept of First Amendment rights, and holds prior consent in abhorrence.

Human rights aspects of free flow. Not only has the United States recently discovered that the First Amendment only applies within its borders, but it has also realized that the Fourth and Fifth Amendments also are not international. A country that exercises prior consent (i.e., censorship) of incoming news may not hesitate to control outgoing news and information. The control of radio, television, and film content is considered an absolute right by many third world nations. Sometimes this is simply accomplished by controlling the national press' reportage in a routine, administrative fashion. On occasion, the control of information takes on some grim aspects. Unofficial domestic news organs are regularly crushed, their publishers sent to jail, their journalists arrested or murdered. Local citizens who give information to foreign journalists are tried for treason, espionage, or "slander of the state." Dissidents who write journalistic pieces for foreign newspapers are incarcerated or simply vanish.[21] In the exercise of some nations' right to control media content, human rights are often expended.

Third World news agency. Several African and Latin American nations are in the early stages of planning a third world news agency. Two issues have emerged: a fear by the established press agencies that once such a news agency is created, some nations may find it expedient to threaten shut-downs of Western news agencies, such as AP, UPI, and Reuters unless the reportage is in

Communications for Tomorrow

line with government policy; and an awareness that the third world could productively use Western telecommunications and computer facilities. The first issue is current in press and media circles. Several administrations, dating back to that of President Kennedy, have spoken on the second issue. Promises for foreign aid have been made, but not carried out.

Sharing the electromagnetic spectrum. The electromagnetic spectrum is a finite resource of communication, whether used for radio, television, voice, or data. The U.S. and the U.S.S.R. have 15 percent of the world population, but use 50 percent of the spectrum. Developing nations are fearful of being shut out in the future, when their growing demand may be preempted by an already occupied spectrum.

In several meetings of the World Administrative Radio Conference (WARC), which allocates spectrum among nations, sharp disputes over "reservations" of both sprectrum and communication satellite parking orbits have remained unresolved.

The decisions of the WARC touch on both the direct broadcast satellite question and on prior consent. The developing nations that fear the U.S.-controlled DBS claim that their concern would evaporate if they are assured spectrum for their own use. That principle is now under negotiation. Under the third world interpretation, it means that spectrum parcels would be reserved exclusively under each nation's control. A solution needs to be devised guaranteeing that the needs of all nations, large and small, are met.

Passive satellite sensing. An immense information advantage enjoyed by the United States comes from its ownership of a vast earth resource intelligence system composed of scanning satellites, communication facilities, large computer processing facilities, and the associated scientific know-how to design and manage the system and to analyze the results. The ERTS and LANDSAT satellites produce vital information about crop conditions, the stocks and movement of fish, the growth and health of forests, and geological formations and deposits. These data are organized in a useful way by various agencies of the U.S. government and can be used to formulate American trade and investment policies.

The U.S. government and, to a lesser extent, U.S. corporations thus are armed with foreknowledge regarding market conditions, and can take advantage of the knowledge. The nations that do not have the analytical capabilites to reduce the data to useful knowledge, thus feel that they are giving away a precious resource (strategic economic information) without receiving much in return.

Lending policies. The traditional areas of foreign aid and lending to the developing countries have been in the transportation and electric infrastructures; in the agricultural, educational, and health sectors; and in heavy industry. Telecommunications has not commanded its share of development funds. Yet the central feature of a modern economy is a tightly integrated production system involving complex flows of commerce between firms, industries, regions, and nations. Each commercial transaction is necessarily accompanied by a transfer of information for negotiation, transmittal of prices, general economic intelligence, and follow-up. It is therefore argued that the development of an information infrastructure is a necessary precondition of development take off, as vital as the transportation and electrical infrastructures.

Several U.S. administrations have pledged technical assistance in the field of communication and information. However, very little of substance has materialized, and the third world is showing signs of impatience with our rhetoric.

Telephony has been construed by the World Bank, the International Monetary Fund, and the Export-Import Bank as generally ineligible for low-interest loans. The justification is that telecommunication projects generate internal rates of return that are competitive, and do not merit subsidy. However, most of the projects have been concentrated in urban centers, with little or no attention to rural telephony. Consequently, almost two-thirds of the developing world is substantially cut off from even basic electronic voice communication with cosmopolitan centers. The gap between urban and rural life can only widen.

Strategic implications of information goods exports. One of the healthiest sources of surplus in the U.S. balance of payments comes from the exports of information goods: computers, telecommunications, and related products. A single order from Saudi Arabia might exceed $1 billion, an order from Iran might top $2 billion. In the 1980s, the outflow of petroleum dollars will partly return to the United States as an inflow of information dollars from the oil-rich third world.

Unfortunately, information resources can be used for military purposes as well as for economic development. A computer that predicts climatic conditions for agrarian planning could also predict the climatic condition for an air, land and sea assault. A computer that monitors and controls the production of a fertilizer plant could accomplish the same purpose for a munitions plant. A microwave telephone system that links commerce between regional economic centers could also link armored units moving rapidly through a country.

The issue is to recognize the dual and joint applications of information resources, and to design a policy regarding their strategic uses.

Human rights implications of information goods exports. The dual purpose of information systems regrettably also implicates our human rights policies. A computer system that modernizes the paperwork of a foreign nation's treasury department can also be used by its police to keep track of dissidents. In less democratic contexts, "keeping track" can take on sinister overtones, resulting in torture or disappearance not only for the dissident but also for friends, colleagues, and family. Computers are efficient instruments for deducing social networks from scraps of information.

Computers can be combined with telecommunication systems in an extremely effective manner. The telephone numbers of known or suspected enemies of the state can be stored in a database. Whenever two dissidents talk to each other by telephone, their converstion can be immediately and automatically tape recorded at the central office. The easy political combination of a minister of communication and a minister of security can make the administrative arrangements trivial.

The U.S., which is by far the largest provider of information technology, thus is faced with a moral dilemma.

Information and technology transfer. The third world has complained loudly about the overwhelmingly one-way flow of cultural information. By 1973, all exports of film and television programs from the United States grossed $324 million.

By contrast, the export of scientific and technical information in 1973 grossed $3,034 million, almost 10 times as much.[22] Scientific and technical information, a derivative of the secondary information sector, is sold directly in the form of royalties, patents, and management and consulting fees. Developing nations, which often lack the human capital needed to manage modern productive systems, have come to rely on the foreign expert. The third world is well aware that this is a new form of economic dependence, vested with the trappings of neocolonialism.

But a critical twist in the technology transfer problem silences the neocolonial criticism: It is to the advantage of the third world to buy U.S. know-how and show-how, because producing that information domestically would be prohibitively expensive. The U.S. scientific and technical information is born of an advanced information-based economy. America's education system is mature, its firms produce vast amounts of management information, and its inventive system is healthy. Information is a natural and continual byproduct that is exportable. The third world's information infrastructure is impoverished. There is a dearth of scientific, technical, professional, and managerial talent; and even that which exists is often trained abroad in European and U.S. universities. The third world, at present, cannot afford to produce its own information.

The long-range economic and cultural effects of technology transfer can be profound compared to the effects of cultural exports. A particularly offensive or insensitive television program may anger the third world viewer or cause some flap in the local press. But the effects vanish after a few days. By contrast, the implementation of U.S. management science—complete with the regalia of profit maximization, an emphasis on capital intensification and managerial efficiency, a system of prices, and creation of financial institutions—these artifacts are enduring. Information is at the leading edge of technology transfer and modernization, and is completely consistent with the shared goal of economic development and improvement of the quality of life. But once the correlates of Western technology are in place, the entire society is inescapably altered.

Problems Between the First and Second Worlds

As we have seen, the North-South framework is rife with information and communication policy problems. The issue of economic equity between the haves and the have-nots is cascaded conceptually and rhetorically into issues of information equity. But information and communication conflicts also occur between the East and West.

Interception of telephone and data communication. An enduring first world-second world problem is the practice of the U.S.S.R. and the U.S. of eavesdropping on each others' microwave communications. The intelligence communities of both countries listen to private conversations of politicians, military leaders, business leaders, and the opposing country's secret police. The private conversations conducted inside a U.S. ambassador's limousine are recorded by the KGB; a conversation between the President and the chairman of the Senate Armed Services Committee is taped and transmitted to Moscow.

Communication satellites, such as INTELSAT, are particularly prey to unauthorized interception, as are all AT&T microwave long lines. Our technologists advise that very little can be done to stop the interceptions, unless we are willing to spend large sums of money for encoding or burying underground all transmission lines. The problem is one of national security, yet its basis and solution are rooted in technology.

Emergency preparedness. The concept of backyard bomb shelters is as stale as the K-rations that were stocked in them in the 1950s. But the thinkers of the unthinkable still worry about

Communications for Tomorrow

national communications in times of emergency. The problem is twofold.

First, an increasing reliance on communication satellites as the key component of both civilian and military communication raises questions about vulnerability. The science-fiction notion of "killer satellites" is now a reality; with a few instructions, an enterprising enemy can destroy all our communication satellites as a prelude to a nuclear strike.

Second, it is not clear that our civil defense communications can handle a crisis requiring universal communications between the President and the people. Some experts fear that with a coordinated sabotage attack, the nation's television, radio, and telephone networks could be rendered ineffective.

Both concerns fall in the realm of unlikely but catastrophic events. Even though the probabilities of occurrence are minuscule, the outcomes would be unconscionable.

Strategic exports. As in the third world, U.S. computer and telecommunication systems are widely sought by the second world. The Soviet Union's command economy is starved for market information, because the price signals do not (as in capitalist systems) embody useful information about demand, supply, or technology.

The Soviets have frequently expressed an interest in establishing a powerful national computer network to compensate for the paucity of market information. But the national security community, acting through the Department of Commerce, consistently blocks the exports of latest generation computers on strategic grounds. A computer that can accommodate the Soviet's economic planning models can also handle a coordinated attack by submarine launched missiles. Again, the twin natures of the information resource are inseparable.

Human rights aspects. During the Helsinki conferences of 1977, the United States clearly stated its position on the international rights of journalists, and on the rights of political dissidents to a fair trial. Unfortunately, we again discovered that the Bill of Rights does not apply east of Ellis Island.

The Soviets impose a different standard on what constitutes a crime. Recently, U.S. journalists who request and receive information embarrassing to the U.S.S.R. have technically been considered as spies. A problem in communication policy is the degree to which the United States can defend U.S. journalists in the U.S.S.R. in their line of duty.[23]

A related problem is the degree to which the official U.S. human rights posture can be blind to the treatment of dissidents who collaborate with U.S. media.

Problems Within the First World

The communication and information problems within the first world are largely ones of trade policy. The stakes are not ideological, inasmuch as they occur underneath the first world's military and economic umbrellas (NATO, Common Market, OECD). But sharp economic clashes have occurred, involving our closest trading partners.

Transborder data flows. U.S. computer manufacturers and software companies have established a strong presence in the European market. The trade volume of the two industries now exceeds $5 billion a year, and could reach $100 billion by the end of the 1980s. European firms are in a quandary, facing aggressive competition from well-financed and scientifically advanced U.S. companies.

Sweden, France, and West Germany have enacted strong laws that protect individuals from unwarranted invasion of privacy. Other countries, including the United States, have no comparable protections. A Swedish firm that keeps its data on foreign soil therefore enjoys a "data haven" not subject to Swedish law. These data havens, flying flags of convenience in the manner of a Liberian-registered ship, offer firms a flagrant loophole through privacy rules.

The United States, by virtue of its mature computer industry, is a particularly attractive data haven. Not only does the U.S. have no relevant privacy laws affecting the conduct of private industry, but it is also a source of cheap and reliable technical support for computing and teleprocessing.

In response to this situation, a convention or treaty to regulate and limit data flows across national borders is being discussed. The deliberations have brought U.S. and European firms into sharp conflict. The former accuse the latter of hiding behind a convention as a sham, where their true purpose is to quash competition. The Europeans accuse the U.S. firms of excessive paranoia, of protectionism, and of a basic insensitivity to the privacy issue.

The transborder data flow issue has a particularly difficult national security aspect. Many European firms and governments store crucial financial data outside their sovereign borders. The question arises: What happens in case of war, terrorist attack, or natural calamity if a firm or government is dependent on another country for the protection of its information and recovery of the information in case of damage or loss? Is it not unreasonable to expect a country to be as diligent and conscientious in protecting foreign data as its own?

Free flow of information. A special case of the free flow of information issue occurs between the United States and Canada. Our television programming is embarrassingly popular in Canada, consistently winning in the ratings battle with domestic programming. Nationalistic feelings, especially in French-speaking Quebec, are running high. A border information battle has started, with nationalists trying to close the border to U.S. television, and almost everyone else in favor of the current arrangement.

Another case of free flow is found in the public television arena. Students of the media have reported that only 3 percent of American television is given to foreign programming, about the same ratio as Chinese and Soviet television give to U.S. programming. The United States, it is alleged, is insulated from the rest of the world. Efforts are currently under way to form a media exchange "bank." Under the scheme, nations would donate and borrow programming, with special "credits" given to needy countries.

Competition on the goods markets. Another trade issue facing the first world is the fierce competition in the consumer goods markets for televisions, radios, citizens band, transceivers, high fidelity equipment, calculators, video tape recorders, and even home computers. The conflict mostly involves the United States and Japan.

Japan, which is starting to call itself "the second information society" (after the U.S.) is a prolific producer of low-cost electronic goods. As with other Japanese products, such as automobiles and steel, the export prices are set with the blessing of the Japanese government. Pricing below marginal cost, internal cross subsidy, and tax leniency are legal devices endorsed by the Japanese government for acquiring foreign markets. The government does not prosecute the same practices which in the U.S. would be subject to Sherman and Clayton antitrust provisions. In fact, the Japanese government actively organizes and disciplines trade cartels as a matter of domestic economic policy.

In 1975, the United States imported $2,076 million worth of radios, televisions, audio equipment, and calculators. The exports amounted to $652 million or less than one-third the imports. In the telecommunication and computer markets, the U.S. holds a clear worldwide advantage. Computer exports in 1975 were $2.2 billion, telecommunications exports were $1.1 billion, and the two industries together generated a $2.4 billion surplus in the balance of payments.

Note that there is nothing inherently wrong with running a deficit. If Japan holds a comparative advantage over U.S. manufacturers in producing information goods, then the U.S. should encourage the importation of the cheaper goods. Our economy would benefit double: Directly, the consumer would benefit from the cheap price, and indirectly, resources would be free to move into a sector of the economy where the U.S. holds a comparative advantage. However, U.S. manufacturers are alarmed at the loss of market share, and would like the government to impose protective tariff or nontariff barriers.

Summary

The United States is transforming from an industrial to an information economy. Twenty-five percent of the Gross National Product in 1967 originated in the production, processing, and distribution of information goods and services. In addition, 21 percent of the GNP originated in the provision of information services consumed internally by noninformation organizations.

The big wheels of social change are rooted in scientific and technological advances. The little wheels are ideological. Revolutionary changes in information technology, involving computers and telecommunications, will cause upheavals at the level of markets, institutions, law, and politics.

Communications policy is concerned with rules regarding the structure and performance of the information industries. Information policy is concerned with the applications of information technologies and techniques as they affect the behavior of industries, governments, and individuals. The need for social judgments about the flow of information, the value of information, and access to information are pressing on the political agenda.

Finally, we should remember that the emergence of the U.S. as an information society is a global matter, and that the baggage of international relations will accompany us into the future.

Notes

1. Daniel Bell, *The Coming of Post-Industrial Society*, (New York: Basic Books, 1973, pp. 126-7.) Daniel Bell's earlier writings on the "post-industrial society," dating back to the Salzburg Seminar in 1959, were reflected in a number of subsequent books. See also Z. Brzezinski, *Between Two Ages: America's Role in the Technetronic Era* (New York, 1970), and Fritz Machlup, *The Production and Distribution of Knowledge in the United States* (Princeton: Princeton University Press, 1962).

2. For a detailed description of 440 occupational types, see Marc Porat, *The Information Economy* (Palo Alto, California: Stanford University, 1976; and Washington, D.C.: Government Printing Office, 1976), Chapter 7, Vol. 1.

3. For a detailed description of the primary and secondary information sectors, see Porat, *op. cit.*, Volumes 1 and 2.

4. The National Income and Product Accounts, the 1967 Input-Output Worktape, the BLS Industry by Occupation matrix, the BEA Capital Flow matrix, IRS data, and private industry data.

5. In June 1977, President Carter sent a reorganization plan to Congress, proposing to create a new National Telecommunication and Information Administration (NTIA) in the Department of Commerce. In April 1978 the NTIA assumed the authorities and resources of the Office of Telecommunications Policy and the Office of Telecommunications (Department of Commerce). The new assistant secretary is the President's principal adviser on telecommunication and information policy.

6. AT&T's 1977 revenues ($36 billion) exceed the GNP of: South Africa ($35 billion), Indonesia ($32 billion), Venezuela ($32 billion), Romania ($31 billion), Norway ($30 billion), Finland ($27 billion), Hungary ($24 billion), South Korea ($24 billion), Greece ($23 billion), Bulgaria ($20 billion), Portugal ($17 billion), Algeria ($16 billion), Libya ($16 billion), Kuwait ($16 billion), Colombia ($15 billion), Israel ($14 billion), and 102 smaller nations. Source: *World Bank Atlas*, World Bank, 1977, pp. 27-30.

7. Technically, the sum of employee compensation (labor) paid to information workers and depreciation taken on information machines in the noninformation industries; PLUS an estimate of the information labor and capital consumed in the management of the information industries; PLUS the information labor and capital consumed by governments. See Porat, *op. cit.*, pp. 175-183.

8. M. Mintz, "Drug Firms Oppose U.S. Price Audit," *Washington Post*, July 9, 1976.

9. Based on an affidavit filed by AT&T on July 11, 1977. The $1 billion covers discovery, pretrial and trial expenses. See also, Stephen m. Aug, "AT&T Ordered to Supply Papers in Antitrust Case," *The Washington Star*, May 3, 1978.

10. Schumpeter, in *The Rise and Fall of Capitalism*, argued that firms go through an almost biological cycle of "creative destruction." They start out small, aggressive, eager to learn and grow. Their scientific and technological edge over established firms provides the wedge into the marketplace. With entry comes the consolidation of market position, and an emphasis on R&D to keep an informational advantage over would-be entrants. Eventually, the market domination earns the firm a place in an oligopoly, and the firm's "knowledge stock" begins deteriorating through inattention. New entrants appear and establish themselves as "high technology" companies, and the established oligopolist begins to deteriorate.

11. In Daniel Bell, *op. cit.*, pp. 82-83.

12. In the manner of the physician, who over-consumes diagnostic evalutions and laboratory tests as a hedge against malpractice suits.

13. Samuel D. Warren and Louis D. Brandeis, "The Right to Privacy," *Harvard Law Review*, Vol. 4, No. 5, December 15, 1890, pp. 194-220.

14. Paul Karger and Roger Schell, "MULTICS Security Evaluation—Vulnerability Analysis," ESD Technical Report 74-193, Vol. 2, June 1974.

15. Although hotly contested by the Federal Trade Commission, which argued in front of the District of Columbia Court that the economic burden is negligible, see "FTC Corporate Patterns Report Litigation," CCH 1977-2, Trade Cases, Para. 61544.

16. This assertion is being researched and documented in the fields of resource satellites (by the National Aeronautics and Space Administration), agricultural commodity information (by the Department of Agriculture) and energy information (by the Department of Energy). Cost benefit ratios of 4.3:1 to 9.3:1 are estimated for the LANDSAT, which gives information on agricultural crops, petroleum and mineral exploration, hydrolgic land use, water resources management, forestry, land use planning and soil management. For a discussion of methodology, see "A Cost-Benefit Evaluation of the LANDSAT," March 1977, NASA Doc #X-903-77.

17. Kenneth Arrow, "Economic Welfare and the Allocation of Resources for Invention," in *The Rate and Direction of Inventive*

Activity: Economic and Social Factors, National Bureau of Economic Research, Princeton University Press, 1962, pp. 607-626. For a review of the subject, see Porat, "The Information Economy and the Economics of Information: A Literature Survey," Stanford University Center for Interdisciplinary Research, Stanford, California, May 1975.

18. The theme is elaborated in Kenneth Galbraith, *The New Industrial State,* Signet Books, New American Library, New York, 1967. See also Peter Drucker, *The Age of Discontinuity,* New York: Harper and Row, 1968.

19. See *New York Times,* "Math a Big Problem for Florida Schools: Failure Rates of 40 to 50 percent is reported," December 6, 1977. The Adult Performance Level Project, reporting on a study for the U.S. Office of Education, concluded that 23 million people—one out of five adult Americans—are functionally illiterate. See Edward Fiske, "Illiteracy in the U.S.: Why John Can't Cope," *New York Times,* April 30, 1978. Similar statistics were developed by the National Assessment of Educational Progress, which estimated that 13 percent of U.S. 17 year-old high school students are functionally illiterate. See Edward Fiske, "Controversy is Growing Over Basic Academic Competency of Students," *New York Times,* April 19, 1978.

20. A "footprint" is the shape of a satellite beam as it is projected on earth. The coverage pattern can be accurately designed by changing the antennae shape and orientation.

21. Article 19 of the United Nations Universal Declaration of Human Rights asserts the right to "receive and impart information through any media." On March 15, 1977, Amnesty International published a list of over 100 journalists imprisoned (often without charges) or disappeared. See Amnesty International "Journalists in Prison," March 15, 1977. See also "Argentina: Repression of the Media Since the Military Coup of 24 March 1976," AI Index AMR 13/05/78. (Excerpt: "In April 1977, *Le Monde* reported that, since the military coup in Argentina, 17 journalists had been killed, 22 had disappeared and 35 imprisoned." In January 1978, *Le Monde* reported that "... 29 journalists had been killed, 40 had disappeared, 70 had been imprisoned, and 400 gone into exile.")

22. Porat, *op. cit.,* Vol. 8.

23. Amnesty International, *op. cit.*

Policy in an Information Society

Appendix Table 1-1 Gross National Product Originating in Primary Information Sector, 1967

	Total Value Added ($ Millions)	Primary Information Value Added ($ Millions)	Primary Information Percent of Total
All industries, total (GNP)	795,388	200,025	25.1
Agriculture, forestry and fisheries	26,733	0	0
Mining	13,886	0	0
Contract construction	36,102	8,527	23.6
Manufacturing	223,729	32,691	14.6
Nondurable goods	90,595	11,762	13.0
Paper and allied products	8,005	1,539	19.2
Printing and publishing	10,718	10,223	95.4
Other	71,872	0	0
Durable goods	133,134	20,929	15.7
Furniture	3,380	528	15.6
Machinery, excluding electric	23,980	3,198	13.3
Electrical machinery	19,959	12,123	60.7
Instruments	5,606	4,309	76.9
Miscellaneous manufacturing	3,305	771	23.3
Other	76,904	0	0
Transportation	32,040	0	0
Communication	17,632	17,609	(a)
Telephone and telegraph	16,024	16,029	(a)
Radio broadcasting and television	1,608	1,580	(a)
Electric, gas and sanitary services	18,429	0	0
Wholesale and retail trade	129,863	16,053	12.4
Wholesale trade	51,802	8,584	16.6
Retail trade	78,061	7,469	9.6
Finance, insurance, and real estate	108,840	41,425	38.1
Banking	11,843	11,731	(a)
Credit agencies, holding and other investment companies	−437	−790	--
Security and commodity brokers	3,582	2,779	77.6
Insurance carriers	7,822	8,826	(a)
Insurance agents, brokers and services	3,344	3,485	(a)
Real estate	82,686	15,394	18.6
Services	86,992	43,021	49.4
Personal and miscellaneous repair services	9,751	853	8.7
Miscellaneous business services	11,919	10,703	89.8
Motion pictures	1,690	1,525	(a)
Amusement and recreation services	3,607	485	13.4
Medical and other health services	21,392	5,754	26.9
Miscellaneous professional services	12,738	12,183	95.6
Educational services	5,446	5,170	(a)
Nonprofit membership organizations	7,527	6,348	84.3
Other	12,922	0	0
Government and government enterprises	95,827	40,699	42.5
Federal	40,559	15,771	38.9
General government	35,865	10,232	28.5
Government enterprises	4,694	3,539	75.4
State and local	55,268	26,928	48.7
General government	49,222	26,928	54.7
Government enterprise	6,046	0	0
Rest of the world	4,510	0	0
Statistical adjustment	802	--	--

Note: Minor discrepancy between the National Income Accounts and the Input-Output Worktape—100% of industry allocated to primary information.

Communications for Tomorrow

Appendix Table 1-2 Gross National Product Originating in Secondary Information Sector, 1967

	($ Millions) Total Value Added	Secondary Information Value Added[a]	Secondary Information Percent of Total
All industries, total (GNP)	795,388	168,073	21.1
Agriculture, forestry and fisheries	26,733	467	1.7
Mining	13,886	1,512	10.9
Contract construction	36,102	13,243	36.7
Manufacturing	223,729	57,880	25.8
Nondurable goods	90,595	21,044	23.2
Food and kindred products	22,340	5,248	23.5
Tobacco manufacturers	3,490	254	7.3
Textile mill products	6,619	1,373	20.7
Apparel and other fabricated textile products	7,816	2,670	34.2
Paper and allied products	8,005	2,109	26.3
Printing, publishing and allied industries	10,718	565	5.3
Chemicals and allied products	16,687	5,266	31.6
Petroleum refining and related industries	7,050	1,337	19.0
Rubber and miscellaneous plastic products	5,626	1,702	30.3
Leather and leather products	2,244	520	23.2
Durable goods	133,134	36,836	27.7
Lumber and wood products except furniture	4,873	1,069	21.9
Furniture and fixtures	3,380	777	23.0
Stone, clay and glass products	6,597	2,035	30.8
Primary metal industries	18,009	4,350	24.2
Fabricated metal products	14,674	4,681	31.9
Machinery, except electrical	23,980	7,259	30.3
Electrical machinery	19,959	3,273	16.4
Transportation equipment and ordnance except motor vehicles	16,417	8,771	53.4
Motor vehicles and equipment	16,334	3,116	19.1
Miscellaneous manufacturing industries	3,305	1,140	34.5
Instruments	5,606	365	6.5
Transportation	32,040	8,115	25.3
Communication	17,632	0	0
Electric, gas and sanitary services	18,429	2,612	14.2
Wholesale and retail trade	129,863	42,447	32.7
Finance, insurance, and real estate	108,840	3,341	3.1
Real estate	82,686	2,764	3.3
Other	26,154	577	2.2
Services	86,992	19,204	22.1
Hotels personal and repair services except auto	14,307	3,740	26.1
Miscellaneous business services	11,919	6,535	54.8
Automobile repair and services	3,889	1,376	35.4
Motion pictures	1,690	0	0
Amusements except motion pictures	3,607	780	21.6
Medical, educational services and nonprofit organizations	34,365	6,773	19.7
Other	17,215	0	0
Government and government enterprises	95,827	18,735	19.6
Federal	40,559	7,693	19.0
General government	35,865	6,357	17.7
Government enterprises	4,694	1,336	28.5
State and local	55,268	11,042	20.0
General government	49,222	9,601	19.5
Government enterprises	6,046	1,441	23.8
Rest of the world	4,510	517	11.5
Statistical adjustment	805	--	--

Note: (a) Includes labor income of information workers and capital consumption allowances on information machines.

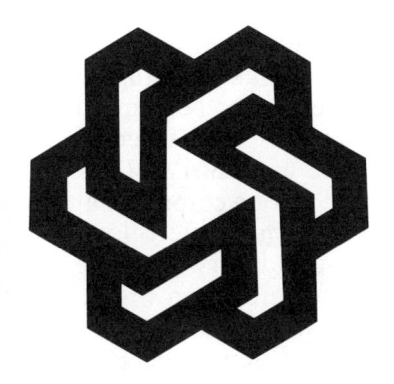

2.
Telecommunications Technology in the 1980s

Walter S. Baer

If technology is indeed a main engine of social change, we might be curious as to what lies ahead. This chapter describes some of the advances in telecommunications technology that can be anticipated during the 1980s. But what follows is intended neither as a technological forecast nor as an assessment of present and future technologies. In a field as rich and rapidly changing as telecommunications, any attempt to predict the technology of 1990 is doomed to failure. One need only look back an equivalent time span—from 1977 to 1964—to see the difficulties inherent in such forecasting.

In 1964, the technologies that seem most significant today were little more than laboratory developments or systems in their early experimental stages. Although communication by satellite had been successfully demonstrated with the Telstar, Relay, and Syncom projects, the first commerical satellite system was still a year away from launch. No one knew how well commercial communications satellites would perform, or for

Communications for Tomorrow

how long they would operate reliably in orbit. Computers were in the midst of a technological revolution from electron tubes to transistors, but computer users still brought stacks of punched cards to a machine room and waited for the results to be printed out on paper. Computer time-sharing was in its first stage of commercial introduction by General Electric and IBM. AT&T was looking ahead to its first installation in 1965 of a commercial computer-controlled, electronic switching system. Cable television had scarcely begun to attract notice as an industry with the potential for far more than retransmission of television broadcasts in areas that lacked adequate broadcast signals. And although the invention of the laser in 1960 had generated new interest in optical communications, practical applications seemed many years away.

This snapshot from 1964 suggests the pitfalls inherent in trying to track the precise path of technological evolution. Still, a look backward suggests some generalizations over the past 13 years that may extend into the future. First, new technological developments have primarily been applied to improve existing communications services rather than to create new ones. Television transmission by satellite across the oceans is one of the few examples of a new communications service introduced since 1964. Most technological advances have been used to reduce costs and improve the performance of services and products that were already offered—such as long-distance telephone calls, data transmission, and color television sets.

A second, related observation is that technological advance in telecommunications has been incremental and often invisible to those outside the field. Microwave and coaxial cable transmission of telephone traffic are two examples of technologies that have shown steady, incremental progress. Of course, over a period of a decade or more, gradual technical improvements can make large differences in the availability, performance, and cost of communications services. Thus, evolutionary changes can appear to outsiders as a communications "revolution." The Marxian view that quantitative changes cumulate to qualitative, revolutionary advances may have more technological than historical examples.

Third, even after new telecommunications technologies have proved economically advantageous, they take a long time to diffuse into widespread use. By 1964, electronic switching, satellite communications, and digital transmission lines were all reasonably well-developed technologically. Their introduction and commercial growth have been determined more by requirements for compatibility with the existing network, by the availability of capital, by depreciation policies, and by regulatory constraints than by their technical "ripeness." A time scale of 20 years or more from commercial introduction to mature saturation seems typical for a system-oriented and highly regulated industry such as telecommunications. Technological advances have been introduced more rapidly in the less regulated industries, such as computing.[1] Consequently, one can expect that the technological developments to be introduced into telecommunications in the 1980s are available in laboratories and experimental systems today. Most have already been widely discussed in the technical literature.

This chapter is based on the published literature, on my knowledge of developments under way in corporate and government laboratories, and on my discussions with many people personally involved in telecommunications technology. However, the conclusions basically reflect my own judgment about the pace and direction of technological developments in telecommunications. And, of course, history will undoubtably prove these conclusions wrong in many respects.

A few additional caveats are in order. The chapter describes technologies which are likely to find application in commercial systems in the 1980s, not those that seem more likely to remain in the laboratory or in the experimental stages. Consequently, the discussion reflects judgments about economic as well as technical feasibility (all costs are estimated in 1978 dollars). Despite their importance, this chapter does not consider regulatory and other policy issues in any detail. Many of these issues are treated in subsequent chapters of the book. Finally, the chapter does not systematically cover military developments or those primarily under way in other countries, except where those technologies seem likely to be applied commercially in the United States by 1990.

Computer and Component Technologies

The "microprocessor revolution," "intelligent terminals," and "teleprocessing networks" are by now familiar cliches. These terms reflect a convergence of communications and computing technologies that, of all technological changes, will have the most profound effects on telecommunications in the 1980s. This technical convergence—the combination of information-processing (computing) and communications functions in new hardware devices and systems—has already begun. Telecommunications systems are extending and integrating their use of computers for storage of information, conversion of electronic signals from one form to another, switching, and network control. Computers increasingly interconnect through communications networks to provide their users with access to more computing power and sources of information.

The growing use of information processing in communications systems is due to the dramatic improvements in cost, performance, and the size of computer hardware. An information-processing system contains three basic units: the computing or processing unit, memory systems for storing information, and input/output units for maintaining external contacts. The costs of the information-processing and storage units have fallen by a factor of three every two years or so for the past 20 years—a total cost decrease of more than 10,000-fold. This means that a few hundred dollars in 1978 buys the equivalent computing power of several million dollars in the mid-1950s. The size of processing and storage units have also shrunk by a factor of about 10,000, while their speed (measured in the number of instructions or calculations processed per second) has increased approximately 50,000-fold. Manufacturers expect these trends to continue at least into the early 1980s.

The Computer on a Chip

Computer processors have gone through three major technnological changes in components—from vacuum tubes to transistors, from transistors to integrated circuits, and from the early forms of integrated circuits to today's large-scale integrated (LSI) circuits. The LSI manufacturing process uses photolithographic and chemical techniques to build tiny regions with different electronic properties on a thin silicon "chip." These regions form the transistors, diodes, resistors, and connecting pathways that make up an LSI device. A computer processor built with LSI technology is called a microprocessor.

The advantages of LSI technology include the high density of components that can be packed on a single chip, greater reliability and less power consumption than individual components, and low incremental production costs. The first microprocessor chip, introduced in 1972, contained about 5,000 components. *(see Table 2-1)* With today's state-of-the-art, about 18,000 components can be placed on a chip that is roughly one-half a centimeter (0.2 inch) square. Such a microprocessor contains about four times as many components as a hand calculator and has more computing power than the largest computers assembled 25 years ago.

Technology in the 1980s

By the early 1980s, continuing improvements in LSI technology should yield commercial logic devices with roughly 60,000 components per chip.[2] These microprocessors will be able to execute four million instructions per second—roughly four times the speed of microprocessors today and 100 times faster than the small computers of the early 1960s. The cost of such devices depends critically on the number produced, but with volume production, they should cost less than $100 by 1980. Smaller capacity microprocessors that now cost $10 to $20, such as those used for programmable calculators and television games, should be available to manufacturers for between $1 and $5.

Today, engineers design new products with the assumption that computer-logic hardware is essentially free. The software is not free, however, and software costs increasingly dominate the expense of building sophisticated logic and controls into other devices. Microprocessors are now incorporated into scales, cash registers, machine tools, and most electronic instruments that cost $1,000 or more. Electronic calculators and home television games are the first consumer products to include microprocessors, but the technology is also being rapidly applied to automobiles and such other consumer items as ovens, washing machines, and dishwashers. And as described below, microprocessors are finding their way into communications equipment of all sorts, including point-of-sale registers, private switching facilities (PABXs), typewriters, data terminals, and even the telephone itself. The microprocessor signifies the demise of the centralized computer controlling the equipment connected to it. In its place, "intelligent" terminals and devices distribute logic throughout a communications network and perform complex tasks under the control of their own, built-in microprocessors.

Memory Devices

Computer processors are useless without information-storage devices, or *memory systems*.[3] Memory systems are of two basic types: primary memory units, which store the processor's instructions and the basic data needed for its operation, and secondary, or mass memory units, which store larger amounts of data.[4] The information stored in a processor's primary memory must be accessible at a speed that matches the processor's own operations. For the past 25 years, magnetic core has been the principal primary memory technology, but microprocessors now use LSI memory chips for primary storage (see *Table 2-2*). LSI memory chips currently cost $1 to $2 per 1,000 bits. At this price, a 16,384-bit memory chip (designated in technical jargon as a

Communications for Tomorrow

"16K" memory) costs about $20 to $30. One such LSI memory chip can store all of the alphanumeric characters from one double-spaced typewritten page or one 24-line graphic terminal display. LSI memory could also be used to store a television picture at the television receiver—a device known as a *frame-grabber*—but the cost of such a device today would be several thousand dollars.[5]

Table 2-1
The Evolution of Microprocessors

Microprocessor Characteristics	1972	1978	Estimated Early 1980s	Estimated Late 1980s
Components per chip	5,000	18,000	60,000	>100,000
Component dimension (millionths of a meter)	10	4	2	1
Speed (million instructions per second)	.1	1	4	>10
Cost per logical unit or "gate" (in cents)	5	1.5	.4	.04

LSI memory costs are dropping even more rapidly than those of microprocessors, so that one expects to see LSI memory chips available in the early 1980s for 30 cents per thousand bits or less. Further projections are more speculative, of course, but most industry sources expect that well before the end of that decade, LSI memory chips will cost less than 10 cents per thousand bits.

Secondary, or mass storage units hold the greater quantities of information that computer systems draw upon for data processing or information retrieval. Names and phone numbers in a telephone directory, payroll accounts, spare-parts inventories, and airline reservations illustrate the kinds of data found in mass storage units. Mass storage need not have the access speed of primary storage, but the memory units should be cheap and reliable.

Costs vary with the mass memory unit's size and speed of access. Magnetic disc memories, which are in common use for airline reservation and similar information-retrieval functions, store more than one billion bits at a current cost of around 4 cents per 1,000 bits, with retrieval times of less than one-tenth of a second. Magnetic tape memories store larger amounts of data at less than one-hundredth of the cost, but with slower access times of several seconds. Consequently, magnetic tape memory units are more appropriate for less urgent tasks, such as payroll preparation or inventory management, than for "real-time" information-retrieval applications.

Technology in the 1980s

Magnetic discs, drums, and tapes are today's principal mass storage media and will remain competitive for many years.

Table 2-2
Digital Storage Technologies

Function and Capacity	Principal Technologies 1978	1985
Primary computer memory units (10,000 to 1 million bits; 1 microsecond access time)	(1) Magnetic core (2) LSI circuits	(1) LSI circuits (2) Magnetic core (3) Charge-coupled devices
Rapid-access mass memory units (more than 1 million bits; .1 second access time)	(1) Magnetic drum (2) Magnetic disc	(1) LSI circuits (2) Magnetic bubble devices (3) Charge-coupled devices (4) Magnetic disc
Large, inexpensive mass memory units (more than 1 billion bits; several seconds access time; $.01 per 1,000 bits)	(1) Magnetic tape	(1) Magnetic tape (2) Optical devices

Magnetic tape cassettes and small magnetic disc units (termed *diskettes* or *floppy discs*) are now manufactured to provide secondary storage units appropriate for small computer systems and intelligent terminals. However, other technologies now under development seem likely to displace these rotating magnetic devices for many applications in the next decade.

Magnetic bubbles and charge-coupled devices, as well as LSI memory chips, are the principal competitors for read-and-write memories. Magnetic bubble devices contain tiny magnetic regions that can be created, destroyed, and moved about in a garnet-like material. Charge-coupled devices store and transfer information as a series of electric charges within a semiconductor chip. These technologies will compete on the basis of cost, access speed, and reliability. Magnetic bubble devices also have an additional advantage of retaining information even if the external source of power is removed. Consequently, bubble memories provide data security in the event of electrical generating failure or other power losses.

Optical readout devices also appear promising for low-cost, slower-access mass storage. At present, optical storage devices are used only for large, specialized applications. But the new optical storage media, such as videodiscs, will have more general uses. Although educational and entertainment programming seem the principal initial markets for videodiscs, they can also be recorded with digital data for computer processing instead of analog signals for video playback. Costs are projected to be competitive with, or lower than, the cost of magnetic tape storage by the early 1980s. RCA and the Digital Recording Corporation, among several companies, appear optimistic about the uses of videodiscs for digital data storage.

Input and Output Devices

The equipment to move information in and out of computer systems is also improving in performance and cost, but at a far slower pace than microprocessors and memory units. Input and output devices dominate the cost of intelligent terminal equipment linked to teleprocessing systems. And as new users of teleconferencing or computer time-sharing services soon discover, many problems remain in effectively communicating with information processing sytems. These human-factor problems will continue to place limits on the widespread acceptibility and use of computer systems throughout the 1980s.

Input devices. The typewriter keyboard or its equivalent provides the usual way to enter data into a computer. Some low-cost systems have tried to develop data-entry procedures using push-button telephones or special equipment with fewer keys than a typewriter. In principle, a 12- or 16-key device with special "shift" or function keys to generate additional characters can handle input requirements. However, these methods have often proved confusing and difficult for the unsophisticated users for whom the low-cost devices were specifically designed.

For some applications, such as pay-television and viewer responses to multiple-choice questions, a simple terminal with fewer push-buttons will serve adequately. Warner Cable is currently experimenting with a five-button terminal for pay-television, alarm services, direct ordering of advertised merchandise, and some simple interactive games. However, this kind of device is too limited for sending messages or for most information-retrieval services.

The typewriter keyboard or its equivalent thus remains the standard input device. Standard keyboards cost product manufacturers about $50. Since typewriter keyboards are a mature electro-mechanical technology that has already passed through many generations of product improvement, costs are unlikely to decrease significantly in the future. Some manufacturers are experimenting with lower-cost, lower-quality keyboards for home computer systems, but their acceptance by consumers remains in doubt.

Other technical approaches for entering information are available at higher cost. These include facsimile scanners, optical character-recognition devices, magnetic-ink recognition devices (used to process bank checks), and code sensors (used for point-of-

Technology in the 1980s

sale transactions). Devices more suitable for inexperienced users include screens and tablets that respond to the touch of a finger, or to a pointer or light pen. Such devices have been used experimentally for many years and are available in military display terminals and other high-cost equipment. They may become practical for low-cost commercial devices in the 1980s.

The most important technical breakthrough in input equipment would be an inexpensive, reliable, voice-recognition device. Most users would find speaking a far easier and more convenient way to issue commands and enter information. Speech-recognition equipment is under development in many laboratories throughout the world, and a few simple systems are now used for such routine operations as stopping and starting luggage conveyors at airports.

The pace of development suggests that more applications of voice recognition for the control of machines and equipment will soon be commercially feasible. A device for voice *identification* (for example, ensuring that a person making a financial transaction is authorized to do so) can be anticipated in the 1980s. A "voicewriter" that can provide acceptable rough draft copy from voice dictation seems a likely prospect by 1990. One cannot yet estimate when voice-recognition devices will be cheap and reliable enough for computer time-sharing or home terminal applications. However, this is one area where rapid technological development could bring about significant expansion and as yet unforeseen uses of computer/communications systems.

Voice and soft-copy output devices. Information is received from a computer system in three basic forms: (1) voice (or other audio) responses, (2) soft-copy (nonpermanent) visual displays, and (3) hard-copy records. Voice response is the cheapest and easiest way to send uncomplicated messages to large numbers of persons, because it requires only the ubiquitous telephone at the user's premises. Voice responses are generated from either prerecorded or electronically synthesized words combined under the computer's control. Equipment of both types is commercially available, and voice response is commonly used today for services that need only a push-button telephone for data entry—computer-based stock market inquiries, credit checks, and changes in telephone listings are three examples. Many new uses will emerge as computer-based mass storage systems become cheaper and accessible to more people. However, voice response has obvious limitations and is less desirable than displays or hard-copy output for most applications.

The simplest kind of soft-copy output is a single-line display of a few numbers or letters using light-emitting diodes (LEDs) or

Communications for Tomorrow

liquid crystals, such as those built into hand calculators. For a few dollars, these displays can be added to an "intelligent telephone" or to any other device containing a microprocessor. Much more versatile is the cathode-ray tube (CRT) which forms the heart of today's TV receiver and the standard computer-display terminal. The CRT's capacity for displaying a changing image or multiple lines of alphanumeric characters is well-known. Small CRTs can now be built for less than $50, but significant cost decreases in the future seem unlikely for this highly mature technology which already enjoys large economies of scale.

The principle technological change in this field in the next decade will be the replacement of CRTs by flat, solid-state panels for both alphanumeric and image displays. Flat display panels do not have the inherent size limitation of CRTs and should therefore lend themselves to large-screen television receivers as well as portable computer terminals. Image quality will also be considerably better than that provided by the current large-screen television-projection devices.[6]

Several different technologies are presently in contention to succeed the cathode-ray tube:

- Plasma (gas discharge) panels that contain an inert gas between two transparent plates. A voltage applied between the plates discharges the gas and produces a bright spot.
- Light-emitting diodes which use the same semiconductor materials as those in calculator and digital watch displays. They emit light when an electric current is applied. Building a large screen with individual LED elements would be too expensive, but methods of producing LED arrays as a single unit are under development.
- Liquid crystals, also used for calculators and watches, inserted as a thin layer between two glass plates. When exposed to an electric field, liquid crystals reflect rather than emit light and hence can be viewed in brightly lit areas.
- Ferroelectric ceramics, materials whose reflective indices for polarized light change with an applied electric field.
- Electroluminescent materials, which emit light when excited by an electric field.
- Magneto-optic films, which deposit on a grid of conducting wires scatter light selectively when current pulses are applied.

Technology in the 1980s

It is too soon to tell which of these technologies will be successful, or when. All are being pursued vigorously in various laboratories. Plasma panels have been manufactured commercially, but the costs are still too high to compete effectively with CRTs. Well before 1985, however, most industry observers believe that solid-state flat panels will gradually begin replacing CRTs in computer-display terminals. Introduction will begin with the high-priced, top-of-the-line equipment and spread gradually to lower-cost displays. By the end of the decade, some portable, battery-powered computer terminals should also be commercially available, but low-cost, portable terminals suitable for the home are not expected before 1990.

Unless there are technological breakthroughs or dramatic cost reductions, flat panels are not expected to make serious inroads into the home television receiver market before 1990. Large-screen, flat-panel television receivers should begin to appear on the market in the mid-1980s, but the costs of such receivers seem likely to remain well above $1,000 throughout the decade. This will place them beyond the reach of the mass market but available for business, commercial, and government users. Like the manufacturers of videoplayers and videorecorders, the producers of large-screen, flat-panel television receivers will face a chicken-and-egg problem in building sufficient consumer demand to realize economics of scale and low-cost production. Thus, CRTs, with their low production costs and large investment in manufacturing facilities, seem likely to retain their place in the television set throughout the next decade. Many new television receivers will contain microprocessors, however, enabling their use as information-display devices.

Many laboratories are also developing holographic and other technologies for three-dimensional displays. Without new breakthroughs, however, there is no indication that three-dimensional displays will be available except for military and special-purpose applications in the next decade.

Hard-copy output devices. Devices that can produce permanent, hard-copy records seem inherently more bulky, expensive, and complex than soft-copy display screens. Hard-copy equipment must store ink and paper, they use more power, and their electromechanical components are noisy and prone to failure. Technological advances will reduce, but not eliminate these problems.

Impact printers, which typically use a mechanical typeface to strike an inked ribbon, will probably remain the dominant technology for hard-copy computer terminals in the 1980s. A number of advances in impact printing have been made in recent years, such as replacing individual type elements with a printing ball, wheel, or pin-like matrix; but it is not clear where the technology goes from here. Although product improvements will undoubtedly be made, costs seem unlikely to drop substantially from present levels.

Nonimpact printing technologies promise greater gains. These include ink jets which are deflected by electric fields to form characters; xerographic and other electrostatic processes; electrochemical processes; thermal processes, possibly including lasers; and photographic processes. Of these, ink-jet printers should gain popularity for high-speed serial printing of characters. New xerographic printers, using lasers to expose the paper, will be used for high-speed character and image (i.e., facsimile) printing. Thermal matrix devices will be used for low-speed, low-cost printing, while some current electromechanical and thermal printers may remain competitive for low-usage facsimile and similar applications. However, thermal and electrochemical devices are limited by their need for special paper and their generally lower quality of print.

Despite occasional newspaper accounts of a facsimile printer using standard paper for under $200, industry sources do not expect such an advance. Some new printers might be available in the next decade at two or three times that price, but even that possibility remains speculative. Generally, forecasts of technological advances for hard-copy printers are less optimistic than those for displays, memories, and other information system components.

Computer Influences on Telecommunications Systems and Services

This review of computer technology trends suggests that computer developmentss will strongly influence the technical evolution of telecommunications systems and the communications services they offer. The following are several specific examples:

The Substitution of Information Processing for Transmission

With the rapidly falling costs of LSI microprocessors and memory units, it has become profitable to make more efficient use of transmission lines by processing information *before* it enters the communications network. This has been done for some years in the transmission of voice conversations over

Technology in the 1980s

undersea cables, where bandwidth is at a premium. People do not talk throughout every second of a telephone call. Electronic equipment at the cable terminals can sense the natural pauses amd gaps in conversations and can interleave segments of other calls within these gaps. Callers think they have a full circuit to themselves, but in fact the channel is shared with others. This is called *time-assigned speech interpolation,* or *TASI.*

Packet switching of data and messages again illustrates the substitution of information processing for transmission bandwidth. If two people have information to exchange, one ordinarily places a call to the other, and a circuit is established between them. They then transmit voice conversations or data between the two points. Usually, however, the circuit is not used to its full capacity. A more efficient approach is to divide the information into a group of bits or "packets" at the terminal. The packets are then shuttled through the network under computer control until they reach their destination and are decoded. Transmission generally takes less than a second, and at no time is there a physical circuit established directly from the sender to the receiver. Packet switching requires a microprocessor and memory at each terminal and a series of computer switches to route the packets along. But even at current prices, packet switching is cheaper than circuit switching for many data applications.

The Trend toward Digital Communications

Computer systems operate with digital bits, whereas telecommunications networks were designed to carry voice signals in analog form.[7] Voice and other analog signals can be coded for digital transmission using pulse-code modulation (PCM) or similar techniques. The LSI technology makes analog-to-digital conversion cheaper, and encourages more direct digital communications traffic. As a result, this technology has accelerated the trend toward installation of digital communications facilities that began in the Bell System in 1960. Still, because of the large existing investment in analog plant, the U.S. switched telephone network will remain a mixture of analog and digital facilities throughout the 1980s.

Increased Mixing of Voice, Data, Message, and Image Communications

After analog signals are converted into digital bits, they can be combined with other digital traffic for more efficient transmission. Interleaving bits from multiple users also provides greater privacy and security, as well as economic advantages. However, it

Communications for Tomorrow

complicates matters for accountants and regulators who want to separate the costs of the various services.

The Integration of Information Processing and Communications Functions

State-of-the-art electronic switches and data terminals already have information-processing capabilities, and equipment built in the 1980s will incorporate increasingly greater computing power. From a technical standpoint, communications and processing functions should often be combined—as, for example, in terminals providing remote access to information files or switches that can be programmed to provide data-processing services. Again, these technological changes will cause problems for regulators, who want a clear division between "data communications" and "data processing." Technology will blur this distinction even more in the future.

Evolution of Communications Terminals

Designers can program a microprocessor for particular applications, rather than build specialized and expensive hardware for every new function. Among other functions, a microprocessor in a communications terminal can convert signals to a form more suitable for transmission, reduce redundant information to save transmission-channel capacity (bandwidth compression), process incoming signals to derive the desired information and check for errors, and serve as a local computing center. Some of these features are described in the following section.

Communications Terminals

This section discusses the evolution of communications terminals and related equipment for sending and receiving information over communications channels. I first describe terminals used primarily in an office or institutional environment, and then those that will be used in the home. One straightforward but oversimplified distinction between the two is that home terminals should cost less than $1,000, whereas office terminals generally are more expensive.

Office Terminals

By the early 1980s, microprocessors will be routinely incorporated into many office typewriters, copying machines, and facsimile devices, as well as into virtually all data terminals and private switchboards (PABXs). These machines will have communications connections to central computers, information data bases, and similar equipment in other locations. Thus, a large number of the business letters, memoranda, and other documents that are now mailed will be sent over communications lines between offices within a building or across the country. Distribution time will decrease from hours (within the same business location) or days (among geographically dispersed locations) to seconds or minutes, and will no longer be a function of distance.

This network of intelligent office machines linked by communications lines is often referred to as the *automated office* or the *office of the future*. Four generic types of devices suggest the evolution of today's office machines into the more powerful, integrated systems of the 1980s. These devices are the word processor, the intelligent data terminal, the intelligent copier, and the computer-based PABX.

Word processors. The evolution of the typewriter into an intelligent word processor is the key step in the development of the office of the future. Word processors contain a microprocessor and associated memory,[8] a keyboard, a printer and/or graphic display, and an interface to the communications network. A draft letter or report typed on the keyboard is stored as a series of bits in the word processor's memory. When the draft has been proofed by the writer, these bits can be transmitted to a central storage location, or they can be sent to a colleague, supervisor, or editor for review. That person can then call up the draft for display on his or her graphic terminal[9] (or, if necessary, receive a printed copy), electronically revise and edit the draft on the graphic terminal, and then send the revised draft back to the writer.

When the draft is considered ready, it can be printed locally at the writer's word processor, sent electronically to a central facility for storage or distribution, or transmitted over communications lines directly to recipients in the next office or at a distant location. This process bypasses not only the U.S. Postal Service, but also the internal mail-distribution systems at both ends. A word processor can also print documents that were prepared in some other location and transmitted to it electronically. Each word processor will have a unique address for sending and receiving material.

The main advantage of an interconnected network of word processors linked to computers, information files, and other devices is that it avoids the costs and delay associated with paper handling. A document is entered on the keyboard only once. It can subsequently be retrieved from digital storage for revision or distribution. Final recipients can view the copy on their terminal displays without any physical document distribution. Paper copies will be made only when and where needed. A word-processing system also allows the writer to incorporate up-to-date information into a document at the last possible moment. For example, one can draft a letter referring to the current month's financial report and then have the most recent data recalled from the computer and inserted into the letter just before transmission.

Precursors of word processors (such as the IBM-MTST) were built more than a decade ago for use in preparing similar letters to multiple addresses. Many varieties of keyboard terminals linked to computing systems are available today and widely used for text editing and information retrieval, as well as for computation. They typically cost $2,000-$10,000, and most do not contain microprocessors. The technical trends ensure that intelligent word processors will be available much more widely and at lower cost in the 1980s. Some manufacturers and consulting groups estimate that over 50 percent of all office documents will be prepared with word processors by 1985.

Intelligent data terminals. Most computer terminals are now used for data entry and retrieval, rather than for word processing. The same technical trends described above will make data terminals less expensive, more powerful, and largely indistinguishable from word processors. The decreasing costs of microprocessors and storage will encourage more data processing at the terminal, without communicating with a central computer. At the same time, computer users will routinely have access to remote information bases and computing facilities.

Over the next decade, more and more work stations will have a graphic or printing terminal available for local processing or storage, for data entry, for text and data editing, and for information transmission. The price of a simple terminal with a keyboard and display, a microprocessor, and associated memory is already below $1,000.

Intelligent copiers. While word processors can print letters, memoranda, and short documents directly at the recipient's work station, large offices will need faster printers for long documents and multiple copies. The evolution of today's copying devices into "intelligent copiers" will occur with the addition of a microprocessor and memory unit, a multiple-font character generator, and an interface to the communications network. By 1990, intelligent copiers are likely to dominate markets presently filled by offset presses and duplicators, photo-composition equipment, and stand-alone office copiers.

Technology in the 1980s

The intelligent copier will still perform its present function—making multiple copies of physical documents brought to the machine. But it will also be able to prepare copies from digital data, using an internal character generator and a laser-imaging system. The type font, spacing, and format of the printed output would be selected by instructions in the incoming data or by the copier's operator. Thus, documents stored in the computer memory could be transmitted to the intelligent copier and printed there for local distribution. Or a memo prepared in one city could be sent electronically to another for printing and distribution.

Intelligent copiers may also incorporate facsimile scanners that use bandwidth-compression techniques to reduce transmission time. Most facsimile devices today scan an entire page and transmit each light and dark element—an inefficient and costly approach. In contrast, an intelligent copier would be able to store the bits representing each element and transmit only the changes between light and dark areas. A further step would be to include logic in the copier, so that it could recognize characters on a printed page and transmit bits describing the character, rather than its individual light and dark elements. This capacity would reduce the data-transmission requirements by a factor of more than ten, as well as improve the quality of the output document. Of course, the need for character recognition would be avoided if the document were initially stored in digital form. Facsimile transmission will be used principally for documents containing handwriting, graphics, or other nonstandard characters.

Intelligent copiers are under development by Xerox, IBM, Hewlett-Packard, and a number of other manufacturers. They should be on the market well before 1980 and comprise the bulk of new copier sales within the decade. In combination with word processors and mass memories, intelligent copiers will replace much of the present physical storage and transmission of documents.[10]

Computer-based PABXs. In the 1980s, businesses will increasingly intermix voice, data, text, and message information on the same communications lines. An intelligent, or computer-based, PABX will control these information flows. A computer-based PABX can have additional logic and memory units built in at little extra cost to provide capabilities for information processing and local storage, as well as the traditional PABX functions of switching, routing, and line control. In addition, computer-based PABXs offer a variety of additional communications services as part of their standard design. These include call-forwarding, conference-calling, and call-waiting (signalling a busy line when another call is waiting). A list of the features presently provided

Communications for Tomorrow

by AT&T's Dimension computer-based PABX is shown in *Table 2-3*.

Computer-based PABXs are available today from AT&T, GTE, Northern Telecom, IT&T, IBM (currently offered in Europe only), and several other manufacturers. More powerful versions will be the standard office communications switching systems of the 1980s.

Table 2-3
Selected Optional Features of the AT&T Dimension PABX

Basic Optional Features	
Attendant position, alphanumeric display	Provides a visual display on an attendant's console of the four symbols used to identify the calling number, etc.
Call-forwarding	Automatically routes to a designated extension either all calls or all calls directed to an extension that is busy or does not answer.
Call-hold	Allows use of a code to hold an ongoing call while originating another call or feature.
Call-pickup	Allows use of a code to answer calls to other extensions within a present pickup group.
Call-waiting	Automatically holds calls to a busy extension while the called party is signalled that a call is waiting.
Outgoing trunk-queuing	Provides automatic queuing of calls when all trunks are busy and automatic ringback when a trunk is available.
Three-way conference transfer	Allows an extension user to dial in a third party while the second party is held; the user can also hang up or drop the third party from the call.
Additional Centrex and Deluxe Business Features	
Automatic callback	Automatically connects an extension to a previously busy number once the line becomes idle.
Automatic identification	Automatically identifies extensions on outgoing calls in order to permit direct billing to extensions for toll calls.
Code restriction	Limits the office and area codes that can be dialed from certain extensions.
Direct inward-dialing	Allows direct dialing to extensions from the dial network without attendant assistance.
Loudspeaker paging	Provides access to voice paging equipment for attendants and extension users.
Remote access to PBX services	Allows a user calling from outside the PBX to access the PBX services via an exchange-network connection.
Tandem tie trunks	Permits a caller at a distant PBX to direct-dial tie trunk calls through the switching system.
Trunk-to-trunk connections	Allows an incoming or outgoing trunk call to be extended via the attendant to another outgoing trunk.

Source: AT&T.

Technology in the 1980s

Interconnection of office terminals. As discussed above, the elements of office communications systems of the 1980s exist today and are used for text editing, information retrieval, and data communications. A few organizations, such as New York's Citibank and the Hewlett-Packard Corporation, are tying terminals together in a deliberate effort to replace paper flows with electronic document storage and distribution. The Hewlett-Packard network now handles more than 20 million messages annually among the numerous company locations.[11] By the early 1980s, interconnection of word processors and other terminals within large commercial and government organizations will be routine.

The next step is to interconnect terminals in different organizations. Many research-oriented institutions already have such links through packet-switched networks, such as the ARPANET and Telenet, or through computer time-sharing services, such as TYMNET. Sending messages among terminals—either as part of a formal, computer-based teleconferencing service or as informal "computer mail"—has emerged as a significant use of these systems (as well as a popular pastime) in the past few years.

As another example, an academic colleague in Los Angeles recently prepared a short paper for submission to a professional journal. The paper was to be sent to an editor in Montreal for review. As is often the case, the author wanted to make some last-minute changes in the draft just before the final deadline date. Consequently, rather than mailing the revised draft to Montreal, he had it keyed into the university's computer system in Los Angeles and transmitted over telephone lines to Montreal, where it was stored and available for display or printout at the editor's terminal. But because the editor was in London at the time, his secretary entered a new address and sent the text via either satellite or undersea cable to a London computer. The editor called up the text for display in London, suggested a few editorial changes, and then transmitted the revised text back to the author in Los Angeles. The editor's suggestions were noted, revisions were made, and the final manuscript was again sent electronically from Los Angeles to Montreal, where it was ready for journal publication. The entire process took less than 48 hours.

Today, this kind of electronic text processing is available to a relatively few sophisticated users of teleprocessing systems. But the number of such users is growing steadily as technology develops less expensive terminals, storage facilities, computers, and transmission channels. By the early 1980s, direct terminal-to-terminal communication will be widely used by most major businesses and by some professionals in their homes. It will have made serious inroads into business use of Telex message services. By the late 1980s, the costs of word processing and message transmission should be sufficiently low to make it attractive to nearly all businesses, large and small.

The costs of electronic message transmission are difficult to estimate because they depend on the number and length of the transmitted message, and on communications tariffs rather than underlying costs. A recent study by Raymond Panko estimates that in 1975, the average cost of sending a 50-word message electronically was about $1.20, made up in roughly equal parts of computer costs, labor costs, communications costs, and terminal costs.[12] Panko projects that the cost will fall to between 25 and 52 cents by 1985. The marginal cost per message will be much lower—perhaps only a few cents—in heavily utilized systems. Even at 50 cents per message, electronic terminal-to-terminal transmission appears a ready substitute for much first-class business mail.[13]

Terminals for teleconferencing. Intelligent terminals in the office will also lead to some substitution of telecommunications for face-to-face meetings. It is by no means clear to what extent telecommunications may reduce travel, either business or personal; but teleconferencing services are expected to grow significantly in the energy-conscious society of the 1980s.

Teleconferencing requires a choice of communications modes: audio only, audio and graphics, messages, or video. Telephone conference calls have been available for some years, of course; but computer-based PABXs and computer-controlled central office switches make them easier to set up and operate. Loudspeaker telephones usually suffice as terminals, although more elaborate audio-conferencing equipment has been designed.

A recent paper by Pye and Williams reports that audio conferencing alone is suitable for 22 percent of business meetings that are currently conducted face-to-face. Adding graphic transmission capability would make teleconferencing suitable for an additional 17 percent of business meetings.[14] Text and other alphanumeric materials can be exchanged using the graphic terminals that will be widely available in offices in the 1980s. Special equipment for transmitting handwriting and other graphics will also be commercially available, but the demand for these services is uncertain.

Teleconferencing by exchanging messages under computer control (computer conferencing) represents a different approach. Participants in a computer conference need not be physically available at the same time; instead, they can enter messages on their terminals and read those of other participants whenever it is convenient. Messages are stored at a central computer, so that a participant can call up the conference file for display or printout at his/her terminal, then respond with individual messages or general comments to all participants. Although computer conferencing presents some problems (particularly for uninitiated terminal users), it is a useful means of communication for busy people at geographically dispersed locations.

Videophone and other switched video services. AT&T's Picturephone was one of the great marketing disappointments of the past decade. The Bell System introduced commercial Picturephone service in 1970, while other telecommunications companies in the United States and abroad hurried to develop their own switched video systems. But at the prices that must currently be charged to recover costs, there has been little demand for video telephone service. Technology will reduce the costs of switched video service over the next decade, but it remains doubtful whether video telephones will achieve commercial acceptance during the 1980s.

The principal technological advance is the availability of inexpensive digital logic and memory units in the videophone terminals, allowing signal processing and storage to reduce transmission requirements. Picturephone was originally designed with a roughly 5-inch-square display that required a bandwidth of about 1 Megahertz (MHz), or a digital data-transmission rate of about 5.6 million bits per second (Mbps).[15] Using bandwidth-compression techniques which transmit only those picture elements that change in time (i.e., the moving areas), Picturephone can be transmitted at 1.5 Mbps without noticeable degradation. This is a significant improvement because it allows transmission over the standard T-1 digital channels installed throughout the telephone network.

Further bandwidth reduction is feasible, but only with some reduction in the quality of the moving image. By 200 Kbps, blurring of the lips and other moving areas becomes noticeable. At 56 Kbps (the transmission rate for a digital voice channel) displaying a continuous moving picture is no longer feasible. Instead, individual frames can be stored and displayed every few seconds, like a series of snapshots. This is sometimes known as *slow-scan* video.

Bandwidth compression requires extensive storage capacity at each terminal. When Picturephone was developed in the 1960s, full-frame storage was too expensive for bandwidth compression to make economic sense. In the 1980s, the cost of LSI or magnetic-bubble storage should be low enough for compression to become economically advantageous. Charge-coupled-device cameras should also reduce the costs of the video terminal.

Technology thus offers a range of potential switched video services with different terminal and transmission requirements and different costs. But the experts do not agree on which, if any, of these services will be marketable. Based on the past experience with Picturephone, some contend that subscribers will demand full television quality and will not accept a smaller, head-and-shoulders display. AT&T is no longer actively marketing Picturephone as a switched video service to individual terminals. Instead, it now offers Picturephone Meeting Service, which uses two-way television transmission among meeting rooms in selected cities. The cost for a three-minute connection between New York and San Francisco is $19.50, which makes the cost of a one-hour video call roughly equal to the round-trip air fare not counting the value of the traveler's time. Other two-way television links between meetings rooms have been established by large companies and government agencies over private lines, but it is not clear that a large demand for this service exists.

Video conferencing could help politicians stay in touch with their constituents and have other important applications in the public sector, but according to Pye and Williams, video adds only 2 percent more to the 39 percent of business meetings suitable for audio and audiographic teleconferencing. And neither business nor residential consumers have yet indicated that they are willing to pay several times the cost of voice communications to see the person at the other end of the call. Consequently, full video service may well remain an expensive, special-purpose service throughout the 1980s.

At the other end of the scale, slow-scan video, oriented toward graphic and document transmission over a standard digital voice channel, may be adequate for most business applications. Slow-scan video terminals could be available in the 1980s in the $1,000 range, and the service would be less expensive and faster than the similar Videovoice service offered unsuccessfully by RCA in the early 1970s. However, the demand for such a service remains uncertain.

Point-of-sale and banking terminals. Several kinds of communications terminals have been designed for remote financial transactions. Among them, the following are all available today:

- Point-of-sale (POS) terminals in retail stores transmit information to a central computer about the goods purchased and the form of payment. With POS terminals, stores can give immediate check and credit authorization, debit or

Technology in the 1980s

credit funds in customers' bank accounts, and maintain accurate sales and inventory records. These terminals generally contain a small (12- or 16-key) keyboard for entering the transaction data, a device to read information from the customer's credit card, a small memory unit, a low-data-rate interface to the communications line, and relatively simple control logic. Credit verification is given from the central facility, either by lighting a lamp on the terminal or by telephone voice response. Telephone companies (e.g., AT&T's Transaction telephones) and a number of manufacturers provide POS terminals.

- Automated teller machines (ATMs) provide such banking services as cash withdrawals or advances, deposits to checking or savings accounts, or transfers of funds among accounts, without a human operator. Processing and communications requirements are similar to those of POS terminals, except that the machines must be heavily armored and have reliable and secure electromechanical devices for handling cash. Consequently, ATMs cost 5 to 10 times as much as POS terminals.
- On-line teller terminals (OLTTs) provide direct access to customer records for financial transactions in banks. The OLTTs are much like office data terminals with graphic displays.

There are no real technical barriers to linking POS and ATM terminals into an integrated electronic funds-transfer (EFT) system that would reduce the flow of checks and credit-card paperwork. With more than 15,000 financial institutions and 6 million retail businesses in the United States, EFT terminals represent a large potential market. Substantial legal and political barriers remain, however, that make it difficult to forecast how quickly consumer-oriented EFT systems will grow.[16]

Home Communications Terminals

Discussing the evolution of communications terminals for the home is much more difficult than projecting the developments in office terminals. The same technologies apply, but costs become a far more critical issue. Consumer demand depends on cost, and cost depends on the production volume. The "chicken-and-egg" cliche applied so often to home communications equipment and services is tiresome, but apt. Yet, as hand calculators and digital watches attest, the introduction problems can be overcome.

The "ultimate" home terminal would be used for a wide range of services—message, voice, image, data, and text. The terminal

Communications for Tomorrow

would have a full alphanumeric keyboard for input, as well as a more easily used key-pad for dialing and frequent instructions. It would provide both hard-copy and graphic-display output. It would store information, send messages to, and receive replies from a distant person or computer. Finally, in John Pierce's words, "the terminal would be as small and light as a portable typewriter. It would cost a couple of hundred dollars and would last forever."[17]

We are still far from building this "ultimate" home terminal, but one can see—cloudily—the technical evolution of at least six generic devices in the home: (1) the telephone, (2) television receiver, (3) typewriters, (4) pay-television terminal, (5) video recorder and player, and (6) the hand calculator, TV game, and home computer.

The intelligent telephone. The 12-key pushbutton telephone can serve as a low-cost intelligent terminal with the addition of a microprocessor and associated memory, a slot for a magnetic card or some other simple data-entry device, and a liquid crystal or LED alphanumeric line readout. Such an intelligent telephone could be available for purchase by home subscribers in the early 1980s for under $200. It would include capabilities for conference calls, repertory dialing, call-forwarding, and other telephone-related features.[18]

Besides telephone-related services, an intelligent telephone would provide access to information and transaction services that do not require extensive graphic or hard-copy output. Credit card and banking transactions from the home are two examples. The Greater New York Savings Bank and a few other institutions now offer these services, using telephone push-buttons for input and voice response for output. However, the necessary data-entry procedures may be too complex for many home subscribers. The addition of a magnetic card reader and an information-processing capability to the telephone would simplify the exchange of information between the home telephone terminal and the central computer. And for financial transactions, a liquid crystal or LED display (similar to that in hand calculators) seems preferable to voice response.

A stock-quotation service providing current prices on individual stocks is another function well-matched to the intelligent telephone and more useful to most investors than the delayed stock ticker offered by cable television systems. The subscriber would call a designated number, enter the appropriate code, and then view the current quotations on the telephone display. In this same way, voice libraries of recorded information—ranging

Technology in the 1980s

from emergency first-aid procedures to a dial-a-joke compendium—could also be automatically accessible by telephone.

Many other information services will require more graphic display capability. A small CRT or (later) a flat-panel display providing 8 to 24 lines of alphanumeric characters could be added to the telephone for under $100. Subscribers could also use a modified television set for information retrieval. However, message transmission generally requires a full alphanumeric keyboard, rather than a 12- or 16-key pad.

The intelligent telephone could provide clock and calculator functions in the home, as well as remote control of other home appliances. It could also serve as the input device for a dedicated home computer system that would, for example, monitor and control energy use. Once the microprocessor, memory, and display are built into a telephone, the marginal cost to use them for other applications becomes quite small. However, the cost of sensors, electromechanical controls, and internal wiring will remain a serious barrier to the use of computers for monitoring and control functions in the home.

Television receivers for data processing. The television set is the obvious display device for home communications and information services. With 525-line resolution, a television receiver cannot comfortably show a full page of printed text. It can, however, display several hundred alphanumeric characters—enough for most message and information services. The television set has a full graphic capability. Most important, it is already in the home, so that the consumer need not invest in a separate display.

The cost of adding an LSI chip with memory and a character generator to a television receiver is about $200 today. With volume production, that cost could easily drop to below $20. The first such application in the United States has been the display of television captions for the deaf.[19] In Britain, the BBC and the British Independent Broadcasting Authority are operating similar experimental systems (CEEFAX and ORACLE, respectively) which display news and other information on LSI-modified receivers. The information is transmitted over spare lines of the broadcast television signal and requires no interactive communications from the home.

Subscribers could request a much richer array of information services via telephone lines or a two-way cable television network. The British Post Office is developing its Viewdata system to display data sent over the telephone lines on modified television sets, using the same format as CEEFAX and ORACLE.

Prospective services include specialized news bulletins, current financial data, and other information ranging from community services to recipe files. Field trials are set to begin shortly, with services available to the public before 1980. Technically at least, a similar system could easily be developed in the United States, with add-on costs to the television set of less than $100 (in volume production). AT&T and other U.S. companies reportedly are working on prototype systems, although they have made no public announcements.

Here, as always, one must be careful in moving from technical possibility to commercial likelihood. Television sets are manufactured for entertainment, not data uses. Those who want information and message services at home may prefer to purchase or rent a separate display terminal rather than to modify the household's primary television set. Professionals who work at home would be especially likely to have their own, dedicated display terminals, either paid for by their employers or justified as a business expense.

A display terminal with a full alphanumeric keyboard would cost considerably more than the add-on cost of a modified television set—probably at least $300, even with mass production. But this equipment could also be used as a home Telex terminal to send and receive messages. While no one knows the consumer demand for a Telex-like service, some observers believe it may be the key to breaking the home terminal "chicken-and-egg" barrier in the 1980s. If so, home terminal-to-terminal messages could displace another significant portion of first-class mail services.

Printing terminals for the home. One might expect that the electric typewriter could form the basis for a low-cost home printing terminal, analogous to the evolution of the office typewriter into a word processor. However, this does not appear economically practical in the next decade. Although new concepts are under development, impact or nonimpact printing terminals for the home, with full alphanumeric keyboard, microprocessor, memory, and communications interface, are expected to cost at least $400 to $500 in the mid-1980s.[20] If rented, the monthly charge would be $20 to $30, depending on the appropriate depreciation and maintenance requirements. Some consumers and professionals who work at home would be willing to pay this much to send and receive messages and obtain hard-copy results, but probably the majority of households would not.

Home facsimile terminals also appear expensive and unlikely to be used for the electronic delivery of newspapers, periodicals, and other bulky documents to the home. The cost of paper and

Technology in the 1980s

other expendable supplies poses an added barrier, in that printing the equivalent of a daily newspaper in the home would require more than 100 square feet of paper every weekday and nearly 400 square feet on Sunday. Newspapers are now delivered to the door for between one-fifth and one-half cent per page. Costs for home delivery would have to go up by a factor of ten before electronic transmission would become competitive.

Cost comparisons for periodicals and catalogs also greatly favor mail over electronic delivery. Moreover, the full-color graphics extensively used in such publications would prove much too expensive for home facsimile. We must await the invention of a cheap, low-power, hard-copy device using ordinary paper (or one that could recycle its own product) before facsimile terminals will be widely used in the home.

Cable and pay-television terminals. In the 1980s, cable television systems may finally develop their long-awaited capacity for two-way communications for pay-television and other interactive services. If so, the cable/pay television terminal would contain the basic logic and communications elements needed to send data and short messages from the home to the cable system studio or headend. In principle, once a two-way terminal is in the home for pay-television, it could be used for other data, information, and message services as well. In practice, however, these services may already be available over telephone lines by the time two-way cable systems become operational. The cable and pay-television industries also have not developed technical standards for terminals that would encourage their use for other services. Consequently, while it is technically feasible to build a cable television data terminal for about the same price as an intelligent telephone—$100 to $200 with volume production—the cable industry may not evolve in that direction. Rather, cable systems may continue to install less-expensive terminals dedicated to pay-television and a few special services.

Video recorders and players. Reports in the 1970s of the birth of a burgeoning new industry for home video players and recorders proved to be somewhat exaggerated. But industry executives and most outside observers now believe that video recorders and players are ready to take hold and will become major consumer products in the 1980s.

Two technologies—videotape and videodisc—are in contention. Videotape systems can record television programs from the home receiver as well as play prerecorded entertainment or instructional tapes. With a camera added, they also make home video recordings. Videotape equipment has been commercially available for over a decade and has gradually improved in performance and price. In early 1978, more than 15 companies were selling home videotape systems at retail prices of $1,000 to $1,300 (some older models were available for $800).

Videodisc systems play prerecorded materials for display on a conventional television receiver. They are only now moving toward commerical introduction after a number of false starts. While several technical approaches were developed in the past few years, the leading contenders today are optical videodisc systems that reflect laser light from marks or depressions on a rotating disc onto a photosensor. The MCA/Phillips "Disco-Vision" system is scheduled to be introduced commercially in the United States in late 1978 at a cost of about $600. Other optical systems will likely follow. RCA has indefinitely delayed introduction of its capacitance-sensing "SelectaVision" system, but at least in its public announcements the company remains committed to this alternative videodisc technology.

It is too early to tell whether videodisc systems will gain broad consumer acceptance. Videotape systems, with their recordings capabilities, are more versatile; but both the hardware and recording media are more expensive to produce. In volume production, videodiscs can be made for a dollar or so per playing hour, while videotape costs are several times higher. However, software prices to the consumer will probably depend more on market factors and the cost of talent than on underlying production costs.

Although hardware costs of both systems should decrease over time, cost reductions are limited by the electromechanical subsystems. As a consequence, cost reductions over a 10-year period may be between 25 and 50 percent. Consumers will not see the much larger cost reductions that have characterized such LSI-based products as home calculators and digital watches over the past few years.

Estimates of video player and recorder penetration by the mid-1980s range from 10 percent or less to nearly 50 percent of U.S. households with color television sets. A 25-percent penetration by 1985 would represent about 18 million units. Whatever the consumer response, most observers agree that the instructional and institutional markets for video players and recorders will continue to expand during the 1980s. Optical videodiscs, with their frame-freezing capability, seem favored for prerecorded instructional programming. They may also prove useful for low-cost mass storage of digital data for computer systems.

Manufacturers are expected to extend and improve these technical systems, rather than to introduce wholly new technologies during the 1980s. Toward the end of that decade, a small

Technology in the 1980s

magnetic disc—or, more speculatively, a solid-state video memory system—might be developed for consumer use. The magnetic video recorders/players incorporating these technologies could then be more readily adapted to information retrieval and other interactive services that require storing a full frame of television picture information at the terminal *(frame grabbing)*.

Current frame-grabbing technologies—storage tubes, magnetic tapes, and LSI memories—are either clumsy or expensive or both. Although solid-state memory costs will decline dramatically, full-frame storage of more than a million bits is still likely to cost the consumer $200 or more throughout the 1980s. Consequently, unless the frame-grabbing capability is built into an entertainment device like the video recorder/player, it seems unlikely to be purchased alone for home communications or information services.

TV games, hand calculators, and home computers. Hand calculators and TV games represent the first wave of LSI devices in the home. The current generation of TV games are programmable; that is, the logic and display instructions for each game are contained on magnetic tape cassettes or plug-in LSI memory chips. These games now sell for $150 to $300, with additional game cartridges costing $15 to $30. Their costs should decrease over time, following the path of earlier game offerings.

Since TV games are already attached to the home television receiver, the next logical step would be to connect them to the telephone or to a two-way cable television network, so that games could be played with remote opponents. Interactive "Pong" or "Tank," or the more sophisticated games now under development, could have great appeal for remote contests between friends or anonymous opponents. When such games have been introduced on a university computer time-sharing system, their popularity has sometimes overloaded the system and led to restrictions on student use. A principal attraction seems to be the ability to cloak oneself in a pseudonym ("Red Baron" and "Hot Dog" seem to be popular handles) and then aggressively but anonymously challenge the world. Both the attraction and the handles are analogous to the CB radio experience in the mid-1970s.

Technically, adding a communications interface and some additional memory to a TV game to allow interaction is not difficult or expensive. However, problems arise in establishing interface standards so that the games from different manufacturers can talk with one another. Overloading local telephone exchanges represents another potential problem. If interactive TV games became popular, their use could saturate telephone central offices that are designed to handle relatively short voice

Communications for Tomorrow

conversations. This would be likely to accelerate the telephone companies' efforts to establish "usage sensitive pricing"—that is, charges for individual local calls—which in turn would dampen game players' demand for telephone communications. The probable response of the cable television industry is less clear. In any event, the development of interactive TV games in the 1980s represents a technical possibility whose likelihood will depend on communications pricing and regulatory responses.

Home calculators and computers will also continue to show large gains in price and performance. Computer hobbyists, according to subscriptions to hobbyist magazines, now number over 60,000. Most hobbyists soon run out of pure computing tasks and turn their attention toward exchanging programs and messages, using data bases, and playing games with other hobbyists. Thus, as their numbers grow, computer hobbyists will want to interact with each other over communications lines. Like professionals who work at home, hobbyists will want their own home communications terminals. How large a market computer hobbyists will represent in the 1980s remains speculative. One analogous group of technological hobbyists are ham radio operators, who currently number about 325,000 in the United States.

Fitting the pieces together—home communications centers. Home communications terminals can evolve through any or all of the devices described above. We already have the technology. As one example, a fully assembled computer with LSI memory, magnetic tape cassette for program storage, CRT display, and standard communications interface was introduced in 1977 at a retail price of under $600. No one knows what consumer demand exists for such a product at that price, but it clearly points toward new offerings at lower prices in the next decade.

A conservative view of the consumer products available by the mid-1980s would include the following: (Price estimates indicate relative prices and should not be taken as literal forecasts.)

- For $100-200, an intelligent telephone or cable television terminal for relatively simple data entry.
- For $50-100, television receiver modifications to display alphanumeric information and data.
- For $300-400, a separate graphic display terminal with full alphanumeric keyboard.
- For $400-500, a printing terminal with full alphanumeric keyboard.
- For $300-500, a video player.
- For $500-800, a videotape recorder.
- For $50 or more, additional computing power and memory.

Technology in the 1980s

Because these devices contain or use many common elements, including microprocessors, solid-state memory chips, pushbutton or typewriter keyboards, and CRT displays, some experts have suggested that they will be integrated into a single home communications center in the 1980s. One such center, as described by Douglass Cater, is affectionately called MOTHER—for Multiple Output Telecommunications Home End Resources. The home communications center would include a large color television receiver, a videotape recorder and/or videodisc player, programmable television games, timer and calculator, appropriate buttons for pay-television and other services provided over the cable system, and a connection to the telephone. A typewriter keyboard and printer would be optional features.

The home communications center would integrate entertainment, educational, information, and message services at a single station. Adults would watch television for entertainment, record programs, pay bills, scan merchandise catalogs and travel brochures, update the social calendar, and do some work at home. Children would use it for computer-assisted instruction as well as for entertainment and games.

This kind of integrated home communications center does not seem a likely possibility for the 1980s for two principal reasons. First, the center would be too expensive—more than $1,000—to buy as a unit. It seems unlikely that many households would be willing to make that kind of investment in electronic equipment. Second, the primary use of a television display in the home will continue to be for entertainment. Most families appear unwilling to have their primary television set diverted for extended periods to other, nonentertainment activities.

A different view is that these home communications devices will be assembled, component by component, like audio equipment. Microprocessors and memories will be cheap enough to be built into several devices, rather than concentrated in a single communications center. Television sets, TV games, calculators, pay-television terminals, video players and recorders, and data displays will, for the most part, be produced separately. But as the decade progresses, an increasing number of these items will be designed for common interconnection. A central control unit will be developed into which individual devices can be connected. It could very well be a part of the television receiver, as most audio device jacks are built into an audio receiver, or it could be a separate, self-contained unit. The end result may not be as elegant and aesthetically integrated as MOTHER, nor as cheap as John Pierce's "ultimate" terminal, but it appears a more practical way of moving toward enhanced communications capabilities in the home.

Communications for Tomorrow

Transmission and Switching Systems

Telecommunications has traditionally distinguished between mass communication and point-to-point communication. Technically, mass communication is much simpler. It involves a few sources sending the same signals to a large number of receivers. The sources—radio and television stations, or cable system headends—are often linked by transmission lines to form mass communication networks. *(see Figure 2-1)*

Point-to-point communication implies a two-way exchange of information between individual senders and receivers. As a consequence, the technical plant for point-to-point communication requires complex switching facilities as well as local distribution links (local loops) and long distance transmission trunks. In the U.S. telephone network, local calls are switched by local central offices.

A long distance or toll call is routed from the local central office to a toll switching office, where it proceeds over toll transmission trunks to the switching office nearest its final destination. A long distance call may pass through several toll switching centers on its way to the destination toll office.[21]

Technological advances affect not only the transmission, switching, and local distribution functions but also the relationships among those functions within the telecommunications network. Satellites, for example, can bypass local loops and switching offices to provide direct end-to-end transmission. Moreover, technological change is blurring the distinction between mass and point-to-point communications, allowing each to provide communications services that were once the exclusive province of the other.

Change has been particularly dramatic for transmission technologies over the past 30 years, bringing the commercial introduction of microwave radio, coaxial cable, and communications satellites. These advances have been reflected in lower prices for long distance calls, but only partially; this is because switching, distribution, and terminal costs have not fallen so rapidly, and because regulators have used long distance revenues to subsidize the costs of local telephone service.

In the 1980s, technology will bring more transmission capacity and more choice of transmission modes. Mass communications will see the development of direct-broadcast satellites and optical fibers for television transmission. For point-to-point communication, new satellite systems and optical fibers will compete with improved versions of today's wire, coaxial cable, and microwave transmission systems.

Technology in the 1980s

Figure 2-1
A Simplified Diagram of the U.S. Telephone Network

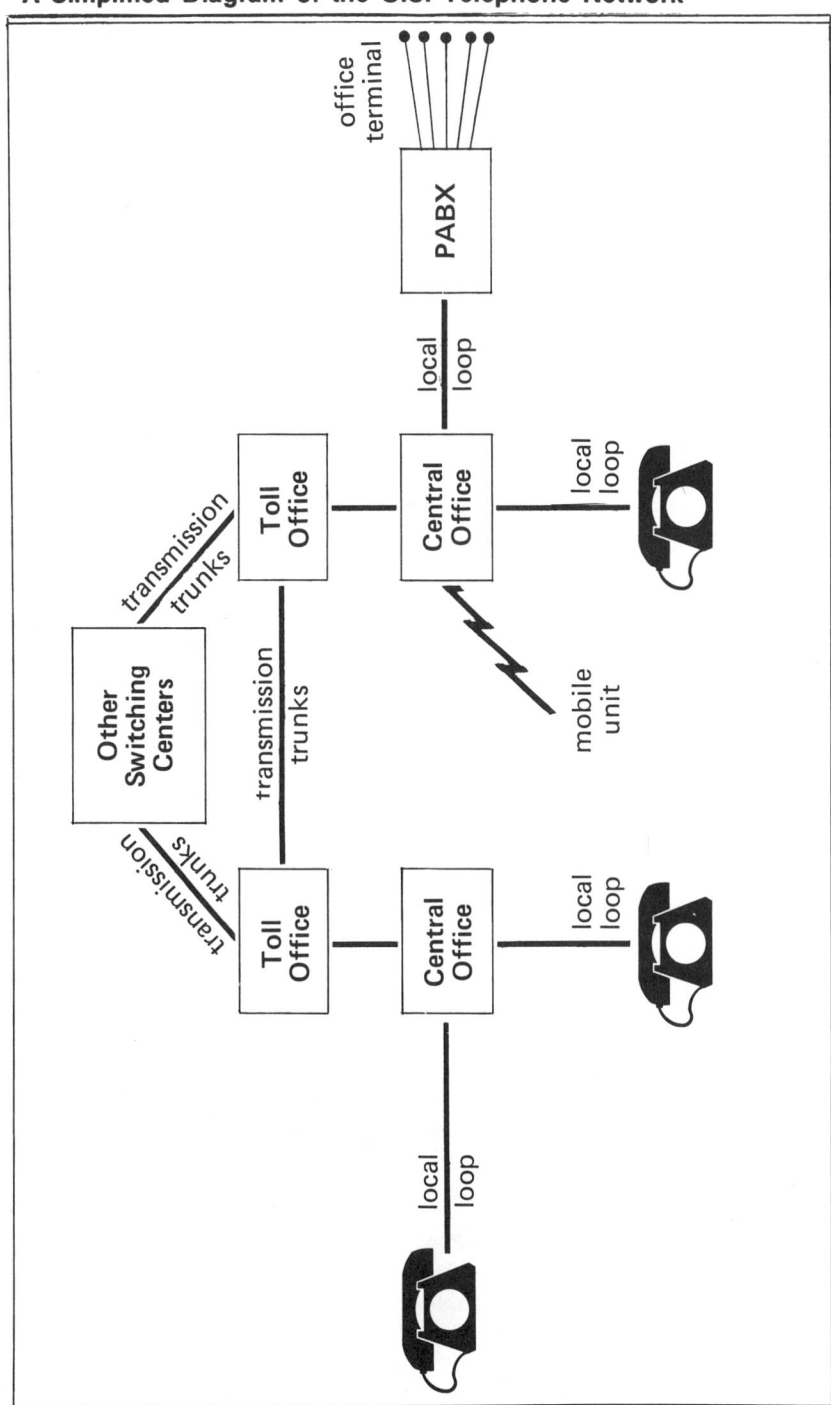

Communications for Tomorrow

Communications Satellites

Since the first commercial communications satellite was launched in 1965, each successive generation of satellites has had greater communications capacity and higher radiated power. More satellite power means that smaller earth stations can be used, a trend that will result in direct rooftop-to-rooftop communications in the 1980s. NASA's space shuttle program will permit the launch of high-powered satellites capable of relaying 100,000 voice circuits or 100 television channels with multiple spot beams. The technology for new services such as direct satellite-to-the-home television broadcasting and satellite switching of mobile communications units will be available in the next decade. However, economic and political considerations make it highly unlikely that the full range of communications satellite technical capabilities will be realized.

Satellite transmission capabilities in the future will be determined more by frequency bandwidth than by power limitations. The frequencies used are set by international agreements and currently fall in the 4 and 6 GHz microwave range. These frequencies must be shared with terrestrial microwave services that are already heavily congested in metropolitan areas.[22] Future satellite systems therefore intend to use the less congested bandwidths available in the 12-14 and above-18 GHz ranges. At these frequencies, satellites can transmit directly to small earth stations within cities.

Other technical advances expected to be incorporated in new satellite systems include:

- More accurate satellite position and attitude control.
- Greater spacecraft power through use of more efficient photovoltaic cells and storage batteries.
- Multiple "spot" beams which will be narrowly focused on the earth's surface to permit greater "reuse" of frequencies in different beams without interference.
- Channel assignment to individual earth stations, as needed, to increase channel utilization.
- Greater use of digital transmission, with switching and signal processing at the satellite.

Commercial satellite systems in the 1980s. At present, four United States and Canadian satellite systems provide commercial communications services in North America, in addition to the Marisat system for services to ships at sea and the INTELSAT system for international communications. These systems are owned by Telesat Canada Corporation, Western Union, COMSAT General (for AT&T/GTE), and RCA. Their technical characteristics are compared in Table 2-4.

Table 2-4
North American Communications Satellite Systems

	Telesat "Anik"	Western Union	RCA	COMSAT General (AT&T/ GTE)	Satellite Business Systems
Operational date	January 1973	July 1974	February 1976	June 1976	Planned for early 1981
Number of satellites planned in orbit	3	2	3	3	2
Channels per satellite	12	12	24	24	10
Channel bandwidth (MHz)	36	36	34	34	43
Channel capacity when used for:					
one-way voice channels	960	1,200	1,000	1,200	1,250[a]
data (megabits per second)	45	45	45	45	43
television channels	1	1	1	1	--
Frequency (GHz)	4/6	4/6	4/6	4/6	12/14
Earth station size (meters)	4.7 - 30	15.5	4.5 - 10	30	5 - 7

Note: (a) Using 32 Kbps per voice channel.

An important new development will be the scheduled launch in 1980 of the first satellite built by the Satellite Business Systems (SBS) consortium, composed of IBM, COMSAT General, and Aetna Insurance. Incorporating most of the technical advances described above, the SBS satellite is designed to operate at 12-14 GHz, with rooftop earth stations 5-7 meters in diameter. The SBS system points the direction for other new satellite systems of the 1980s.

The INTELSAT consortium will expand its international communications capacity through a new generation of satellites (INTELSAT-V) introduced in the early 1980s. Unless political factors change, the INTELSAT system seems likely to remain oriented toward large earth stations tied into national telephone systems, even though some users might prefer small earth stations located at their sites.[23]

Broadcast station and cable system networking. Distributing television programs to commercial networks, the public television system, and independent stations was seen in the 1960s as an initial application for a U.S. domestic satellite system. In recent years, however, AT&T and other carriers have reduced their tariffs for terrestrial video transmission, so that the television networks have not moved to satellite service. Satellites now primarily interconnect cable television systems. Home Box

Office, the nation's largest pay-television syndicator, leases circuits on the RCA satellite system to distribute pay-television to about 100 earth stations serving cable television systems throughout the country. Another firm, Southern Satellite Systems, distributes programming from an independent television station in Atlanta, Georgia, to cable systems in the Southeast.

More regional and national television networking via satellite should occur in the next decade as satellite capabilities expand and the costs of earth stations drop. In early 1977, the Federal Communications Commission authorized the use of 4.5 meter earth stations instead of the 9 meter stations previously used for television reception. This has reduced earth station costs from roughly $100,000 to under $40,000, making satellite reception economically feasible for most cable systems with more than 1,000 subscribers. The Public Broadcasting Service has contracted with Western Union for satellite distribution of three television channels to more than 150 public television stations around the country. By the mid-1980s, building satellites with the capacity to relay 100 television channels with multiple spot beams at 12 GHz or above should be technically feasible. However, it remains doubtful whether a demand for this many television channels will materialize.

Direct satellite broadcasting (DBS). The trend toward higher satellite power and smaller earth stations leads to the concept of direct television transmission from the satellite to a rooftop antenna. This is known as *direct satellite broadcasting.* DBS would bypass terrestrial television transmitters and cable television networks by providing national distribution of multiple television channels directly to the home. Consequently, DBS is highly controversial politically—in the United States because of concerns about preserving local television service, and in other countries because of concerns about television transmission across national borders without government control. The choice of whether or not to adopt direct satellite broadcasting will therefore be based more on political than on technical or economic factors.

There is no question that DBS is technically feasible. The principle issue is the cost of the rooftop satellite receivers in comparison with over-the-air broadcast antennas and cable distribution systems. The cost of a satellite receiver depends on the size of the antenna, which depends, in turn, on the transmission frequency and the power of the satellite. Transmission at 2.5 GHz (suggested for some educational channels) might require a 3 meter antenna, for example, while transmission at 12 Ghz could use a .75 to 1.5 meter antenna. Today, at volumes of 10,000 or so per year, the installed cost of a 1.5 meter antenna would be around $1,000. Of course, manufacturing costs (but not installation costs) would fall with mass production. A recent study for the National Research Council reported that a 1.3 meter antenna capable of receiving 12 color television channels transmitted at 12 GHz could be installed for about $250 if mass produced in the millions.[24]

Even at $250, a satellite receiver would be far more expensive than an ordinary broadcast antenna and considerably more than the average cost per household to build a 12-channel cable television system. Moreover, it is not clear what advantages direct satellite broadcasting would have for U.S. households in metropolitan areas. Rural Alaska could certainly utilize satellite broadcasting; but in other U.S. rural areas, such a system would seem to be less cost-effective than extending over-the-air television service with translators. Proponents of direct satellite service to rural areas usually cite the ability to carry instructional television, video teleconferencing, and other subsidized noncommercial services once the basic system is in place.

Direct broadcast satellite systems seem more economically attractive in developing countries and in nations with widely dispersed, low-density populations. Canada, Brazil, Indonesia, Australia, India, and the Soviet Union, among others, are interested in the concept. Japan is pursuing the technology most aggressively, with a direct satellite broadcast experiment scheduled to begin in 1978, even though Japan seems well-suited for conventional broadcasting and cable television distribution. The Japanese may well foresee a large export market for direct broadcast satellite terminals in the 1980s.

Point-to-point satellite services. The use of satellites for point-to-point services should expand significantly when systems designed for small earth stations are introduced in the 1980s. Satellite Business Systems plans to use earth stations installed on office rooftops, each with sophisticated computers for signal processing and channel assignment. Such terminals would today cost more than $300,000, but SBS expects their cost to drop below $200,000 by the early 1980s. Other point-to-point satellite systems, such as that proposed by the Public Service Satellite Consortium, could be designed for lower cost earth stations with considerably less communications capacity.

SBS has tailored its system for large communications users who can justify the cost of dedicated earth stations. Voice, data, facsimile, video, and other traffic will be combined into a single digital bit stream at the sending terminal, transmitted to the satellite in short bursts under computer control, and switched at the satellite for transmission to the proper receiving terminal. Customers will use only as much bandwidth as they need for the time they need it. End-to-end service costs will depend on the volume and type of communications transmitted, but the costs are expected to be below those now charged by other carriers (SBS has not yet filed tariffs). Based on the SBS cost projections filed with the FCC, a large user's average cost to send a page of text from one office word processor to another will probably be between 1 and 10 cents, independent of the distance between sending and receiving terminals.

The SBS system should be particularly attractive to customers located far from the large earth stations operated by other satellite systems. The system should encourage more data, message, and text transmission among geographically dispersed offices and industrial plants. Moreover, unlike AT&T's present competitors, the SBS system will directly link end users, bypassing the terrestrial telephone network with its local loops and switching hierarchy. AT&T would also be able to offer its own roof-to-roof satellite service if it chose and were permitted to do so. Thus, technically at least, the stage is set for end-to-end competition among AT&T, SBS, and other satellite carriers in the early 1980s.

One particularly interesting application of satellite communications is that now used by the *Wall Street Journal.* Each evening, the next day's newspaper is composed in New York City and transmitted digitally by satellite to three regional printing plants for simultaneous printing and distribution. During the next decade, other publications are likly to use this technology to achieve rapid printing and national distribution of their products.

By the late 1980s, even more powerful satellites launched by the space shuttle could be used to link vehicle-mounted or hand-held mobile transceivers. Regional or national mobile communications networking by satellite therefore appears to be a technical, although not necessarily cost-effective, possibility. However, such a development would require substantial spectrum reallocation in the UHF or another region. One other possible application of satellite technology in the 1980s would be to monitor large-scale, geographically dispersed sensor arrays that could detect forest fires, floods, and even oil spills. Although this again appears technically feasible, the economics of large sensor systems remain questionable. These and other applications could be tested by experimental satellite programs in the early 1980s.

Technology in the 1980s

Optical Fiber Transmission

Technical developments in lightwave communications. The technology for transmitting signals with light over glass fibers has matured so rapidly that lightwave communications has been brought from laboratory development to commercial introduction within five years. Optical fibers are smaller and lighter than coaxial cables or wire pairs. They offer increased transmission capacity and avoid electromagnetic interference with other signals. Optical fibers are already in use in military systems and are undergoing commercial tests in computer, telephone, and cable television systems. They appear likely to replace copper wires and coaxial cables for many new high-capacity data and video links in the 1980s.

All of the components required for lightwave communications are in a rapid state of technological advance. Losses and distortion within the glass fiber itself have been reduced sufficiently so that the fiber can carry more than 100 megabits per second—enough for 1,500 voice conversations or 2 television channels—more than 7 kilometers (4 miles) without amplification. The fiber's usable bandwidth remains limited by light dispersion, but it, too, is improving remarkably. Projections for the 1980s suggest that a single, hair-sized fiber will be able to carry half a billion bits per second. However, in many applications it may be more economical to use additional fibers in parallel, rather than to push a single fiber to its capacity limits. Given their small size, a number of lower-capacity fibers can be run together in a cable to fill virtually any transmission need.

Both light-emitting diodes (LEDs) and solid-state injection lasers are available as sources for lightwave communications. Today, LEDs are less expensive, more reliable, have longer lifetimes, and are better suited than lasers for analog signal modulation. However, LEDs have greater usable bandwidths. Thus, as their lifetime and reliability improve, injection lasers should become the dominant light sources in the next decade. Silicon photodiodes provide inexpensive and reliable detectors, and more sensitive detectors (avalanche photodiodes) are available for high-performance applications. Repeaters that can be directly integrated with the fiber transmission line have been built experimentally and should be commercially available by 1980. Low-loss splicing techniques, which had previously posed a barrier to practical fiber installations, have also been developed. The technical trends therefore suggest several commercial uses of optical fiber transmission in telecommunications systems of the 1980s.

Optical fiber transmission in the telephone network. The first widespread applications of optical fibers in the telephone network will be for video transmission and for digital trunks connecting switching offices within metropolitan areas. The small size of optical fiber cables is particularly attractive in urban areas, where underground duct space is limited. The standard data rates for interoffice trunks (1.544 Mbps, known as a *T-1 rate*, and 44.7 Mbps, known as a *T-3 rate*) are technically feasible today for optical fiber transmission.

AT&T successfully field-tested optical fiber transmission at 44.7 Mpbs in Atlanta during 1976. Further tests are proceeding in Chicago (AT&T), in Long Beach, California (GTE), and outside of London (ITT), among other places. By 1980, industry sources expect the cost of optical fibers to fall from the present level of $1 per meter to around 10 cents per meter. This would make fiber optics cheaper than the special video cables now installed by the telephone carriers, and should lead to their rapid adoption for video transmission in metropolitan areas. As costs fall, optical fibers will also become competitive for the T-1 digital trunks now carried on wire pairs. Optical fibers could begin replacing these copper wires in the early 1980s and are likely to be the preferred choice for interoffice trunking by 1985.

Long distance transmission requires higher data rates. Field tests in England achieved rates up to 140 Mbps in 1977. Bell Laboratories is developing a 274-Mbps optical fiber system capable of carrying more than 4,000 voice channels. Such a system might well be competitive with other long distance transmission links by 1985, and so that we may see the installation of long distance optical fiber transmission lines in the late 1980s. However, for the following reasons, the Bell System may not have a great need for additional long distance terrestrial facilities at that time: (1) Bell's long distance traffic may not be growing as rapidly in the 1980s as in previous decades due to competition and some saturation of demand. (2) New satellite systems may be able to handle much of the increased traffic. (3) AT&T plans to expand the capacities of its present coaxial and microwave systems on existing rights of way. Consequently, although optical fiber systems may be cost-competitive with other new terrestrial facilities, the demand may not warrant their construction. Optical fibers appear likely to be feasible for some undersea cable installations by the mid-1980s.

Technology in the 1980s

Microwave, Cable, Wire-Carrier, and Waveguide Transmission

Significant improvements are also under way in the "conventional" microwave, coaxial cable, and wire-carrier transmission systems. New single sideband (SSB) equipment can double or treble the capacity of existing microwave links, and higher frequency microwave systems are available as well. Consequently, present microwave routes can be upgraded in capacity at relatively low cost, because new rights of way will not be needed. The capacities of coaxial cable transmission systems have also improved. New systems can carry more than 13,000 voice channels per coaxial pair, or as many as 132,000 channels along each route. The installation of new repeaters can double the capacity of the digital T-1 links carried on wire pairs in metropolitan areas. And higher capacity digital carriers (T-2, carrying 96 voice converstions, or 6.3 Mbps) have been introduced into the urban network.

As outlined above, these improvements may limit the extent to which optical fibers are needed for long distance transmission. Together with the developments in optical and satellite systems, these improvements also signal the demise of millimeter waveguide transmission, which was considered among the most exciting new transmission technologies a decade ago. Waveguide technology has advanced considerably in the past ten years, but it appears doubtful that it will compete in the 1980s with other transmission systems.

The Trend toward Digital Transmission

Although microwave transmission systems are likely to remain analog because of bandwidth limitations, most new guided-wave transmission systems in the 1980s—wire carrier, coaxial cable, and optical—will be digital. This reflects not only the expected growth in digital data communications, but also the following general advantages of digital transmission.

- Efficient integration of voice, data, video, and other services.
- No cumulative signal degradation with increasing distance.
- Better error control.
- Prospects for increasing transmission efficiency through signal processing and coding techniques (bandwidth compression).

Communications for Tomorrow

- Likelihood of large cost reductions in LSI digital circuitry.
- Opportunities for greater privacy and security.

These advantages hold for digital transmission of voice and other analog signals as well as data.

Digital and packet networks. Still, the bulk of the existing telephone plant is analog. As a result, the telephone carriers and their competitors have developed separate, digital facilities to serve data users. AT&T, by adapting a section of existing analog microwave links for data (called *data under voice,* or *DUV*) and by other techniques, now provides end-to-end digital service to more than 50 cities. Bell plans to expand the Dataphone Digital Service (DDS) network as quickly as the FCC and state regulators allow. Competitors also hope to expand their present facilities and to introduce new satellite and terrestrial digital transmission links in the next decade. Throughout the 1980s, however, digital transmission networks will be separate from AT&T's switched-voice network, which will remain a hybrid of analog and digital facilities.

Packet networks, described previously, will be widely used in the 1980s for transmitting digital data and messages among terminals and computers. Once in place, these networks can be used to transmit voice messages as well. Packet services are now available commercially in about 80 cities. Like end-to-end satellite communications, packet services are designed for business data users who can afford the required processors at each terminal. At present commercial prices, a large user's average cost to send a page of characters or 20,000 bits runs between 5 and 25 cents; the marginal cost per page is about 1 cent. Prices should decline in the future and will probably be competitive with satellite transmission, although this will be determined more by market than by technological factors.

The telephone carriers presently lease digital circuits to other firms that provide packet services, but there is every indication that AT&T and other carriers plan to offer packet services of their own. Bell's Transaction Network Service (TNS) for credit checking and point-of-sale communications represents a clear step in this direction. If AT&T is successful in establishing TNS on an intrastate basis, it will surely try to generate other data services over these facilities, as well as extend them to form a national packet network. Many observers expect that AT&T will announce plans for such a Bell Data Network in the next year or two. Electronic funds-transfer systems also will use the packet-switching concept over digital facilities.

Toll-call signaling (CCIS). Another innovation, known as Common Channel Interoffice Signaling (CCIS), illustrates the growing use of digital transmission in the switched-telephone network. CCIS transmits the information necessary to route a call through the toll network over digital circuits separate from those used for the call itself. At present, call-control information is transmitted as analog tones in the voice-frequency range, giving the familiar series of "beeps" heard when one places a long distance call. With CCIS, call-control information can be transmitted at high data rates, enabling toll calls to be set up much faster on the switched network. This is vitally important for data transmission that sends short bursts of pulses between computer terminals.

Besides using toll circuits more efficiently, CCIS brings several additional advantages to the telephone carriers. It encourages use of the switched network for both voice and data. A separate circuit for call-signaling deters unauthorized use of the toll network by those who have assembled their own call-signaling devices. CCIS also carries the number of the calling party throughout the network as it sets up a call, thus making it possible to offer a number of new services. An intelligent telephone, for example, could display the caller's number while the phone was ringing, or provide a different tone for preselected "important" callers.

CCIS would also allow a nationwide firm to establish a single, toll-free "800" number and have calls routed automatically to the nearest local office. Another commercial service would be to provide a business customer with a list or computer tape with the names, addresses, and phone numbers of everyone who had called. Commercial subscribers could then use this information for, among other things, preparing advertising mailing lists. Such a service raises questions of privacy, however, especially if unauthorized parties could obtain such lists.

Finally, CCIS provides the technical means to offer other data services over its high-speed digital links, similar to those provided by the packet networks described above. The Bell System has made a major commitment to CCIS and expects to have it operating among large switching centers by the early 1980s.

Privacy on digital networks. Despite issues of privacy that may arise from new developments such as CCIS, digital communications in general offer technical prospects for increased privacy and security. Combining pulses from voice, data, and other services into a single digital bitstream makes it more difficult for amateur eavesdroppers to discern individual conversations or messages. Of course, a professional wiretapper with a computer could sort out information on a digital channel, but this would require training and dedication, as well as some sophisticated equipment.

Digital signals can also be coded or encrypted more easily than analog signals. IBM has developed a mathematical data encryption procedure that provides considerable security protection and is relatively easy to implement on a communications channel. This algorithm has been accepted by the National Bureau of Standards and seems likely to be generally adopted for commercial encryption. An LSI chip to implement the IBM algorithm is relatively inexpensive—$50 or so in large quantities—but the additional electronic circuitry needed to build encryption into a communications terminal may cost several hundred dollars or more.

Many business and government users seem willing to pay this price to increase communications privacy and security, but it is clearly too high for use in a home telephone. Also, although encryption and other techniques can be used by the telephone carriers on digital long distance channels, these techniques would not protect subscribers' local loops, where eavesdropping can most easily occur. When digital channels finally reach the home, a less expensive encryption circuit could be offered as a plug-in module to an intelligent telephone. Of course, it would be useful only for communicating with others who had similar decoders and appropriate keys. There is no indication that residential telephone (or cable television) subscribers would be willing to pay substantially more for greater assurance of privacy, but digital encryption is technically feasible and could be offered as an added service by communications carriers.

Switching Technologies

Computer-controlled switching. The major switching innovation in the past decade has been the introduction of computer-controlled switching in local central offices, often called *electronic switching.* The Bell System's No. 1 ESS is an example of a computer-controlled switch; other telephone companies and independent manufacturers produce comparable equipment. Computer-controlled switching has not meant a change in the basic switching mechanism itself, which relies on physically making and breaking electrical contact between circuits—a technique still oriented toward analog voice communications. Rather, computer control makes such electromechanical switching faster and cheaper in large telephone exchanges, and far more adaptable to changing calling patterns.

Technology in the 1980s

To the telephone subscriber, computer-controlled switching brings the introduction of pushbutton telephones, faster call connections, and the availability of new services, such as those listed in *Table 2-3*. To the telephone carriers, computer-controlled switching means lower switching costs per line, more flexibility of service, and opportunities to earn new revenues. Besides offering new calling features, computer-controlled switches permit inexpensive automatic recording and billing of both toll and local calls. This makes usage-sensitive pricing of local calls considerably more attractive to the telephone carriers.

If the trend toward usage-sensitive pricing continues, the telephone carriers will have more incentive to introduce (or encourage others to introduce) information-oriented services, such as those being developed by the British Post Office for its Viewdata system. Thus, automatic local-call billing, made possible by computer-controlled switching, seems a key to introducing a wide array of information services accessible over the telephone network. The conversion of some 20,000 telephone central offices from one switching technology to another is a formidable task involving an investment of tens of billions of dollars. AT&T has programmed the introduction of computer-controlled switching over 40 years. As of early 1978, 13 years after the first No. 1 ESS switch was introduced in the Bell System, only about 25 percent of subscriber lines were connected to computer-controlled switches. It will be the mid-1980s before half of the Bell System's lines are served by computer-controlled switches, and well into the 21st century before the conversion is completed.

The introduction of computer-controlled switching in the independent (non-Bell) telephone companies that serve predominantly small towns and rural areas has been even slower. Computer-controlled switching exhibits economies of scale, so that costs per subscriber are higher in small exchanges. This has led to the concept of a single computer-controlled switch handling several small exchanges via remote links. The Rural Electrification Administration and several independent companies are actively pursuing this approach. Still, the majority of non-Bell subscribers will probably not be served by computer-controlled switches during the 1980s.

Digital switching. Even as telephone central offices have begun converting to computer-controlled, electromagnetic switches, a different approach to switching has been developed that more fully integrates computer technology and is more compatible with digital communications. In digital or "time division" switching, the conducting paths connecting external circuits are not physically opened and closed. Instead, the data pulses associated with a given incoming signal are identified and sorted onto the correct outgoing circuit. In essence, a digital switch acts as a timed electronic gate, opening and closing under computer control.

Some of the newer computer-controlled PABXs employ digital switching, and digital switches (No. 4 ESS) are beginning to be installed in the toll telephone network. By 1990, the Bell System plans to route more than 75 percent of toll calls through digital switches. In 1977 the first commercial digital switches were introduced into a few independent telephone local exchange offices. However, no plans have been announced to bring digital switches to AT&T local central offices, even though many experts outside the Bell System believe them to be technically superior to the computer-controlled electromechanical switches scheduled for installation over the next 25 years.

Local Distribution

Broadcast radio and television, telephone wire pairs, and CATV coaxial cables are today's principal modes of local distribution. Each of these modes will evolve technically during the 1980s; but the distinction between voice and other narrowband services delivered over the switched telephone network, and broadband services provided on a cable distribution network, is likely to continue. Substantial expansion of mobile two-way communications should also occur throughout the decade, but portable telephones are likely to remain too expensive for most households. Optical fiber distribution will be introduced principally for commercial and government applications. Although integration of the cable and telephone networks may become technically feasible, it seems unlikely to be implemented before 1990.

Telephone Wire Pairs

The copper wire pairs running from telephone central offices to subscribers have been well-engineered to carry voice conversations and other analog signals—otherwise known as POTS, for Plain Old Telephone Service. No changes in metropolitan area local loops are needed to handle POTS. However, where underground ducts are already filled, or in rural areas with long local loops, more electronics may be placed between the subscriber and the central office to cut down on the number of necessary wire pairs and to improve service (e.g., to provide single-line rather than party-line service).

Local loops designed for POTS will also carry digital data at speeds that can fill a CRT display with text in a few seconds—

Technology in the 1980s

fast enough for information and message services to the home. Higher speed services up to 56Kbps can be distributed on special wire pairs that are engineered for digital data. In a business or institutional environment that makes heavy use of voice, data, and other communications, the technical trend is to concentrate the traffic at the subscriber's premises and transmit it digitally to the telephone central office. This is a principal function of the computer-based PABXs that will be commonly used by commercial subscribers in the 1980s. Some computer-based PABXs will connect directly to a digital switch in the toll network to avoid analog switching at the local central office.

Coaxial Cable Television

Cable television has evolved over the past 30 years to a reasonably mature technology for distributing television signals one-way from a central source (the headend) "downstream" to multiple subscribers. State-of-the-art cable systems can carry 30 to 35 television channels on coaxial cables attached to utility poles or installed in underground ducts. About 12 million households currently are cable subscribers; forecasts for 1985 range from 20 to 40 million.

Technical advances in coaxial cable technology in the next decade will be modest and evolutionary. Systems will be designed for greater reliability, lower distortion (and, hence, improved signal quality), and lower cost. Increasing the number of channels carried per cable is technically feasible, but not likely to be a major design emphasis, since few cable systems use the 30 or more channels they now can carry. Consumer demand for high resolution television, in conjunction with projection television or large flat-panel displays, would call for additional downstream bandwidth and perhaps digital television transmission to the home. However, neither seems likely to be a major factor in cable system design during the next ten years.

New cable systems in metropolitan areas will typically have several distribution points, or hubs, each with its own cable distribution network to subscribers. The hubs will be interconnected by microwave, coaxial cable, or (within five years) optical fibers and will be linked to other cities by satellite or terrestrial transmission. As a consequence, more cable systems will be able to provide channels from other cities, as well as nationally syndicated pay-television and other programs unavailable on broadcast television.

Portions of the cable bandwidth can be reserved for local video channels, computer data, or any other form of communications service. If the system is carefully engineered, some digital data can be carried along with the analog television signals on the same cable. Few cable systems today are designed to carry data, but the urban systems of the 1980s probably will be.

The technical capability for two-way cable communications—that is, for sending signals "upstream" from the subscriber to a hub or headend—was developed in the last decade but has been relatively little used. Two-way cable systems can reserve a certain bandwidth (usually 6 to 30 MHz) for upstream communications, using electronic filters to keep these signals from interfering with the downstream flow. Carrying signals in both directions on a single cable is not without problems, including added distortion, added cost, and sensitivity of the upstream signals to noise introduced at each television receiver. But it is technically feasible to send data, voice, and video signals upstream from the subscriber to a hub or headend.

Other technical possibilities for two-way cable communications include looping the cable back to the headend, or installing a separate wire or cable for the upstream link. Most new urban cable systems are designed for eventual two-way operation with an added investment of 15 to 30 percent. However, a more practical approach for many proposed interactive services is to use the telephone lines for the return link from the subscriber to the headend.

Cable systems have not yet found the combination of new services that would justify the added cost of two-way transmission or switching capabilities. In some metropolitan areas, cable companies may be able to compete effectively with the telephone carriers for the distribution of data and closed-circuit video services to institutional subscribers. But to date, the principal new service to the home is pay-television. More than one million households now subscribe to pay-television on cable. Their numbers are forecasted to double by the early 1980s. Broadcast pay-television is now available in Los Angeles and is scheduled to begin operation in several other U.S. cities by 1980.

Most cable systems now provide pay channels at frequencies not used for their basic services and install a special converter for each pay-television subscriber. This arrangement makes it relatively easy for subscribers to obtain their own converters and get the pay programming for free. Consequently, cable systems are moving rapidly to install electronic "traps" that can filter out the pay channels for those who do not pay to receive them. Alternatively, an "addressable tap" can be installed that allows pay programming to pass through upon a signal from the headend. Systems also can scramble or code the pay-television signals and provide a home decoder. Decoders are more expensive than frequency converters, but the decreasing cost of logic is bringing their cost down to below $100. Broadcast pay-television systems must use decoders, since all viewers in the area receive the same broadcast signals.

Pay-television subscribers today pay a fixed monthly charge for the special channels. This avoids the need for identifying and recording when subscribers are watching a particular pay program. However, per-program charging seems desirable from both the industry's and subscribers' standpoint. Several technical approaches to per-program charging are feasible and have been tried with moderate success. One approach is to record (on a card or magnetic strip) at the subscriber's terminal all pay programs watched, and then to have the subscriber periodically mail the record for billing purposes. An alternative is to have the subscriber dial a special access code by telephone, which then produces a signal from the headend to the subscriber's addressable tap or home terminal that permits viewing of the pay program.

Neither of these approaches requires two-way transmission on the cable system. However, two-way cable communication makes pay-television transactions much simpler. The subscriber can simply push a button to view a pay program, sending a message to the headend for billing purposes. Alternatively, a computer at the headend can routinely poll each subscriber's terminal to determine whether or not a pay program is being watched, and then record the result automatically.

Each of these technical approaches has advantages and disadvantages. The reduced costs of LSI logic and memory should favor installation of more sophisticated terminals in the home for pay-television billing. However, the fact that pay-television can be implemented relatively inexpensively on a one-way basis inhibits the introduction of two-way cable services.

Mobile Communications

Among local distribution modes, mobile communications stand to gain most from advances in LSI technology. Mobile systems of the 1980s will make extensive use of microprocessors and associated LSI circuits for signal processing and control functions in order to use the limited available frequency spectrum more efficiently.

Mobile communications include several different kinds of services:

- *Paging service*, requiring only one-way transmission from a base station to a portable paging unit. Most paging units today sound a tone or otherwise signal the user to call the paging service for a message. Advances in LSI circuitry will allow pocket pagers to record or display a brief message (e.g., the name and telephone number that the paged party is to call), as well as reduce the cost of paging receivers substantially below their present level of about $250.
- *Citizens band (CB) radio*, "party line" broadcasting channels over which any user can send and receive messages. CB radio has proved immensely popular in the past few years, but equipment sales have slowed recently, and there are signs of consumer saturation. Although CB radio channels have been used for data transmission (e.g., between computer hobbyists), the high congestion makes CB radio technically less desirable for such uses than two-way dispatch or mobile telephone services.
- *Dispatch and mobile telephone services*, requiring two-way communications between a fixed base station and a mobile unit, or between two mobile units. Dispatch services for taxis, ambulances, delivery vans, and other vehicles have traditionally been distinguished from mobile telephone service because they transmit very short messages, and consequently more users can share a given bandwidth. Technically, the two services are quite similar and can be provided on the same physical system. Mobile telephone base stations connect with the "wireline" carrier network to transmit calls between mobile users and fixed telephones.

In addition, telecommunications technologies are applicable for such other mobile functions as vehicle location, identification, and control.

The growth of mobile telephone services has been severely limited by the restricted frequencies made available by the FCC. Older mobile telephone units had fixed assigned frequencies which generally could not be shared by other users. Newer units share a block of channels, known as Multi-Channel Trunked Systems (MCTS), and automatically search for a free channel on which to transmit. Still, an MCTS transmitter broadcasts throughout its mobile service area (MSA), thus denying that frequency to other users and consequently limiting the total number of mobile units within the MSA. Current FCC guidelines recommend only 40 mobile telephone subscribers per channel on a 5-channel MCTS system.

Technology in the 1980s

Technology in the 1980s

Cellular mobile systems. One technical approach to expanding the number of mobile users is to reassign or "reuse" frequencies within a mobile service area. This is done by dividing a large MSA (typically covering several thousand square miles) into smaller cells ranging from one to eight miles in radius. Each cell contains a low-powered base station transmitter, with transmitters in adjacent cells operating at different frequencies. Frequencies used in one cell can then be reused in other nonadjacent cells within the MSA.

A cellular system can accommodate roughly 3 to 50 times the number of mobile units as the conventional multi-channel trunked system, depending on cell size and the frequency-assignment scheme. The smaller the cell size, however, the more elaborate the processing capability required at the base stations and mobile units to allocate of channels and "hand-off" units from one frequency to another. Fortunately, computer-controlled telephone switches and LSI-equipped mobile units can adequately handle such complexity at a cost roughly comparable to that of MCTS systems—about $1,500 to $2,500 per subscriber.

In 1974, the FCC authorized an additional 115 MHz for mobile communications, of which 40 MHz was allocated to the telephone carriers for development of cellular mobile systems. AT&T plans to test its first cellular system in Chicago in 1978 and to begin regular commercial service by 1980. The Bell system estimates that it can have cellular systems operating in 25 urban areas by 1985.

Digital packet radio. An alternative new technology for mobile communications is digital packet radio. In a packet radio system, a mobile unit transmits bursts of digital pulses using the entire mobile frequency band, rather than transmitting continuously over an assigned frequency channel. The pulses are passed along by fixed repeaters to a base station, where the information is either transmitted to the receiving mobile unit in the same MSA or sent out over the wireline network. The advantages of a packet mobile radio system, like the advantages of digital communications in general, are the system's ability to integrate voice and data communications efficiently, its flexibility in handling varying demand patterns, and its extensive use of LSI digital circuitry, which promises large cost reductions over the next decade.

Packet radio would be particularly suited for mobile data transmission. For example, the address of an ambulance or taxi pickup could be displayed in the vehicle, instead of communicating with the driver by voice. Similarly, a newspaper reporter could transmit story copy directly from the scene of the news event. With packet radio, a portable computer terminal would serve as an office. The principle disadvantage of the packet radio system is that, without bandwidth compression, digital transmission of voice conversations represents an inefficient use of the limited spectrum available for mobile communications. As bandwidth compression and coding techniques become less expensive in the 1980s, packet radio will become more competitive for analog voice transmission.

Developing either cellular or packet mobile systems would allow mobile telephone service to grow from its present level of about 200,000 users to perhaps several million subscribers by 1990. Although costs should come down somewhat from present tariff levels, mobile users probably will still pay several times the cost of regular telephone services. As a result, two-way mobile telephony in the 1980s will remain a service principally for business and professional users. Consumer demand may open up additional CB radio channels, but broad expansion of other mobile services to the general public would require even more spectrum reallocation and therefore appears more than a decade away.

Optical Fibers

The spectacular advances in optical communications technology over the past few years make optical fibers attractive for local distribution, as well as for long distance transmission of video, voice, and data services. This development will occur first among business and other institutional users. Optical fibers are already in use for high-speed computer-to-computer links in data-processing centers. They should soon also be practical as well for distributing video and high-speed data within an office building or commercial center.

In factories, optical fiber links are particularly good for protection against electrical interference, for eliminating damage to terminals due to damage to the link, and for personal safety. Providing the internal communications links to a rooftop satellite terminal appears to be another application. By the early 1980s, optical fibers may also provide the digital local loops from a computer-based PABX to the telephone toll network, or to a competitive transmission system.

The introduction of optical-fiber local loops to home telephone subscribers appears to be further away. Wire pairs are adequate for voice and low data-rate communications. Consequently, even though optical fibers might be less expensive than copper wires for new installations within the next decade, the lower cost would not justify the expense of removing existing local loops

Technology in the 1980s

(unless the cost of copper increased to a point where it became profitable to recycle telephone wires). Moreover, wire pairs carry direct electrical current to the home, providing power for the telephone, ringing voltages, and several other control functions. Since optical fibers are nonconducting, these functions would still have to be provided with conducting wires, or by using the subscriber's electric power, which would make telephone service more vulnerable to power failures. The conclusion is that optical fibers are unlikely to replace telephone wire pairs for residential local loops until a demand arises for switched broadband services to the home. Picturephone represents one such possibility, but there is no indication that this service will have arrived by 1990.

Optical fibers will be used for cable television distribution in the 1980s, although not as quickly or pervasively as some enthusiasts have forecast. Costs are still high (though dropping rapidly), and intermodulation distortion limits the capacity of present fibers to two or three analog television channels. However, within three to five years, optical fibers could be competitive with coaxial cable and microwave links for interconnecting cable system headends, distribution hubs, studios, and satellite receivers. TelePrompTer successfully demonstrated such an application in New York City in 1976. Within the next decade, optical fibers seem likely to supplant coaxial cables as trunks and feeders in some new systems, particularly those installed underground. The choice to use optical fibers rather than cables will be made on the basis of installed cost, rather than on the fiber's potential for increased capacity. And these initial installations are likely to be on conventional, entertainment-oriented, basically one-way television distribution networks.[25]

A conservative scenario thus shows optical fibers in the 1980s gradually spreading from cable system headends and hubs to trunk lines and then toward the final link to the home. By 1990, a small but growing fraction of the nation's television distribution plant will be optical. Eventually, it will prove economically feasible to bring a switched, two-way optical system to residential subscribers. At that point, a single integrated network for the distribution of television as well as telephone services may be preferable to maintaining separate systems. On the other hand, the advantages of having competitive links to the home may overshadow any cost savings from integration.

The decision whether the communications system of the 21st century will have a single superhighway or two parallel expressways to the home will undoubtedly be made on political as well as economic grounds. The technology will be available to support either approach.

Communications for Tomorrow

Notes

1. Had the mechanical calculator industry been regulated, we might still be waiting for delivery of electronic gate-to-electromechanical gear converters, rather than buying $9.95 electronic calculators at the supermarket.

2. The number of components per chip is several times greater for LSI memories, with their regular, repetitive arrays, than for microprocessors. The technological improvements expected during the next five years include the manufacture of larger chips, higher production yields, and the use of electron beams and x-rays to form even smaller components. LSI technology has advanced so rapidly that component density now is limited by the wavelength of light used to expose photolithographic patterns on the chip, and by the additional interconnections needed as the number of components increases. The size of the "wires" between components rather than the components themselves has become the limiting factor. See Ivan E. Sutherland, Carver A. Mead, and Thomas E. Everhart, *Basic Limitations in Microcircuit Fabrication Technology*, The Rand Corporation, R-1956-ARPA, November 1976.

3. The common unit of information is the binary digit, or bit. A single bit can be represented by the on-off switching of an electrical current in a simple telegraph system. As in telegraphy, letters, numerals, and other characters can be formed by a succession of on-off signals, or bits. Conventional telegraph systems have used five bits for each character, with a "shift" character to indicate whether a letter, numeral, or other character is being transmitted. This data code is still used for Telex transmission. Other forms of data transmission and storage typically use a seven- or eight-bit code to designate alphabetical and numerical (alphanumeric) characters, without need for a shift character. The "ASCII code" is the U.S. standard version of a seven-bit-per-character data code. In a teleprocessing system, an extra bit is often required for error correction purposes. Thus, eight bits become a standard unit—a "byte"—for computation and data storage.

4. An additional distinction is drawn between memory units whose data remain unchanged (e.g., the basic instruction set for a microprocessor, or a reference file of statistical data), and those whose data is changed or updated regularly (e.g., the stored results of a computation or a file of current stockmarket prices). The former are known as read-only memories (ROMs); the latter, as read-and-write or random-access memories (RAMs). ROMs are generally less expensive but less versatile than RAMs.

Technology in the 1980s

5. Using a data code with eight bits per character, a page of typewritten text with 24 lines, 80 characters per line represents 15,360 bits (15.4 kilobits). Facsimile scanning of that same page requires nearly 1 million bits (1 megabit). Storing each of the 310,000 picture elements in a standard U.S. television frame would require more than 2 megabits. A 200-page printed book contains about 4 megabits; the *Encyclopedia Britannica*, roughly 100 megabits.

6. The 525-line U.S. television standard limits the image quality of any large-screen television display. Consequently, unless new standards are introduced, consumers may not find large-screen television displays particularly attractive for home viewing.

7. Digital signals must be converted (modulated) to analog form before transmission over an analog channel, and demodulated before entering another digital system. The conversion devices are known as modulator-demodulators, or modems. Conversely, coder-decoders (codecs) are used to process analog signals for transmission over digital communications facilities.

8. In some word processing systems, individual units will share logic and memory functions, similar to minicomputer time-sharing systems.

9. Both professionals and their secretaries are assumed to have a terminal in the office of the future. Today's experience suggests that most professionals without prior exposure to computing systems (including those over 40) resist using a graphic terminal directly and hence will require printed materials. This seems primarily a generational problem; within ten years, most newly hired professionals and support personnel will have had some prior exposure to word processing. Word processing systems also will change the relative work roles of professionals, secretaries, and typists—a topic beyond the scope of this paper.

10. Another approach in current use is to use microfiche or other microforms to reduce physical storage requirements. This has advantages in storing handwriting, images and other graphics, and is currently less expensive than digital storage for most applications. Major disadvantages include problems of updating microforms and in transmitting them over communications lines. With memory costs declining at 30 to 40 percent per year, and with an increased emphasis on electronic document transmission, one expects to see digital storage chosen over microform storage for more systems in the 1980s. One also expects to see more integration of information processing and microform systems, as, for example, by maintaining an index of microform records on the computer system.

11. Henry D. Taylor, Jr., "HP-Communication System," *Proceedings of the National Telecommunications Conference*, Los Angeles, December 1977, pp. 21:6-1 to 21:6-5.

12. Raymond R. Panko, "The Outlook for Computer Mail," *Telecommunications Policy*, Vol. 1, No. 3, June 1977, pp. 242-253.

13. Cost comparisons are difficult to make and interpret, since electronic message transmission will substitute for intracompany document distribution and telephone calls as well as the mails. Relative labor costs (currently estimated to be $6 to $8 per letter) are more critical than transmission or postage costs, as is the importance placed on speed of delivery. Given these uncertainties, estimates of first-class mail lost to electronic transmission by 1985 range from 20 to 60 percent.

14. Roger Pye and Ederyn Williams, "Teleconferencing: Is Video Valuable or Is Audio Adequate?" *Telecommunications Policy*, Vol. 1, No. 3, June 1977, pp. 230-241.

15. Frequencies and bandwidths are expressed in cycles per second, or hertz (Hz), and in multiples of thousands of hertz (kilohertz or KHz), millions of hertz (megahertz or MHz), and billions of hertz (gigahertz or GHz). Analog voice channels require a nominal 4 KHz bandwidth; U.S. television signals occupy 4.6 MHz within a nominal 6 MHz channel bandwidth. Digital data transmission is commonly expressed in bits per second (bps), kilobits per second (Kbps), and megabits per second (Mbps).

16. EFT systems already exist for funds transfer and settlements among banks. They include the FEDWIRE system which links the Federal Reserve Banks and their member commercial banks in each district; the BANKWIRE teletype network, which interconnects several hundred U.S. banks; the CHIPS network, which provides more comprehensive funds transfer services among New York City banks; and the SWIFT system for international funds transfer among 400 member banks in Europe and North America. See, for example, R. A. Hall, "Money Movement Transfer Systems," *Proceedings of the National Telecommunications Conference*, Los Angeles, December 1977, pp. 02:1-1 to 02:1-4. These interbank networks should grow rapidly in the next several years and provide a base of experience for further expansion into consumer EFT systems in the 1980s.

17. J. R. Pierce, "The Outlook for Communications," undated.

18. These features can also be provided by computer-controlled switches in telephone company central offices. In some cases, the sharing of equipment located at the central office may result in lower service costs. However, telephone companies may have incentives to install logic at the central office, which

they control, rather than place it at the terminal, which is subject to competition. Consequently, telephone companies may not be enthusiastic about implementing these services with intelligent telephones.

19. Another early application of LSI logic to U.S. television sets is for automatic picture tuning.

20. A single-line strip printer, as is contained in some hand calculators, can be added to any terminal device for about $50. It appears to have limited utility, although some users prefer it over a liquid crystal or LED display for records of transactions and short messages. A receive-only matrix printer for full-page printing would cost about $200.

21. To give some indication of its size and complexity, the Bell Telephone System now includes approximately 120 million telephones, 80 million local loops, 150,000 PABXs and key systems serving 30 million telephones and other terminals, 10,000 local central offices, and 1,000 toll switching offices connected by 1 million toll transmission trunks.

22. Interference may also arise between beams directed to or from different satellites. Current regulations require satellite spacing at approximately 3 degree intervals in their orbits around the earth. Thus, only about 25 satellites can be assigned to use the same frequencies over the United States, which covers a 75-degree arc. However, the capacity available is still quite large. A 500 MHz satellite bandwidth at 12 GHz could accommodate 220,000 telephone channels and 100 television channels with multiple spot beams. Improved technology could also permit closer satellite spacing in the 1980s.

23. International communications are not considered further in this chapter. For an account of possible future developments, see Ithiel de Sola Pool and Arthur B. Corte, "Implications of Low-Cost International Non-Voice Communications," unpublished report, Cambridge, Massachusetts, Center for Policy Alternatives, M.I.T., September 1975.

24. S. Metzger, "Possible Use of Satellite Transmission for Direct Broadcast TV in a Future Metropolitan Communications System," in *Telecommunications for Metropolitan Areas: Near Term Needs and Opportunities*, Committee on Telecommunications, National Research Council, Washington, D.C., 1977.

25. The availability of optical fibers will bring renewed interest in the concept of a switched television distribution service. In a switched system, television signals are distributed from a central source to a local distribution center serving some tens to hundreds of subscribers. Each subscriber has a signaling link to the distribution center and can select a television channel

with a special dial or pushbutton terminal. The subscriber's signal then switches the appropriate channel at the distribution center for transmission to the home. This concept was developed in the 1960s by Rediffusion, Ltd., using twisted wire pairs, and by Ameco, Inc., using coaxial cables for the link from the distribution center to the home. It was never successful in the United States because of high cost and the large number of wires required. Any switched system designed for the 1980s could include the capacity for services other than television, and two-way communications among local distribution centers. It remains doubtful, however, that this kind of switched system would be cost competitive with conventional tree or hub systems for television distribution, even with low-cost optical fibers. Some early tests of the concept may well occur, especially if the experiments using optical fibers for two-way communications that are scheduled in Japan during 1978-1980 prove successful.

Selected Bibliography

Communications Technology—General

Committee on Telecommunications, *Telecommunications for Metropolitan Areas: Near Term Needs and Opportunities*, National Research Council, Washington, D.C., 1977.

Douglass D. Crombie (ed.), *Lowering Barriers to Telecommunications Growth*, Office of Telecommunications, U.S. Department of Commerce, Washington, D.C., November 1976.

James Martin, *Future Developments in Telecommunications*, 2d, ed., Prentice-Hall, Inc., Englewood Cliffs, New Jersey, 1977.

James Martin, *Telecommunications and the Computer*, 2d, ed., Prentice-Hall, Inc., Englewood Cliffs, New Jersey, 1976.

J.R. Pierce, "New Trends in Electronic Communication," *American Scientist*, Vol. 63, January-February 1975, pp. 31-36.

Ithiel de Sola Pool, "International Aspects of Computer Communications," Telecommunications Policy, Vol. 1, No. 1, December 1976, pp. 33-52.

Telecommunications Research in the United States and Selected Foreign Countries, A Preliminary Survey, Vol. I and II, Panel on Telecommunications, Research Committee on Telecommunications, National Academy of Engineering, Washington, D.C., June 1973.

Telecommunications and Society, 1976-1991, prepared for the Office of Telecommunications Policy, Arthur D. Little, Inc., Cambridge, Massachusetts, June 22, 1976.

Thompson, G.B., "Convivial Interactive Mass Communication Media," *Proceedings of the International Communications Conference*, June 1976, pp. 9: 3-4.

Computer and Component Technologies

A. Barna and Dan I. Porat, *Introduction to Microcomputers and Microprocessors*, John Wiley & Sons, New York, 1976.

John W. Bremer, "Hardware Technology in the Year 2001," *Computer*, Vol. 9, No. 12, December 1976, pp. 77-85.

T.A. Dolottam et. al., *Data Processing in 1980-1985*, John Wiley & Sons, New York, 1976.

David A. Hodges, "Trends in Computer Hardware Technology," *Computer Design*, February 1976, pp. 77-85.

Proceedings of the IEEE, Special issue on microprocessor technology and applications, Vol. 64, No. 6, June 1976.

Science, Special issue on electronics, March 18, 1977.

Scientific American, Special issue on microelectronics, September, 1977.

I.E. Sutherland, C.A. Mead, and T.E. Everhart, *Basic Limitations in Microcircuit Fabrication Technology*, R-1956-ARPA, The Rand Corporation, Santa Monica, California, November 1976.

J. Warren, "Personal and Hobby Computing: An Overview," *Computer*, March 1977, pp. 10-22.

Computer/Communications Terminals

V.G. Cerf and A. Curran, "The Future of Computer Communications," *Datamation*, May 1977, pp. 105-115.

R.R. Daynes, *Videodisc Technology Use Through 1986: A Delphi Study*, Navy Personnel Research and Development Center, San Diego, California, December 1976.

Edward Dickson and Raymond Bowers, *The Video Telephone: Impact of a New Era in Telecommunications*, Praeger, New York, 1973.

Paul Green and Robert Lucky (eds.), *Computer Communications*, IEEE Press, New York, 1975.

R.W. Hough, *Teleconferencing Systems: A State-of-the-Art Survey and Preliminary Analysis*, Stanford Research Institute, SRI Project 3735, May 1976.

Institute for the Future, *Development of a Computer-Based System to Improve Interaction Among Experts*, Menlo Park, California, 1974.

Raymond R. Panko, "The Outlook for Computer Mail," *Telecommunications Policy*, Vol. 1, No. 3, June 1977, pp. 242-253.

Raymond R. Panko and R.U. Panko, "An Introduction to Computers for Human Communication," *Proceedings of the National Telecommunications Conference*, Los Angeles, California, December 1977, 21:1, pp. 1-6.

Proceedings of the IEE, Special issue on digital signal processing, Vol. 63, No. 4, April 1975.

Proceedings of the IEEE, Special issue on man-machine communications by voice, Vol. 64, No. 4, April 1976.

Roger Pye and Ederyn Williams, "Teleconferencing: Is Video Valuable or Is Audio Adequate?", *Telecommunications Policy*, Vol. 1, No. 3, June 1977, pp. 230-241.

Raymond M. Wilmotte, "Technical Frontiers of Television," *IEEE Transactions on Broadcasting*, Vol. BC-22, No. 3, September 1976, pp. 73-80.

Technology in the 1980s

Raymond M. Wilmotte, *Technological Boundaries of Television*, Vol. 1-3, Prepared for the Federal Communications Commission, Washington, D.C., November 1974.

Transmission and Switching Systems

Applications of Satellite Business Systems for a Domestic Communications Satellite System, Vol. II, "System Description," Filed with the Federal Communications Commission, Washington, D.C., December 1975.

J.A. Arnaud, *Beam and Fiber Optics*, Crawford Hill Laboratories, Academic Press, Holmdel, New Jersey, 1976.

J.E. Burtt et al., *Technology Requirements for Post-1985 Communications Satellites*, Summary Report, Prepared for Ames Research Center, Mountain View, California, October 1973.

Burton I. Edelson, "Satellite Communications Technology," *The Journal of Astronautical Sciences*, Vol. XXIV, No. 3, July-September, 1976, pp. 193-219.

R.L. Gallawa, *User's Manual for Optical Waveguide Communications*, Office of Telecommunications, OT Report 76-83, Washington, D.C., March 1976.

R.L. Gallawa, "Optical Waveguide Technology for Modern Urban Communications," *IEEE Transactions on Communications*, Vol. Com-23, No. 1, January 1975, pp. 131-142.

Detlef Gloge, *Optical Fiber Technology*, The Institute of Electrical and Electronics Engineers, Inc., New York, 1976.

I. Jacobs, "Telecommunication Applications of Fiber Optics," *Proceedings of the National Telecommunications Conference*, Los Angeles, December 1977, 06:3, pp. 1-3.

J. Palmer and Y.F. Lum, "Communication Satellites: The Future," *Telesis*, March/April 1974, pp. 162-167.

Pritchard, W.L., "Satellite Communications—An Overview of the Problems and Programs," *Proceedings of the IEEE*, Vol. 65, No. 3, March 1977, pp. 294-307.

Proceedings of the International Switching Symposium, Institute of Electronics and Communication Engineers of Japan, Kyoto, Japan, October 1976.

C.A. Siocos, "The Use of Satellites in Broadcasting," *Journal of the SMPTE,* Vol. 84, February 1975, pp. 61-70, 135-150.

N.E. Snow and N. Knapp, Jr., "Digital Data System: System Overview," *The Bell System Technical Journal,* Vol. 54, No. 5, May-June 1975, pp. 811-832.

L.S. Stokes, *Technology Forecasting for Space Communication,* Executive Summary, Prepared for Goddard Space Flight Center, Hughes Aircraft Company, El Segundo, California, June 1973.

Local Distribution

A Review of Land Mobile Radio, Office of Telecommunications Policy, Washington, D.C., June 1975.

J. Frey and A. Lee, "Technologies for Land Mobile Communications," Working Paper No. 11, Cornell University Technology Assessment Project, Program on Science, Technology, and Society, Ithaca, New York (undated).

D. Gillette, "Innovation in the Exchange Area," *Bell Laboratories Record,* October 1975, pp. 361-367.

IEEE Transactions on Communications, Special issue on interactive broadband cable systems, Vol. Com-23, No. 1, January 1975.

Robert E. Kahn, "The Organization of Computer Resources into a Packet Radio Network," *IEEE Transactions on Communications,* Vol. Com-25, No. 1, January 1977, pp. 169-178.

J.F. Lester, "What's Happening to the Loop Network," *Proceedings of the National Telecommunications Conference,* Los Angeles, California, December 1977, 33:1, pp. 1-4.

L.J. Lukowski, "Loop Plant Evolution," *Proceedings of the National Telecommunications Conference,* Los Angeles, California, December 1977, 33:2, pp. 1-6.

Technology in the 1980s

E. Nussbaum, "Emerging Technologies in Local Exchange Communication Networks," *1975 IEEE Intercom Conference Record*, April 1975, pp. 5/6:1-6.

L.R. Pamm, "Transactions Network: Data Communications for Metropolitan Areas," *Bell Laboratories Record*, January 1977.

J.L. Stern, "Business Data on CATV," *TV Communications*, January 1975, pp. 18-27.

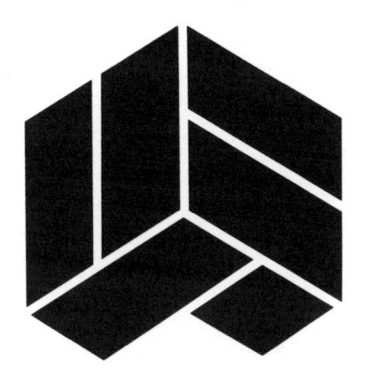

Part 2.
Communications Industry Structure and Regulatory Boundaries

3.
Boundaries to Monopoly and Regulation in Modern Telecommunications

Leland L. Johnson

Development of the Industry

In pursuing the mandate of the Communications Act of 1934 "...to make available, so far as possible, to all the people of the United States a rapid, efficient, nationwide, and worldwide wire and radio communication service...," three propositions have played prominent roles:

(1) Only one entity can serve a single geographic area most economically.
(2) Cost averaging, affording cross-subsidization among routes and services, would contribute to rapid development of a nationwide, easily accessible system.
(3) A single entity should have overall responsibility for providing end-to-end service to ensure efficiency and reliability.

127

With respect to the first, attempts by competitors to string parallel lines through cities and villages, as well as over long distances, were regarded by the industry and most customers alike as patently wasteful. Economies of scale in both local and long-distance transmission, under which the cost per unit of service declines as the number of units increases, led to the commonly held view that the telephone industry was a "natural" monopoly able to serve the public at a lower unit cost than would be possible if several firms attempted to compete. This view gained credibility in the early decades when telephone companies struggled to gain control over particular geographical areas by installing duplicative facilities. While some cities (Los Angeles, for example) ended up being served by two telephone companies, each retained its own exclusive local service area.

Today, the Bell System provides about 83 percent of the nation's telephones covering about 41 percent of its land area, while 1,600 independent telephone companies serve the remainder. The Bell System, interconnecting with independent telephone companies, holds a monopoly of message toll telephone service; Western Union holds a monopoly in public telegraph service; and, until recently, only the two competed for private-line services.[1] Interstate services are regulated by the Federal Communications Commission, while intrastate services are regulated by individual state or municipal regulatory bodies.

In previous decades there was little challenge to this market and regulatory structure because common carrier services were few and relatively homogeneous. The message toll network served individual subscribers using the ubiquitous black telephone. Private-line services encompassed four basic submarkets: telegraph private lines, used largely by newspaper wire services; telephone private lines, principally used by large businesses; audio and video private lines, used by broadcasters; and audio and video private lines, used by television networks.

The second proposition was that rapid development of a nationwide "wire and radio" system would be facilitated by permitting common carriers to average their costs in setting rates to customers. Under protected monopoly, profits from high-density routes would be used to support unprofitable services to sparsely populated, rural areas. With the requirement that they serve all classes of customers under "fair and equitable" terms, the carriers were permitted to price individual services largely as they chose, subject only to the overall constraint that total profits not exceed a "fair" return on their investment.

The third proposition was that common carriers should have integrated end-to-end service, including transmission and terminal facilities. Only if a single company had responsibility (or companies suitably interconnected subject to a single set of technical standards) could "technical integrity" be ensured. If responsibility for the system were divided among numerous suppliers, the industry argued, each would attempt to make cost and technical tradeoffs favorable to itself, but perhaps harmful to the network and to other users by causing interference, dangerously high voltages, and other problems. Thus, because telephone companies preferred to lease rather than sell terminal equipment to their customers, in most cases they use equipment manufactured by their own subsidiaries and they have sought to prohibit customers from attaching their own terminals to the network.

Guided by these three propositions, the telecommunications system has grown rapidly, with the Bell System and many independents enjoying strong financial positions. Although Western Union has suffered reverses as a consequence of a long-term decline in message telegraph service, vigorous development of the industry, financed by private capital, stands in contrast to some other regulated industries, such as the airlines, with their history of large federal subsidies and frequent financial difficulties.[2]

The Changing Environment

In recent years, dramatic advances have called into question all three propositions. Although the boundaries of monopoly were drawn earlier around the entire industry—with competition between Bell and Western Union for private-line service being the sole exception—new technologies and service demands have brought a marked shift. Perhaps the most significant development is the digital computer.[3]

> The emergence of the commercial computer began in 1951 when UNIVAC, the world's first electronic digital computer, was delivered to the U.S. Census Bureau. By 1962, over 10,000 computers were installed in the United States and, by the mid-1960s, about 30,000 were in place. Today, it is estimated that there are over 200,000 digital computers installed throughout the world. In just 25 years the computer has become a vital and integral part of the U.S. economy and has virtually revolutionized the manner and efficiency of performing business transactions, analyzing and storing information, performing scientific calculations, controlling complex manufacturing processes, and automating office functions.

Communications for Tomorrow

The computer has generated demands for services quite different from those of the conventional telephone. With connection of scattered remote terminals to central computers and time-sharing of computers among multitudes of users, the requirements for data transmission cannot be met by a network designed for voice communications.

For example, the voice network's analog transmission involves an infinite number of amplitudes, in contrast to digital signals that have only two levels (on and off). Within a wide range of "noise" levels, distortion in digital signals can be eliminated during transmission so that an exact replica is received at the destination. But exact replicas of analog signals cannot be produced. Although the human ear can tolerate this distortion in an analog system, it increases the error rate for data transmission, requiring costly corrective measures.

Moreover, computers use communications links in short bursts of a few seconds or less, while a typical telephone call continues for minutes. Just to set up a call on a switched telephone network takes 10 to 15 seconds—a small time requirement for a telephone call, but a large one for computer use.

The voice network also is limited because it cannot maintain full duplex operation, that is, transmit in both directions simultaneously at all times. Because of the need for echo suppressors and limitations imposed by the two-wire portions of the telephone work, a delay—generally imperceptible to the human ear—occurs when the direction of conversation is reversed. But this delay poses difficulties in efficient use of high-speed computers.

Another advance, commercialized after World War II, is terrestrial microwave transmission. A microwave network permits communication over long distances by use of radio relay towers located about 20 miles apart to provide the line-of-sight operation required in the microwave portion of the radio spectrum. This technology is radically superior in reliability and cost per channel to the earlier narrowband wire network. Among other things, its broadband capability permitted the first coast-to-coast transmission of television signals. Economies of scale afforded by microwave technology helped to maintain the basis of natural monopoly; but microwave links can be easily built at low cost for specialized use as well, as elaborated later.

Development of the communications satellite has attracted extraordinary public attention. One satellite can handle the equivalent of a dozen or more television channels or some 5,000 telephone calls over distances of thousands of miles (the only constraint being that transmitting and receiving earth stations be within line of sight of the satellite). Eliminating the need for numerous microwave relay stations, satellites offer possibilities of lower unit costs over long distances. Although they also have economies of scale, large and rapidly growing demands for long-distance transmission permit economic use of several systems side by side, as discussed later.

With other advances afforded by such developments as the transistor, integrated circuit, and minicomputer, the proliferation of applications for remote data processing and other specialized communications services requires a wide variety of terminals. Those for verifying credit cards are different from those for handling airline reservations systems, for text editing, for graph-plotting, or for high-speed facsimile transmission. High-speed operation is important in some applications, but not in others. Light-weight portability is required in some cases but can be sacrificed in others.

These advances raise questions about whether new and quite different networks are required in addition to the conventional telephone network, and whether these need to be supplied by a single, regulated monopolist. For not just economies of scale for basic services are at issue, but the economies of delivering specialized services as well, services quite different from the relatively homogeneous ones discussed earlier. With the increasing heterogeneity of the terminal market, nascent terminal manufacturers and others have questioned whether the needs of society are best met by having responsibility for end-to-end service in the hands of one supplier, or whether it is preferable to have the customer provide terminals and other similar equipment and plug them into the telecommunications industry's transmission network, just as the customer provides appliances and plugs them into the electric power industry's network.

The Emergence of Competition

Long-Distance Transmission

The first major break with tradition came in the FCC's "above 890" decision in 1959.[4] Previously only railroads and other "right-of-way" companies had been permitted their own communications links. But the flexibility and low cost afforded by microwave technology encouraged many other groups to seek systems. After lengthy inquiry, the FCC decided that authorizing private point-to-point microwave systems did not pose problems in efficient use of the limited radio spectrum; that enough spectrum space was available for all needs in the foreseeable future; and that neither Bell nor Western Union would suffer adverse economic effects from private systems. About 190,000 route miles of privately owned microwave have been constructed since 1959.

In response to increasing needs for channels engineered for burgeoning computer uses, the FCC granted permission a decade later to Microwave Communications, Inc. (MCI) to provide data and other specialized services between St. Louis and Chicago. Despite counter assertions by Bell and other carriers, the Commission concluded that the new services would meet an important public need and would not unduly duplicate existing facilities.

In 1971, the Commission broadened possibilities for competition in private-line markets with its "specialized common carrier" decision to permit additional carriers to enter the market:[5]

> Permitting the entry of specialized common carriers would provide data users with the flexibility and wider range of choices they required. Moreover, competition in the private line market was expected to stimulate technical innovation, the introduction of new techniques, and production of those types of communications services which would attract and hold customers.

The FCC deliberated for years about whether to permit only existing common carriers to use satellites for domestic services, or allow Comsat—the nation's chosen instrument for international satellite service—to serve the domestic market as well or open entry to firms outside the common carrier field. It concluded that here, too, competition would serve the public interest best, and it adopted an "open skies policy" in 1972 to permit entry by any firm meeting certain qualifications.[6]

As a consequence of these decisions, a number of specialized carriers entered the market: the Data Transmission Corporation (Datran), the first to provide an all-digital data transmission service; an expanded MCI Communications Corporation and its several affiliates; the Southern Pacific Communications Company, and United States Transmission Systems. Among satellite carriers are Western Union, the American Satellite Corporation (currently leasing Western Union channels), RCA Globcom, and AT&T.

In addition, three "value-added" carriers—Telenet, Graphnet, and Tymshare—lease channels from other carriers to provide "packet" switched services. By temporarily storing data and transmitting them in packets over whichever route is open between origin and destination at that instant of time, these carriers enhance efficient network use.

Of course, the future of these markets cannot be predicted reliably. One firm, Datran, has gone bankrupt, and the first to apply for value-added service—Packet Communications—failed to begin operation for lack of capital. At the same time, large and well-financed firms are planning to enter. Notably, the FCC recently has approved applications by IBM, Comsat General Corporation, and Aetna Life and Casualty to provide satellite links among scattered plants of large business firms by using small earth terminals located on their premises, thereby bypassing entirely the existing telephone network.

Terminal Markets

In line with the third proposition about end-to-end service and technical integrity, Bell tariffs prohibited interconnection of customer-supplied equipment until the mid-1950s. Previously, even shoulder rests attached to the receiver, dial locks, and plastic covers on phone books were forbidden.[7]

Despite the telephone industry concerns about integrity of the system, the FCC gradually has forced Bell to liberalize tariffs on customer-provided equipment interconnected with the telephone network, starting with the Hush-a-Phone decision in 1957 and the Carterfone decision in 1968.[8] In response to these early decisions, AT&T filed tariffs permitting customers to connect their own terminals to the network if carrier-supplied protective devices were placed between the terminals and the network and, in some cases, if carrier-supplied network control signaling units were included.

Subsequently, the equipment that users obtained from the telephone company in some cases was identical to that which others purchased or leased directly from the manufacturer. Yet the telephone company did not employ special protective devices when it provided this equipment, but required customers to do so when they sought to connect the same device. Moreover, installation and monthly charges were assessed for these connecting arrangements—a substantial additional cost not incurred by telephone company customers.

Judging this practice to be unwarranted discrimination, the FCC established a federal-state joint board proceeding in 1972, under which an FCC registration program was initiated in 1975 for ancillary data and extension telephone equipment. Under the program, suppliers either can certify the equipment itself (e.g., a telephone or an answering set) or use registered protected circuitry in the equipment. After certification, terminals can be connected to the telephone network without carrier-supplied arrangements. In 1976, the registration program was expanded to include main telephones, PBXs, and key telephones. The registration program is now in effect.

The telephone companies—particularly the Bell System—have opposed the registration program in the FCC's Dockets 19528 (dealing with the technical aspects of interconnection) and 20003 (dealing with the economic effects of competition). The FCC maintains that even before its policies and procedures were modified, about 1600 independent telephone companies—many of them small co-ops serving rural areas with a wide variety of commercially available equipment—satisfactorily interconnected with Bell's network. Moreover, certain large businesses and government organizations (railways, pipelines, other public utilities, the National Aeronautics and Space Administration, and the Department of Defense) were permitted interconnection without specific protective equipment.

Despite delays of litigation—which has now been resolved in favor of the FCC's registration program—the "interconnect market" involving sale or lease of terminal and related equipment directly to users has been growing with an impressive array of offerings. Office switchboards, automatic dialers, answering devices, hand sets, key sets, data terminals and facsimile devices are among the many examples.

Issues of Public Policy

With growing competition in private-line and interconnect markets, a number of issues arise:

- Effects on technical integrity of the network.
- Cross-subsidization between monopoly and competitive services.
- The proper scope of competitve entry.
- Distributional effects on classes of users.
- Adequacy of current institutional arrangements.
- Effects on innovation and new services.

Technical Integrity

Only scattered and anecdotal evidence indicates possible harm or difficulties experienced in the network as a consequence of attaching customer-provided equipment. A recent evaluation of filings in the FCC's Docket 20003, dealing with the impact of

Monopoly and Regulation in Telecommunications

competition in the telephone industry, concludes that "no persuasive evidence...was presented to substantiate Telco assertions that interconnection will result in network harm."[9] Nor has any telephone user group complained about interference it has experienced as a consequence of interconnection by others.

Although one easily can conceive of particular terminal devices that would cause harm, the standards promulgated under the FCC's registration program, combined with earlier Bell interconnection arrangements, apparently have sufficed to protect the network. Bell's own standards were adopted in the registration program. In laying down standards for certification, the Commission asked Bell to set forth standards it thought necessary to protect the network. The Commission adopted those standards, and required that they be met by both Bell and outsiders to guard against the possibility of Bell establishing impossibly high standards to freeze out interconnect suppliers.

However, AT&T and other carriers argue that continuing interconnection will affect the network through accumulated interface. Although adding a single customer-supplied terminal may contribute little interference, the carriers contend that as the build-up continues, aggregated interference will become significant, and it will be difficult or impossible to identify which terminals are the offenders. This view has been espoused by one AT&T official who likens the problem to "environmental pollution and conservation, where large numbers of small deviations result in intolerable general degradation which affects all."[10]

However, if we are to take seriously the danger of "pollution," then we must be alert to the possibility that among the potential offenders are the 1600 independent telephone companies that connect all manner of terminal equipment to the nationwide network and whose continuing growth, according to the argument, could contribute to aggregated interference. Should we not therefore seriously entertain the possibility of cracking down on independent telephone companies, especially small rural co-operatives?

In short, given the long experience of many users successfully interconnecting with the nationwide network and the apparent success of the FCC's registration program, the issue of maintaining technical integrity is the easiest to handle among all those discussed in this paper.

Cross-Subsidization

When a service, or combination of services, does not generate sufficient revenues to cover its cost, the deficit can be covered by a variety of means, including a subsidy from taxpayers, contribution by stockholders or, in the case of cross-subsidies, by users of other telecommunication services. Cross-subsidization is probably the most fundamental problem in telecommunications policy for, if it did not exist, one seriously could entertain the notion that the entire field be opened to competition. Particular portions of the industry would move toward monopoly or toward competition depending upon economies of scale, the nature and level of demand, and other factors as expressed in the marketplace, rather than as estimated in paper studies and adversary filings.

The danger of permitting a regulated monopolist who is guaranteed a fair rate of return to compete with outside firms is that the regulated firm may use revenues from its monopoly markets to price its competitive services at less than the additional cost of providing them. As well expressed by one government representative:[11]

> I think we must begin by acknowledging the presence of a real dilemma. The established carriers have a multiple-service capability. Let them respond in an unrestrained fashion to new entry into selected markets, and they may easily draw on high earnings for monopoly services to subsidize unreasonably low or predatory prices for competitive services. As you see, this is the other aspect of the problem of cross-subsidy. But if you impose restraints that effectively keep the carriers out of the competitive markets, then you are condemning existing plant and facilities to less than fully productive use; and you are also—very importantly—foregoing one of the chief benefits of competition, which is to stimulate greater flexibility and innovation and responsiveness to customer needs on the part of the established carriers.

Were it possible to separate easily the costs of each service, the FCC simply could require that total revenues for each service just cover the total costs of each (including a fair rate of return on investment devoted to each), in which case cross-subsidization would not arise. But how does one compute the underlying costs of a particular service? The costs that can be identified directly with the service, such as a teletypewriter used for data communication, can be accounted for without great difficulty. However, a serious stumbling block arises because services share so many telecommunications facilities. The basic telephone instrument itself is used for local exchange, intrastate toll, and interstate toll. Microwave towers contain radio relay equipment for a variety of data, audio, and video needs. The problem, then, is to devise satisfactory ratemaking criteria to ensure that each service covers at least its own directly identifiable cost and, at the same time, to enable all services collectively to generate sufficient additional revenue to cover common cost facilities and to provide the firm with a more or less "fair" rate of return on investment.

The FCC generally uses a "fully distributed cost" (FDC) approach under which these costs are allocated to various services roughly in accordance with their relative use of the facilities in question. Thus, if a common-cost facility is used 40 percent of the time for service A and 60 percent of the time for service B, its cost would be allocated accordingly.[12] The difficulty with this approach is that the total cost estimated for a particular service, including an FDC allocation of common cost, does not necessarily measure the cost burden on society of that service. If the monopolist is required to price at levels high enough to cover this fully distributed cost in competitive markets, the danger of "umbrella" pricing arises with the possibility of market entry by competitors whose costs of providing the service are *higher* than those of the regulated monopolist.

The danger of umbrella pricing arises in a different form when the Commission simply prohibits the firm from offering a particular service—for example, the Commission requirement that Bell *not* use communication satellites for its private-line services for three years in order to give competitors an opportunity to gain a foothold. Such practices immediately raise fears of cartel-type situations similar to those that have arisen in the transportation field where markets in effect are carved up among firms by regulatory decree through erection of barriers to entry and restrictions on price competition.

In contrast, one could argue that the monopolist shoud be permitted to reduce prices so long as it is able to cover the *additional* cost generated by its competitive service—the "incremental cost pricing" approach along the lines advocated by the Bell System. Bell argues that, to avoid predatory pricing, tariffs should be set to cover at least the long-run incremental costs of its competitive services—that is, the additional costs taking into account the time required to adjust the scale and nature of long-lived capital facilities. With these costs serving as a floor, Bell's provision of the competitive service would not constitute a burden on its monopoly markets because basic telephone users would pay no higher rates than they would were Bell not permitted to offer the competitive service. In fact they would benefit to the extent that the competitive services contribute *anything* to covering common cost that otherwise would be borne entirely by other users.

However, many regulators see two main problems in incremental cost pricing. First, common costs would be allocated roughly in accordance with the differences in price elasticity of demand for the various services. Services having a demand relatively unresponsive to price changes (presumably basic telephone services) would bear a relatively heavy portion of common costs, while services having a demand highly responsive to changes in price (presumably competitive private-line services) would bear a small portion of common costs. To put it more crudely, common costs would be allocated in accordance with what the market would bear. This approach strikes many people as unfair, especially for services provided by regulated monopolies. The basic telephone user not able to enjoy the fruits of competition would be left picking up the lion's share of the tab while others would escape by paying only their own incremental costs. Although recognizing that, under certain conditions of "competitive necessity," a reduction in rates for particular services should include an apportionment of common costs roughly in accordance with the relative use of common-cost facilities.

Second, long-run incremental costs are extraordinarily difficult to estimate. Among other problems, they involve forecasts of revenues and costs subject to wide error. Complications arise insofar as elements that appear at first blush to be common costs may be affected by offering additional service. For example, a future expansion in capacity to accommodate private-line service might involve installation of additional radio repeaters on a microwave tower used simultaneously for message toll and other services. The radio repeaters themselves are direct costs easily attributable to the private-line service. The microwave tower is a common-cost facility. However, were it necessary to strengthen the tower because radio equipment and antennas were added, a directly attributable cost also would arise from this additional requirement—a requirement difficult to estimate because the future demand for private-line service, as well as for other services sharing the tower, is subject to uncertainty. Moreover, the growing multiplicity of services over the route in question might be satisfied by a newer technology (such as coaxial cable) whose costs and physical characteristics are in doubt.

Whatever are the merits of the incremental-cost approach, its pursuit would leave the Commission dependent upon whatever data and estimates are provided by Bell. In the words of a former chairman of the FCC:[13]

> ...incremental costs methodology leaves the consumer completely unprotected from the telephone company pricing decisions

which, inadvertently or intentionally, favor large-volume, private-line users to the detriment of the public telephone subscribers. Incremental pricing is based on the estimates of future costs. Obviously, such predictions are difficult for any regulatory body to intelligently review since they are based on estimates rather than on actual data. If the telephone industry is allowed to selectively price its services incrementally, the Congress would be guaranteeing that shortfalls in those estimates of future costs would be borne by the public telephone rate payer. In other words, telephone rate payers would pay the costs of company errors or predatory pricings of competitive services.

Recognizing the dilemma of umbrella and predatory pricing, the Commission has gone through a long exercise in its private-line case (Docket 18128) to define better criteria for allocating common cost. In fact, it has delineated and examined no fewer than seven different ways of allocating costs in accordance with fully distributed cost criteria! At this writing, the FCC is working with Bell to implement so-called Method No. 7 which moves part of the way toward a long-run, incremental-cost approach. However, problems of umbrella pricing and predatory pricing persist. Part of the difficulty is that it is not enough to make estimates for a broadly defined market such as all of Bell's private-line services; together, but one also must investigate sub-markets in direct competition with specialized carriers and interconnect companies as elaborated below.

As a consequence, the FCC has been greatly troubled in responding to tariff applications by various competitors. For example, several years ago Bell moved away from average-cost pricing, as a competitive response to MCI, by filing a "hi-lo" tariff under which high-density routes competitive with MCI would bear lower tariffs than those elsewhere in the country. After many months of litigation, the tariff went into effect in mid-1976 but subsequently was disapproved by the FCC on grounds that Bell had not supplied sufficient information to show its reasonableness. In response, Bell filed a "multiple schedule, private-line" tariff now being examined by the FCC to determine its consistency with the fully distributed cost methodology developed in Docket 18128.

The problem of cross-subsidy arises also in cases where Bell is both competitor and a supplier to outside groups. Perhaps the best example is development of the electronic funds transfer services (EFTS). This field is not now subject to common-carrier regulation but is served by a number of firms—banks, credit card institutions and other financial entities—that lease lines from common carriers and provide packages of terminals, computer facilities and software for end-to-end service. However, Bell is offering a competitive service, Transaction Network Service (TNS), which is a switched data communications system with terminals designed to accommodate high-volume, short-duration, inquiry-based applications such as credit and check authorizations. Bell is offering TNS in the states of Minnesota and Washington, with plans to submit tariffs to other state commissions, and may introduce an interstate service as well. One of the objections to TNS is that competitors are dependent upon Bell's "transparent" network consisting of dial-up and lease lines in offering their own services. They fear that Bell, being both a competitor and a supplier, will provide the transparent network to them under increasingly unattractive terms as an anticompetitive measure, while at the same time offering its packaged TNS service at predatory prices.[14]

The specialized common carriers' dependence on Bell for access to local loop and switching facilities poses similar problems. Indeed, starting from the MCI decision in 1969, controversy has raged about conditions under which Bell should be required to provide interconnection by specialized common carriers to reach customer premises. After much court action and strenuous efforts by the FCC, agreement finally was reached in 1975 among MCI, other carriers, and Bell about technical and financial terms of interconnection arrangements. During this process, MCI filed an antitrust suit in 1974 against Bell, charging that it "had attempted to monopolize and unreasonably restrain trade in the business and data communications markets in violation of federal antitrust laws."[15]

In such cases, evaluating Bell's underlying costs to determine whether it is competing "fairly" is a monumental task. As yet more offerings emerge in response to continuing technological advance and growing demands for service, uncertainty and delays from lengthy administrative processes and litigation will likely become increasingly serious if the FCC's policy of encouraging "full and fair" competition is to be served.

The Scope of Competitive Entry

Compounding the problems of predatory and umbrella pricing, serious difficulties emerge in defining the scope of appropriate competitive entry. The Commission's decision in the specialized common carrier case was premised on the notion that a clear distinction existed between the public switched telephone network, generally deemed to be a natural monopoly, and private-line services, where competition might provide a feasible alternative. However, some services offered by the specialized carriers are taking on progressively more of the characteristics of message toll services. As noted by the former chairman of the FCC:[16]

> ... I frankly have been disappointed that some new carriers—while preaching the virtues of diversity, innovation and specialized offerings in the private-line market—have attempted to establish what to me are clearly basic message services. In my opinion, these attempts represent something of a breach of faith with the Commission and should be precluded by definitive administrative rulings.

As one example, the FCC concluded that MCI's "Execunet" service introduced in 1974 was essentially a switched public telephone service similar to Bell's offering, and that the service therefore should be withdrawn. MCI is seeking approval for its switched private-line service, while the Southern Pacific Communications Company is seeking authorization for its "Sprint" service. The FCC's disapproval of Execunet has been overturned in the courts and, at this writing, the way is clear for specialized carriers to offer switched services.

Issues also have arisen about the extent to which the scope of regulation should be expanded to include previously unregulated activity and the extent to which common carriers should be permitted, if at all, to enter unregulated markets. These are well illustrated by concerns about the extent to which the computer industry should be regulated and the appropriate role of common carriers.

With rapid advances in computer technology, apprehension was widely expressed in the 1960s that one or a few giant "computer utilities" would emerge to dominate the data processing industry. If construction of few very large computers to handle the nation's data processing needs became economic, then issues of monopoly power would arise.

The question also arose whether common carriers already supplying data communications should be permitted to compete with the several hundred existing service bureaus and other nonregulated firms in data processing.[17] Taking positions similar to those above, data processors were concerned that telephone companies could use revenue from basic telephone services to underprice their competitive data processing markets. (A Department of Justice antitrust consent decree in 1956 already had prohibited Bell, but not other telephone companies, from offering unregulated services, including data processing.)

After lengthy inquiry, the Commission ruled in 1971 that data processing, as distinguished from data communications, would remain *unregulated.* Any common carrier would be permitted to provide data processing services only through a separate corporation that "must maintain its own books of account, have separate officers, utilize separate operating personnel, and utilize computing equipment and facilities separate from those of the carrier for its data processing service offerings."[18]

However, problems are posed by "hybrid" services that by their very nature involve both data processing and data communications. The FCC ruled that hybrid services involving primarily data processing, with message-switching as an integral, but incidental feature, would remain unregulated. But hybrid services that are "essentially communications...warrant appropriate regulatory treatment as common carrier services...."[19]

Subsequent developments appear to have justified the decision not to regulate data processing, for continuing technological advances, including microprocessing, have pushed toward decentralization rather than toward centralization of computer "power." Economies of scale of large computer facilities have not been as great as had been thought earlier, and many users have found relatively small computers economic. Thus the computer "utility" and the problems of regulating it have not emerged.

At the same time, the distinction between data processing and data communications is becoming increasingly blurred. Some hybrid services cannot be categorized neatly as "primarily" one or the other. For this reason, the FCC has reopened its search for ways to delineate improved criteria to determine whether particular services should be regulated.

This problem is particularly troubling for Bell which, given the antitrust consent decree, is not permitted to engage in unregulated data processing or in sales of data processing equipment—limitations that may pose increasingly pressing issues of public policy.

For example, controversy has arisen over whether Bell's Data Speed 40/4 is a data processing or a communications device. In 1976 the FCC's Common Carrier Bureau decided that it was a data processing device and therefore not a legal Bell tariff offering. The Commission subsequently reversed the Bureau's decision, concluding that the 40/4 is a communications device and is therefore a legal Bell offering. At this writing, the decision has just been affirmed by the court of appeals.

Another example is Bell's Transaction Network Service. Providers of EFT services assert that Bell's TNS involves data processing and therefore should not be permitted. (Because an interstate tariff for TNS has not been filed, the FCC has not yet addressed this issue; presumably the states in which tariffs have been approved do not regard TNS as data processing.)

Thus, questions of defining the scope of entry both of competitors into traditional monopoly markets and regulated monopolists into traditionally competitive markets has generated much litigation, uncertainty, and costly delay. Again the basic problem is cross-subsidy. Were we reasonably confident that predatory pricing of umbrella protection would not arise, then regulated monopolists could compete on a "full and fair" basis in competitive markets and free entry could be permitted into previously monopolistic ones.

The Shift in Burden Among Users

An issue emphasized by Bell, the independent telephone companies, and state regulatory commissions is that competition in the interconnect and specialized common carrier markets will erode revenues of the existing carriers, so that users of basic telephone services will be forced to pay higher rates, allegedly contrary to the mandate of the Communications Act of 1934 regarding rapid and efficient development (and presumably maintenance) of a nationwide "wire and radio" network. In its Docket 20003, the FCC undertook a broad examination of potential competitive effects and found that estimates of such harm depend critically on the underlying assumptions made by various parties. For example, Bell claims that basic exchange rates might rise by as much as 70 percent, based on two assumptions: a) that the company is so impotent in the face of competition that it would lose *all* the revenue contributions from both its competitive and monopoly services (apart from basic exchange service), with competitive private-line suppliers taking over the whole private-line market and eroding the market for Bell's WATS and message toll service as well because of high assumed cross elasticities of demand; and b) that none of Bell's common costs associated with these services would be reduced. Thus, basic telephone users would be forced to bear the whole common-cost burden.

Others filing in Docket 20003, such as the new specialized carriers, assume that the private-line market will remain a small part of the total, that competitors will make only partial inroads and that, with continuing overall growth of the telecommunications industry, the basic telephone user will be little affected.

Some evidence suggests that basic telephone service *subsidizes* private-line and terminal services rather than being subsidized by them, so that basic telephone users might be even better off, rather than worse off, with competition. With respect to this possibility, the FCC concludes that basic telephone services may be subsidizing private-line services because on a fully distributed cost basis, private-line services have been earning a lower rate of return than have basic services. However, because of the arbitrary way in which common costs are allocated, as discussed earlier, this is insufficient evidence to determine whether private-line service is a burden on basic services. Again, so long as private-line services are covering their long-run incremental costs—and we cannot determine this despite the enormous records in the FCC's Dockets 18128 and 20003—they are not being subsidized by basic users in the sense that the latter would be better off in the absence of private-line services.

In assessing the impact of competition on the distribution of costs and benefits, several considerations are salient:

The extent to which cross-subsidy exists. As discussed previously, this is very difficult to estimate because of uncertainties in projecting costs and revenues and problems of allocating common costs.

The cross elasticities of demand between competitive and monopolistic services. As the price of private service falls, we would expect users to substitute private-line service for WATS and message toll. But the strength of the substitution effect is unclear. The extent to which competitors potentially can make inroads into Bell's business depends in part on this factor.

The extent to which both competitive and monopolistic services continue to grow. This will depend in part on continuing innovation and increases in demand as a function of overall economic growth.

The price elasticity of demand for basic telephone service. Suppose rates were to rise by 20, 40, or 60 percent: Would this affect telephone usage substantially? Would it reduce accessibility to telephone service in rural areas substantially? More generally, would universality of service be compromised seriously? Without tolerably good estimates of the price elasticity of demand, answers are necessarily conjectural. Unfortunately, despite the voluminous filings in Docket 20003, we still do not have good estimates of the price elasticity of demand. However, even if the elasticity were high, the possibility remains of encouraging telephone hookups by expanding usage-sensitive pricing for local service. At a low "life-line" rate, the subscriber could have a telephone, and beyond some minimum number of local free calls, he would be charged for additional ones. Indeed, whether or not competition were to place an increased burden on the telephone user, usage-sensitive pricing may increase the efficiency with which the telephone system is used substantially.[20]

Modifications in separations and settlements arrangements. Part of the difficulty some perceive in shifting burdens to the basic telephone user as a consequence of competition is the nature of "separations" arrangements for dividing costs between state and federal jurisdictions, and of "settlements" arrangements under which Bell compensates independent telephone companies to cover costs of their plant used with Bell's toll facilities. In cooperation with state public utility commissions, the FCC is examining prospects for modifying these arrangements to reduce the impact of competition on basic telephone users.

The extent to which cost reductions generated by competitive telecommunications services are passed on to other groups. Business firms are allegedly the primary beneficiaries of competition; but they reasonably would be expected to reduce prices to consumers by some portion of competition-induced price reductions in telecommunications services in the same manner that cost reductions for other things they buy would trigger price reductions to their customer. This "second round" effect generally has been ignored as insignificant because, in the popular view, firms seeking to maximize profits would be expected to keep any cost reductions for themselves. Yet, in accordance with economic theory, even a monopolistic firm would be expected to pass a portion of cost reductions forward and, depending on conditions of demand and cost, even could pass on an amount *equaling* the cost reduction.

All in all, we simply do not have the information to pin down tolerably well the potential effect of competition on basic telephone users. The existence of a strong nationwide system, combined with possibilities of usage-sensitive pricing, suggests that "universality" of service will not be compromised. At the same time, if existing public telephone carriers are expected to provide reliable basic telephone service to all areas of the country as their "public service" obligation, while competitors are free to pick and choose their routes and services, long-term growth of competition eventually could force a substantial increase in costs borne by basic telephone users in outlying areas. Continuing assessment of these potential distributional consequences is important, as new evidence accumulates, to face squarely such questions as whether an increase of, say, 20, 40 or 60 percent in local telephone rates in some areas would compromise national objectives seriously, whether national objectives themselves should be reappraised, and whether losses to certain groups such as basic telephone users would be more than offset by gains to others.

Adequacy of Current Institutional Arrangements

Questions frequently are raised about the adequacy of current institutional arrangements in coping with the many problems discussed above. In particular, the split between federal and state jurisdiction has posed difficulties. Fearing that competition will force an increase in intrastate telephone rates, state utility commissions generally have opposed the FCC's decisions and continue to reflect the philosophy embodied in the three propositions noted at the beginning of this paper. Questions have arisen whether FCC decisions on interconnecting customer-provided equipment to the telephone network have preempted state action. Decisions of the North Carolina Utilities Commission and the Attorney General of Nebraska, if carried out, essentially would have prevented interconnection of customer-provided equipment to telephone systems in those states. In its Telerent decision, the FCC concluded that it did have preemptive jurisdiction over terminal equipment—a decision that has been upheld by the courts.[21] With these developments, it is not clear what legislative changes need to be made in federal and state jurisdictions, or what other measures need to be taken.

Much concern is being expressed about the adequacy of the Communications Act of 1934, and attention is being devoted to possible amendments. As I have argued elsewhere, the Act does seem to have served the public well, precisely because its broadly worded mandate has permitted the FCC and other interested parties wide latitude to introduce and interpret evidence in accordance with the particular facts and circumstances at hand.[22] Perhaps the Act could be amended to provide better guidance for the future, as discussed briefly later. But one must be concerned that amendments not be worded so narrowly that they unduly constrain the Commission and other groups in taking appropriate action in light of circumstances that cannot be foreseen now.

Yet other changes might be proposed such as reducing the fragmented oversight of Congress with respect to portions of the industry, or strengthening the capability in the Executive Office of the President. Some such institutional changes, which cannot be explored here, may be merited.[23] But the important point is that leading problems in the industry do not stem from institutional factors or from the absence of a "national policy" on telecommunications, but rather from the extreme difficulties of

Monopoly and Regulation in Telecommunications

establishing workable ground rules for moving toward "full and fair" competition between regulated monopolists and their challengers. No amount of rhetoric, no wholesale reorganization of telecommunications responsibilities in either the executive or legislative branches, by themselves will cope with these hard problems.

Innovation and New Services

What has competition actually accomplished in stimulating innovative activity and new services? Carriers have argued strenuously that competitors are mainly cream-skimming, with services essentially duplicating those already available. One cannot assess accurately this early what the effects of competition have been or are likely to be in the near term, despite the time and money poured into litigation in response to the many issues discussed above.

Moreover, one must recall that Bell has had decades of experience and billions of dollars for R&D. In 1973 Bell Laboratories alone spent about $420 million in comparison with approximately $30 million of *total* revenue earned by specialized terrestrial common carriers in 1975. With such disparities in size, one cannot expect to see rapid changes in industry behavior at this early state of competitive entry.

Yet from a wide range of firms, we have seen an impressive array of new terminal devices that otherwise likely would not have emerged. The FCC has compiled a long list of manufacturers of "non-traditional" terminal devices used with the switched network—facsimile equipment, computing typewriters, small business computers and peripherals, answering devices, dictation equipment and intercom equipment, to name a few.[24]

Specialized carriers also have offered a range of services that have met with satisfactory consumer response. Among other things, users have a wide choice of service qualities including data transmission error rates. The Satellite Business Corporation mentioned earlier clearly will provide innovative services if, in accordance with its application, the system permits "each customer with geographically dispersed locations to combine voice, data, and image communications into a single, integrated private-line switching network."[25]

Notably, Bell has responded to this competition. It has established a new marketing department to better mesh new services and customers' needs. Its "New Dimension" PBX, its "com-key" system, and its use of standard telephone circuits for transmitting data under voice are all responses to competitive forces.

147

In summary, with such a short history of competition, and so much of it consumed in litigation, the identifiable benefits of competition are as large as one reasonably could expect.

Possible Solutions

In light of the preceding issues, several possibilities for modifying regulatory policy have been suggested from time to time to which we shall now turn. Throughout we must remember that no solution is without its pitfalls, and that the principal objective must be to devise policies that show promise of working better than others in a highly imperfect world.

Separate Subsidies for Competitive Services

Suppose all competitive services of the existing common carriers were spun off into arms-length subsidiaries as a response to problems of cross-subsidization. This approach is attractive for two reasons. First, by requiring the parent firm to make available facilities and services on a nondiscriminatory basis to its own subsidiary and to outside competitors, the parent firm would be dissuaded from both predatory pricing and umbrella pricing. If it leased facilities or sold services to its own subsidiary at prices below incremental cost, the parent firm also would have to offer the facilities or services to competitors on the same terms. With assurances of nondiscriminatory access by all, the FCC might be more inclined to move away from fully distributed cost criteria so that the parent firm would not be forced to pursue umbrella pricing in offering facilities and services to its subsidiary and to outsiders. In short, as long as competition exists between the subsidiary and outsiders, the parent firm would be under pressure to price its competitive services at long-run incremental costs—which is the pricing strategy that Bell has advocated all along.

Second, dangers of uneconomic construction of duplicate facilities would be reduced. Competitors would build their own facilities or not depending upon how their own costs compare with the cost of leasing facilities from Bell. Presumably, if particular facilities show strong economies of scale, it would be in the interest of both Bell and society for Bell to build them and to offer to lease them to outsiders at levels sufficiently low to discourage construction of duplicative facilities. If economies of scale are in fact large, facilities would tend to be offered on a monopoly basis and services on a competitive basis—as dictated not by the FCC or Congress but by market forces.

Monopoly and Regulation in Telecommunications

However, apropos of the foregoing discussion of the scope of entry, where does one draw the line between parent company and subsidiary operations? Today one can differentiate fairly clearly between local loop facilities, with their strong natural monopoly characteristics; terminal equipment, where competition is playing a progressively stronger role; and long-distance service which falls some place in between. But new technologies and growing customer needs either may add to or subtract from economies of scale. Fiber optics, digital transmission, and electronic switching might increase natural monopoly barriers to entry in some markets, while sophisticated terminals, high-speed facsimile and packet switching may eliminate them in others. Indeed, the development of fiber optics—perhaps eventually permitting a wide range of voice, data, and television services to be brought economically into the home—would place common carriers in direct confrontation with cable television operators.

Although local loop facilities currently have strong monopoly characteristics, packet radio eventually could become a strong competitor for linking telephones with the nationwide network while providing the additional advantage of mobility to the user. Satellites working with small earth stations located on customers' premises for business and other institutional uses, such as those planned by the Satellite Business Systems consortium, also would provide a competitive alternative to some existing local exchange facilities. Yet, eventual use of millimeter-waveguide technology, permitting high-capacity, long-distance transmission through communications "pipes," and continued development of high-capacity coaxial cable, may increase economies of scale in long-distance transmission further, thereby discouraging competition in that area.

The critical consideration is that competition may appear and disappear in various quarters depending upon the pace and direction of technological advances combined with evolving customer needs. Suppose, then, a separate subsidiary were set up for private-line services. Should it include all of Bell's private-line services or only those for which competition has emerged? Should the subsidiary operate over the whole nationwide network or, should it be confined to competitive routes? If competition arises for local exchange services in a few cities, should Bell be permitted or required to respond by setting up a separate subsidiary in response? And if the competition subsequently disappears, would the subsidiary be abolished? The shifting winds of competition and monopoly may make it mechanically difficult or impossible to establish, expand, contract or abolish subsidiaries as a competitive response of the monopoly carrier.

However, if this problem can be solved, the subsidiary approach would be an attractive way of dealing with problems of common-cost allocations and of avoiding undue duplication of facilities subject to economies of scale.

Deregulation of Terminals and Line Brokerage

Because manufacture of terminals exhibits few economies of scale, and because there is already vigorous competition, many groups outside the existing common carriers advocate that services be unbundled with terminals sold or leased separately. Terminals could be deregulated except for the enforcement of technical standards. This appears to be an attractive idea except for the problem of what to do with Bell. Is it to be permitted to compete in the unregulated terminal market? If so, then modification of the antitrust consent decree should be required. The problem of cross-subsidization would arise insofar as Western Electric selling in both regulated and unregulated markets could cross-subsidize across product lines—although this problem would not be as serious in multiproduct manufacturing as it is in long-distance and local transmission where common-cost components play such a major role. To guard against the problem of cross-subsidy, one could advocate that Western Electric be spun off entirely from the Bell System and be required to deal at arm's length with all customers on a nondiscriminatory basis. Such a course brings us to issues being confronted by the Justice Department and its current antitrust suit against AT&T that cannot be explored here.

Another problem in deregulating terminals lies in the jurisdictional separation between federal and state authorities. Much equipment is not tariffed with the FCC, but with state regulatory agencies, because they serve intrastate rather than interstate markets. While the Telerent decision afforded federal preemption in cases where a state action might impede interstate service as, say, by prohibiting a particular type of equipment that does not meet state technical standards, this decision does not forbid states from regulating the price of communications equipment as part of an intrastate tariff. Thus deregulation of terminals would require a good deal more centralized authority. Otherwise we might end up with a patchwork in which some states might decide that certain kinds of equipment should be regulated, while others might opt for deregulation. If deregulation of terminal markets appears desirable, perhaps it is here that revision of the 1934 Act would be appropriate.

Finally, questions arise about the regulation of line brokerage. Suffice it to say that those who lease lines from common carriers and combine, divide and resell them—perhaps in combination with terminals and computer software for packaged services—have little reason to be regulated. Again the question is the proper role of Bell and other regulating companies in serving these markets.

A Carrier's Carrier

One possible way of coping with problems of cross-subsidy is to convert Bell into a carriers' carrier that would sell services on its lines and switching equipment to retail carriers at wholesale on a nondiscriminatory basis. One should be troubled about this approach, however, because it is so similar to the FCC's Authorized User decision regarding Comsat's sale of international satellite services. Comsat was prohibited by the FCC from dealing at the retail level with ultimate users and was limited to selling only to existing common carriers. Because these carriers have incentives to employ their own facilities rather than to lease from Comsat to maintain or increase their rate bases, and because restricting Comsat in providing services reduces competitive pressures against existing carriers, the Authorized User decision has been widely criticized. The situation would be somewhat different here because Bell would be reselling presumably to *competitive* unregulated users such as EFT providers who then would serve the public. But again the problem is where to draw the line between competitive and monopolistic services. Suppose a firm wants to compete with the carriers' carrier in providing certain "wholesale" services? Or suppose certain ultimate users would prefer to deal directly with the wholesaler, feeling that the wholesaler can do a better job in terms of price and service quality? Unfortunately, we are back to where we started.

Better Criteria for Decisionmaking

Of course, it would be highly desirable to have better regulatory tools for decisionmaking based on criteria that make outcomes in the public interest as automatic and predictable as possible. But how do we translate the general goal into anything operational? As emphasized above, issues of cross-subsidy are particularly pressing; if we could come to grips with them satisfactorily, then we could move rather straightforwardly to expanded competitive entry, including deregulation of terminals and brokerage. But establishing criteria for allocating costs among services, as discussed previously, is extraordinarily difficult. The FCC has made progress in developing a fully distributed cost methodology that improves past procedures in moving closer to long-run incremental costs; but much more will need to be done to provide the basis for guarding against either umbrella or predatory pricing for competitive submarkets.

As one palliative, should the burden of proof be put on the monopolist faced with competitive entry, by amending the Communications Act of 1934? Perhaps that would help to stimulate competition; but the critical point is that, regardless of who has the burden of proof, it is very difficult to demonstrate on paper that the proposed tariff for a service is "fair" or that one firm can perform better than another—especially in the case of a new service. It is not clear that the delays, costs, and uncertainties of administrative processes and litigation we have seen in the past would diminish under these circumstances.

Concluding Remarks

All this is not to end, however, on a note of pessimism. On the contrary. Much progress has been made in the past 20 years in critically examining the three propositions noted at the beginning of this paper, introducing changes in regulatory policy and permitting an infusion of fresh ideas and approaches. We already have seen an impressive level of innovation, given the short time involved, and the future looks bright. Despite its strident opposition to competition, the Bell System probably will continue doing well financially (it is earning record profits today) and basic telephone users will continue to be well served.

The issues discussed above will continue to nag us, for there simply are no ready solutions. But there are possibilities for improvement that, although not perfect, may take us well down the road. In the first place, deregulation of the terminal market seems warranted. The weight of evidence suggests that there are no strong economies of scale in the manufacture of equipment; suitable technical standards probably can be maintained so far as present evidence suggests; and the problem of cross-subsidy in Western Electric multiproduct lines is more tractable than in other areas of telecommunications. Second, the possibility of requiring Bell to offer competing services through an arm's-length subsidiary is appealing if somehow we can cope with the problem of shifting boundaries between monopoly and competitive services. Third, although cost studies are time consuming and lead to frequently contested results, scope exists for improved cost methodology moving toward incremental cost principles for pricing competitive services. Even crude approximations of these costs as the basis for rate making would be useful in widening the potential scope of competition while avoiding the dangers of umbrella and predatory price response.

In contrast to experiences elsewhere in the economy, with continuing rapid growth in telecommunications, there is little prospect of competition leading to disruption of the industry's labor force, or to gross underutilization of capital equipment. Nor do we need to be concerned about prospects of massive federal subsidies required to prop up a declining or sick industry. Expressed differently, it is much easier to introduce competition into a dynamic, growing industry than into a static or declining one.

In short, in comparison with other economic activities, the telecommunications industry is doing well and there is good reason to expect that it will continue to prosper. Those who seek to address societal "crises" with their analytic tools should search elsewhere—as in the fields of energy and delivery of health care, where problems are far worse than anything discussed here.

Notes

1. Mobile telephone service also is provided competitively. About 500 radio common carriers scattered about the country offer two-way mobile service and one-way paging service in competition with telephone companies. An excellent history of the telephone industry from which portions of this paper are drawn is contained in *Report by the Federal Communications Commission on Domestic Telecommunications Policies*, submitted to the Subcommittee on Communications, House Committee on Interstate and Foreign Commerce, Washington, D.C., September 27, 1976, Attachment B.

2. The only notable government subsidy to the telephone industry consists of low-interest loans by the Rural Electrification Administration to small, independent telephone companies. An additional implicit subsidy arises because common carriers, as well as other users of radio spectrum space such as television stations, obtain rights free of charge despite the fact that spectrum is a scarce resource.

3. *Report by the Federal Communications Commission on Domestic Telecommunications Policies, op. cit.,* Attachment B, pp. 15-16.

4. *Allocations of Frequencies in the Bands Above 890 MHz,* 27 FCC 359 (1959).

5. *Report by the Federal Communications Commission on Domestic Telecommunications Policies, op. cit.,* pp. 74-75.

6. Domestic Communications Satellite Facilities, 35 FCC 2d 844 (1972) (second report and order in the Domsat proceeding). See also *Network Project v. F.C.C.* 511 F. 2d 786 (District of Columbia Cir.) 1975 (affirming FCC grants of authority pursuant to Domsat decision).

7. It was not until 1975 that in one such case involving plastic covers, the North Carolina Supreme Court overturned a prior state PUC ruling that this tariff restriction was valid.

8. At the time of the Hush-a-Phone decision, Bell's tariffs barred telephone customers from connecting to its lines any device—no matter how innocuous—which was not furnished by itself. The Hush-a-Phone was a cup-like device placed on the telephone handset to facilitate privacy in telephone conversations in crowded offices. Upon court remand, the FCC ruled that any blanket prohibition upon customers' use of devices without discriminating between harmful and harmless ones is unreasonable. The "Carterfone" is a customer-supplied device to connect mobile radio systems with the nationwide telephone system. The FCC found that the Carterfone met an unfilled communications need and did not affect the telephone system adversely. *Hush-a-Phone v. U.S.*, 238 F. 2d 266 (District of Columbia Cir.) 1956; Carterfone, 13 FCC 2d 420, reconsideration denied, 14 FCC 2d. 571 (1968).

9. Robert F. Stone, Mark A. Schankerman, and Chester G. Fenton, *Selective Competition in the Telephone Industry*. T&E Inc., Cambridge, Massachusetts, 1976, p. 23. For a critical appraisal of the T&E study and other studies in Docket 20003, see Stanford Research Institute, *Analysis of Issues and Findings in FCC Docket 20003*, April 1977.

10. Henry M. Boettinger, AT&T Director—Corporate Planning, as reported in *Telecommunications*, February 7, 1977, p. 14.

11. Statement by John Eger, Acting Director of the Office of Telecommunications Policy, before the Subcommittee on Communications, House Interstate and Foreign Commerce Committee, Washington, D.C., November 10, 1975.

12. In principle, common cost can be allocated in many ways, as by relative time use of common-cost facilities, by relative revenues of various services or, as in the Postal Service, by relative weight or volume. The one characteristic of all fully distributed cost approaches is that they do not take into account price elasticities of demand for various services as a criterion for allocation.

13. Letter from Richard E. Wiley, Chairman of the FCC, to Lionel Van Deerlin, Chairman of the House Subcommittee on

Communications, commenting on the "Consumer Communications Reform Act of 1976," September 23, 1976.

14. An extensive study of the EFTS industry, including competitive relationships with the Bell System, has been completed by the National Commission on Electronic Fund Transfers, *EFT in the United States*, Washington, D.C., October 1977.

15. MCI Communications Corporation, *1975 Report to Stockholders*, Washington, D.C., June 6, 1975.

16. Richard E. Wiley, address before the United States Independent Telephone Association, Dallas, Texas, October 16, 1975.

17. "Data processing" involves using the computer for storing, retrieving, sorting, merging and calculating data whereby the *content* of the information is changed. Data communications involves "message switching" with computer control transmission of messages where the content remains unaltered.

18. Final Decision and Order, Docket No. 16979, 28 FCC 2d 287.

19. *Ibid.*, at 278.

20. Bridger M. Mitchell, *Optimal Pricing of Local Telephone Service*, The Rand Corporation, Santa Monica, California, R-1962-MF, December 1976.

21. *Telerent Leasing*, 45 FCC 2d 204 (1974), affirmed sub. nom., *North Carolina Utilities Commission v. F.C.C.*, 537 F.2d 787 (4th Cir.) 1976, on rehearing, 552 F.2d 1036 (4th Cir.) 1977.

22. Leland L. Johnson, *Domestic Common Carriers and the Communications Act of 1934*, The Rand Corporation, Santa Monica, California, P-5798, April 1977.

23. A discussion of possible institutional changes in light of some of the problems discussed is contained in U.S. House of Representatives, Committee on Interstate and Foreign Commerce, *Fundamental Changes Needed to Achieve Effective Regulation of Communications Common Carriers*, 94th Congress, 1st Session, U.S. Government Printing Office, Washington, D.C., 1975.

24. *Report by the Federal Communications Commission on Domestic Telecommunications Policies, op. cit.*, Attachment B, p. 50.

25. Satellite Business Systems, application before the FCC, December 1975.

4.
International Telecommunication Regulation

Henry Goldberg

Of the areas the FCC regulates, none seems more arcane than the regulation of common carrier communications services between the United States and other nations.[1] Currently, the principal issues involve complex questions of planning and constructing transoceanic communications facilities.

For new facility construction, investment decisions are based on forecasts of demand made in ten-year segments. Undersea cables are promoted by United States carriers and their foreign counterparts—the governmental ministries of posts, telegraph, and telephone/telecommunications (PTT). Satellite facilities are promoted by the congressionally created Communications Satellite Corporation (COMSAT)—and sometimes by the FCC. Executive branch agencies fight over turf and also stalk the FCC. The Congress presses the FCC and executive branch agencies for a coherent policy, yet shies away from making legislative changes in the existing statutory policies—policies that were created and frozen in 1927, in 1943, and in 1962. The FCC involvement in facility planning confuses, and sometimes enrages, the American and foreign carriers.

In addressing the issue of new facility construction, several proposals exist including those that emphasize government ownership, mandatory merger of all U.S. international carriers, permissive merger divestiture, competition, open entry, and both satellite and cable circuits.

Against this backdrop, observers and analysts conclude that the regulatory structure for international communications is in shambles and that the final crisis is either upon us or around the next corner. Since this has been the state of affairs in international communications for 30 or 40 years, some historical perspective is needed.

A 1951 report by the President's Communications Policy Board, "Telecommunications: A Program for Progress," offered the following analysis.[2]

> Proposals for merger of American companies providing cable and radiotelegraph services have provoked vigorous debate ever since radio emerged as a practical means of international communications. The traditional American policy against monopoly has affected this debate throughout. During the years immediately following the first world war, it was a chief concern of the government—the Navy Department in particular—that the well-established cable companies should not be allowed to hamper the full development of radio as a medium for telecommunications. Hence arose the obstacles to the ownership or control of radio companies by cable companies later embodied in Section 17 of the Radio Act of 1927 and in Section 314 of the Communications Act of 1934.
>
> A related policy introduced in 1943 calls for the separation of companies doing overseas business in record communications from domestic record communication companies.
>
> Within this broad framework, however, proposals have persistently recurred during the last 20 years for mergers of American communications companies. Fundamental to this problem is the possibility offered by radio of providing, with relatively small capital outlay, circuit capacity exceeding the normal requirements of international communications. This raised difficult economic questions of cost of service, and the future profitability of cables in the face of radio competition. There are some who have suggested that cables are now obsolete, but considerations of reliability and security point to the necessity of retaining cable service.

Several of the companies have asked permission to merge in the hope of avoiding deficits. From time to time, some government departments have favored consolidations for reasons of national defense, conservation of radio frequencies, or for other reasons, while other government departments have opposed consolidation. Some of these agencies have shifted their positions from time to time of the desirability of one or another form of merger. At no time have all the interested executive agencies been in agreement on this issue. As of May 1950, this was still the case.

The move for merger in the field of international record communications has never been able to win complete congressional support because of traditional resistance to monopoly. Numerous hearings have been held by committees of the Congress, but no legislation has resulted. Either the case has not been strong enough, or prevailing international situations have delayed consideration of the various proposals. In the meantime, however, Congress has approved mergers of telephone companies and of domestic telegraph companies, and permits the domestic telephone companies to operate in the international field.

Both the title of the 1951 report and its analysis sound startlingly fresh to those familiar with the regulation of international communications. This is particularly so when one realizes that the issue of facility construction—particularly the authorization of "competing" telegraph cable and radiotelegraph circuits to the same foreign points—was the critical issue in the 1940s; even reaching the Supreme Court in 1953 in the *RCA* Case.[3]

The following excerpts from the Supreme Court decision make the current controversy on communications facilities seem like *deja vu*—except the players have switched sides.

From 1934 until 1939, when radiotelegraph was just emerging from its infancy, the Commission generally denied applications for circuits to countries already served by other American radiotelegraph carriers. From 1939 to 1942 the Commission generally granted applications for new circuits, regardless of whether the points involved were served by an existing radiotelegraph circuit. From 1942 to 1943 an affirmative policy of authorizing duplicating American circuits (a 'duplicate circuit policy') was followed as a war measure at the behest of the Defense Communications Board. From 1943 until 1945, also as a war measure, the reverse course (a 'single circuit policy') was followed at the behest of the Board of War Communications (the successor of the Defense Communications Board).[4]

RCAC asks us to uphold the Court of Appeals decision on another ground, that the grant of authorization to MacKay would violate Section 314 of the Communications Act, which forbids common ownership, control or operation of radio and cable in international communication whose purpose or effect may be substantially to lessen competition, restrains commerce of unlawfully to create a monopoly. We cannot agree. There has been in recent years a considerable shift of international telegraph traffic from cable to radio, a shift strongly accentuated in some countries, including Portugal and the Netherlands, where the overseas correspondent of American companies is a government-controlled monopoly which strongly advocates radio transmission. RCAC, in the two instances before us, is the beneficiary of this discrimination against cable transmission; with negligible exception, it has a monopoly of radio traffic to these countries.[5]

This historical perspective suggests that international communications regulation has long been a baffling subject. Although the traditional legal, technological, or policymaking yardsticks serve analysts fairly well in other areas of the FCC's regulatory scheme, they are difficult to apply to international communications. Often it is even difficult to know what is happening in the field; not to mention determining the significance of the known events. International communications and its regulation are simply different. However, if one avoids gazing at the legal and technical trees and, instead, focuses on the economic forest, one can discern a pattern in the regulatory structure that is not entirely alien.[6]

The pattern is one of regulated oligopoly. The international communications industry includes one dominant firm offering international "voice" communications, only five providers of international "record" communications, and one technological "monopolist" who supplies satellite circuits to the other carriers. In most industries that have experienced rapid technological change, the technological advances give rise to new firms, that compete with the existing firms. With a regulated oligopoly, the advance of technology also encourages potential competitors, but the new firms are absorbed into the existing industry and regulatory structure, where they coexist, rather than compete, with the existing firms. Professor Bruce M. Owen has described the phenomenon that, for purposes of this paper, is considered an hypothesis.[7]

Since the object of embracing new firms and technologies within the regulatory fold is precisely to prevent competition, the regulatory agency must take measures to thwart the natural instincts of the members of the industry. In this role, the regulatory agency becomes in effect a "cartel manager." The role of cartel manager has two functions—to maintain industry profits at a satisfactorily high level, and to prevent encroachments by one firm on the market shares of its rivals. Such encroachment can take place in a number of ways—by price cuts, by variations in the quality of service, or advertising, and by investment decisions. The regulator must set up rules and procedures for preventing any of these forms of competition, or else suffer a 'lack of stability in the industry. Both of these cartel management functions are presumably at variance with the role of the regulator as protector of the consumer. Accordingly, the objective of regulation in such cases is often stated as protection of consumers from destabilizing competition,' and particular emphasis is placed on that group of consumers (usually rural) who would in fact be disadvantaged by the demise of one competitor.

Historical Review of International Telecommunications

The Development of Radio Communications

For many years, the sole transmission medium for transoceanic communications was telegraph cable. Between 1866, when the first transatlantic cable was completed, and the late 1920s, when regular telephone service by high frequency radio became available between the United States and Europe, several transatlantic cables were constructed and operated by private foreign companies—predominantly British—authorized by their governments to provide telegraph service. Their only contact with the United States government arose out of the need to receive an American cable landing license. Similarly, the two principal domestic telegraph companies—the Western Union Telegraph Company (Western Union) and the Postal Telegraph Company (Postal)—owned and operated cables and received landing licenses or permits from those countries where their cables reached.

The first technological innovation occurred with the development of international radio communications. Early radio communications and facilities were controlled by the British, through the Marconi Company. However, following World War I, the United States determined that, in the national interest, the British domination should be broken and an American company was created to control radio facilities. The Radio Corporation of

America (RCA) was formed in 1919 to perform the functions that had been performed by the Marconi Company. In 1927, the American Telephone and Telegraph Company (AT&T) also offered a commercial voice communications service between the United States and Europe.

This new technology led to "intermodal" competition in the services that were offered in the international telecommunications market. Thus, the telegraph cable was used for record communications, or hard copy, and the high-frequency radio facilities of RCA and AT&T were used for voice communications, or telephone. While this separation of service had a technological basis in international communications, it reflected a legal division of the communications services market already present in the domestic communications industry.

The domestic industry was divided between voice and record communications as a result of the competitive relationships among AT&T, Western Union, and Postal. In 1879, as part of a settlement of a patent dispute between AT&T and Western Union, the two companies agreed not to compete in each other's markets. In 1909, AT&T acquired control of Western Union and this dichotomy was ended. However, in 1919 the Justice Department forced AT&T to divest itself of its interest in Western Union, thereby reestablishing the traditional voice and record dichotomy in domestic common carrier communications.

Since RCA did not have a legal monopoly over radio technology, several new common carrier firms exploited the business opportunities created by the development of international radio communications. In addition to Western Union and AT&T, a number of international record carriers (IRCs) evolved to provide overseas telegraph services either by cable or by radio facilities. Such IRCs included All America Cables, Inc., and Mexican Telegraph Company. The radio telegraph carriers included Tropical Radio Telegraph, RCA, and a number of subsidiaries of International Telephone and Telegraph Company (ITT), which was also the parent company of Postal. During this early period, the various carriers of telegraph communications began to compete with each other using high frequency radio signals and telegraph cables in serving the same overseas points. Moreover, some telegraph cable IRCs added to this "intermodal" competition, as they used high frequency radio circuits as backup facilities. ITT became the most significant company using both radio and cable facilities, as the result of acquisition of existing radio and telegraph carriers, such as McKay Radio and Telegraph, Commercial Cable Company, and All America Cables and Radio. Also, as noted in the earlier quoted RCA opinion, the FCC's policy on authorization of competing radio circuits to the same international point underwent many shifts.

The activities of the competing IRCs also began to undercut the separation of services into domestic and international spheres. While the IRCs were interconnected at their "gateway cities" by Western Union and Postal, several IRCs that employed high-frequency radio facilities internationally also used them to transmit their international messages within the United States. The radio facilities of RCA and the ITT companies were also used, to a limited extent, for domestic telegraph services. However the competition resulting from the development of radio ended with new legislation adopted in 1943.

Ironically, the 1943 ossification of the international industry occurred not as a result of a conscious adoption of a policy for international telecommunications, but as the fallout from the remedy applied to the financial collapse of the domestic telegraph industry.

A Policy for International Communications

In 1943 the two domestic telegraph companies were on the verge of financial disaster. Merger was viewed as the only way to assure continued provision of domestic telegraph service. In Section 222 of the Communications Act, Congress exempted the merger from the antitrust laws, but also conditioned it in four ways to prevent possible anti-competitive abuses that might result from Western Union's domestic telegraph monopoly.

First, Section 222 required Western Union to divest itself of its international telegraph operations, a divestiture that was not completed until 1963, with the formation of Western Union International (WUI) as an IRC.

Second, the Congress required Western Union to distribute among the IRCs all international record communications originating in the United States and bound for overseas points, as long as the customer did not designate a particular IRC. Section 222 authorized the FCC to create a formula to distribute this traffic. As of 1943, the formula froze each IRC's market share and required a carrier whose share of routed traffic increased to relinquish an equivalent amount of unrouted traffic so that its overall share would remain the same. Naturally, the formula discouraged the IRCs from competing with each other for routed traffic and removed rates as a competitive factor in the industry.

Third, the Congress drew a rigid demarcation between international and domestic traffic, to carve out the permissible sphere for Western Union operations. Section 222 defines domestic and international record service, allowing Western Union to provide the domestic service but not the international service.

However, as a practical matter, since Western Union was the only IRC already serving Mexico and Canada, domestic service was defined to include these countries as well as the continental United States. All other locations including Hawaii, Puerto Rico, and the Virgin Islands, were designated as international points under Section 222. In addition, the Section 222 traffic distribution formula adopted pursuant to Section 222 encouraged the FCC to limit the IRCs activities to "gateway cities," thus precluding them from threatening Western Union's domestic record service in the "hinterland."

Fourth, Section 222 authorized the FCC to prescribe division of revenues between the IRCs and Western Union for Western Union's message delivery from other American locations to the IRCs facilities in the gateway cities.

Ten years after the Congress froze the international communications industry, the Supreme Court seemed to place its imprimatur on it in *FCC v. RCA Communications.* All that was lacking to solidify the regulatory, legislative, and Supreme Court determinations in favor of a non-competitive industry structure was a technology that more nearly lent itself to natural monopoly than the high-frequency radio circuits used for international voice communications. In 1956 the gap was closed with the joint AT&T-British installation of TAT-1, the first undersea voice-grade cable.

Although created to provide reliable telephone communications on an intercontinental basis, the voice-grade cables were capable of transmitting record as well as voice communications. Initially, the FCC allowed the IRCs to lease circuits in the voice-grade cables. However, the distinction between voice and record services was maintained on the new, single facility. AT&T used the facility to provide traditional voice services and the IRCs used the facility to provide their traditional record services. The development of alternate voice data service in the late 1950s presented the FCC with its first major blurring of the distinction between these two services. In 1959 the FCC determined that AT&T could provide voice and record service—alternate voice data (AVD)—to the Department of Defense (DoD) over its undersea cables. While the FCC authorized both AT&T and the IRCs to provide AVD to DoD, the latter shifted its telegraph leases from the IRCs to AT&T. This action could have been the death knell for the IRCs. Their competitive position vis-a-vis AT&T would have been perilous unless the FCC protected a portion of the international communications market as the exclusive domain of the IRCs.

The FCC took this step in 1964. In that year, the TAT-4 decision established the symbiotic relationship between the IRCs and AT&T, which exist today. The FCC rejected an application of an ITT subsidiary and granted the cable application of AT&T. However, the Commission required AT&T to offer ownership shares—called indefeasible rights of use (IRU)—to the IRCs, based on a formula reflecting each carrier's use. Moreover, the FCC prohibited AT&T from providing any new alternative voice-data services. Thus, the FCC established the precedent of dividing the market for international services and also ruled out the possibility of competition between cable paths. There would be only one cable path and it would be operated jointly by the "competitors." Although the IRCs lost any substantial say in future facilities development and investment, the TAT-4 decision gave them the right to include in their respective rate bases their proportionate ownership shares in cable facilities. They also assured themselves the right to an exclusive segment of the international communications market, one that would be protected by the FCC.[8]

The development of the voice-grade cable technology also made a profound change in another set of relationships, which has proven to be critical to international communications. Unlike the old telegraph cables, which were owned by a single company or government, the new cables were jointly owned, not only by United States carriers but by their foreign correspondents. The foreign correspondents obtained rights in cable ownership and use that were similar to those of the IRCs. Whether or not the cable landed in their country, foreign telecommunications organizations that wanted access to the cable could affiliate to purchase an ownership interest in the form of an IRU.

Therefore, a single, high-capacity, voice-grade undersea cable is an amalgam of a variety of United States and foreign ownership interests. With the exception of a few circuits that are wholly-owned, the American organizations own the cable from its domestic point of origination to its midpoint in the ocean. The foreign correspondents jointly own the cable from that point to the cable's destination on the foreign shore. To a degree unknown at the time these new ownership relationships were created, the new submarine cable technology would give the FCC authority to affect the economic and national policy interests of a large number of PTTs. The complications and foreign policy implications of this expansion of authority were not felt fully until the FCC attempted to become involved in planning international communications facilities. This involvement initially occurred when the Commission began to exercise its responsibilities under the Communications Satellite Act of 1962.

The Creation of a Satellite Monopoloy—COMSAT

The development of communications satellite technology for international use provided an opportunity for an alternative, and possibly a competitive means of transmission—an opportunity that had not been present since the development of high-frequency radio in the 1920s.

The creation of the Communications Satellite Corporation (COMSAT) even offers some interesting comparisons with the creation of RCA in the 1920s. Like RCA, COMSAT was created to develop a single technology. While RCA was created to wrest control over radio technology away from the British, one Congressional objective for COMSAT was to assure the United States' dominance in space technology. Moreover, in both instances, there was a concern with the national interest in manufacture and control of the technology. There the similarities end. Unlike RCA, COMSAT was created by statute and the legislative policy contained in the 1962 act, as well as the regulatory response to that policy, was to preclude any possibility that there would be competition between satellite technology and cable technology in international communications. Satellite technology and COMSAT, as the "chosen instrument" for using that technology, would be thoroughly integrated into the industry structure for international communications that had been established in 1943.

When COMSAT was created by the Communications Satellite Act of 1962, it represented a new departure in public institutions. COMSAT essentially was created as a private corporation, charged with accomplishing national purposes. The structure of the company was somewhat unwieldy as a result of bitter congressional and public debate that surrounded passage of the Communications Satellite Act. On one side were the advocates of integrating satellite technology into the private communications industry. On the other side were advocates who supported public ownership, because of the government's investment in the development of space technology. The resulting compromise cast COMSAT as a private, profit making corporation, whose shares would be privately owned. However, the private ownership was to be divided equally between AT&T and the IRCs on one hand and the general public on the other. The board of directors would be a mixture of carrier-elected directors, an equal number of directors elected by the other private stockholders, and three directors appointed by the President.

The federal enabling legislation and the presidentially appointed directors were not the only features of COMSAT that cast it as public-private hybrid. Despite the private ownership, COMSAT was charged with achieving the national policy objectives expressed in the legislation. COMSAT was intended to serve the policy of establishing a commercial communications satellite system, as part of an approved global communications network which would be responsive to public needs and national objectives, which will serve the communication needs of the United States and other countries, and which will contribute to world peace and understanding. Moreover, the 1962 act divided governmental supervisory authority for COMSAT between the FCC and the Executive Branch.

In addition, the 1962 legislation laid the foundation for a series of regulatory decisions that made it unlikely that COMSAT and the new satellite technology would be used to compete with the existing carrier and cable technology for a share of the international communications market.

One such regulatory decision involved ownership of earth stations that send and receive signals from satellites. The 1962 legislation gave the FCC a choice among a variety of earth station ownership configurations. Initially, the FCC chose to allow COMSAT to own and operate earth stations without the participation of other carriers. Moreover, the Commission indicated that the point of interconnection between COMSAT and the international carriers would be at gateway cities, allowing COMSAT to control the link between the earth station and the gateway. When the IRCs and AT&T objected, the FCC reconsidered its decision and ruled that the point of interconnection should take place at the earth station rather than the gateway city. Later, when faced with competing applications for new earth stations, the FCC selected a consortium arrangement for earth station ownership. Ownership was divided equally between COMSAT, with 50 percent, and the IRCs and AT&T, which divided the other 50 percent. COMSAT operated the earth stations for the consortium of earth station owners.

Another regulatory decision in the same mold resolved the interpretation of the "authorized user" provision of the 1962 act. This issue also involved the larger question of COMSAT's role in international communications. The 1962 legislation authorized COMSAT to deal with "authorized users" for providing the communications satellite services. However, the legislation did not define the term "authorized users." This critical definition was left to the FCC.

After considering the matter for a year, the FCC in June 1966 ruled, in effect, that COMSAT should be a "carrier's carrier." The Commission found that, as a matter of policy, COMSAT should be limited to providing satellite circuits only to international communications carriers. Thereafter, with certain limited exceptions, COMSAT has been prevented from dealing directly with the ultimate users of satellite services.

As a corollary to the Authorized User decision, the FCC adopted a composite rate policy for international communications. That is, irrespective of whether cable or satellite technology is used to serve the customer's needs, the carrier's rate to the customer is the same. Any cost savings derived from the use of satellite circuits are intended to be reflected in the carrier's overall rates and passed on to the customers.

At the same time that the FCC was making the regulatory decisions that integrated the role and functions of COMSAT into the international industry structure, it was making similar decisions to integrate satellite technology.

The first time that cable technology and satellite technology came into conflict was in a number of applications to extend facilities to Puerto Rico and the Caribbean. AT&T and ITT applied to construct a cable from the mainland to the Virgin Islands via Puerto Rico. In December 1966, the FCC, seeking to carve out a portion of the market for satellite services, approved the cable and one earth station. In doing so the FCC was responding to what it interpreted as a congressional commitment to promote satellite technology. Balancing this statutory policy was the FCC's recognition that the United States' carriers had a strong interest in cable, in some instances because they manufacture cable and, in all instances, because investment in cable facilities could be part of their rate bases. Moreover, the FCC found a need for diversity of facilities to assure continuity of service in case either the cable or the satellite failed.

Once the FCC had authorized the construction of both cable and satellite facilities, logically it found it necessary to decide how the traffic would be divided between the two types of facilities. The FCC sought and received commitments from the carriers to use satellite circuits. The precedent established in the Puerto Rico case later became the FCC's proportional fill and "reasonable parity" policies. These policies were used to govern allocation of international communications traffic between cable

International Telecommunication Regulation

and satellite facilities, as new satellites and new cables were constructed to serve the burgeoning international communications market across the North Atlantic.

Therefore, for the past 10 years, the regulatory history of international commercial communications has been characterized by an increasing FCC participation in facilities planning and traffic division, in order to further its cable and satellite policies and to preserve a competitive status quo in the international industry.

The most dramatic effect of the FCC's activities has been the resulting strain on international relations. As noted previously, the development of voice-grade cables brought foreign telecommunications entities in the FCC's regulatory sphere. The foreign telecommunication organizations have found the yoke of regulation to be confining and irritating. More particularly, they have objected to the FCC's decisions concerning the construction and activation of new transatlantic communication facilities. These specific complaints involve the way the FCC controls the investment in new facilities and the distribution of traffic between cable and satellite circuits.

Prior to 1970, the foreign telecommunications organizations negotiated the introduction of new facilities and would plan for their use with their counterparts, the American international carriers. Agreements would be reached which the foreign telecommunication entities believed were final agreements. However, when their American counterparts filed Section 214 applications with the FCC, reflecting these agreements, the FCC imposed the proportional fill and other requirements intended to foster the development and use of satellite technology. After the foreign entities had experienced such difficulties in connection with FCC authorization of both the TAT-5 and TAT-6 submarine cables, they sought a change in the process for planning international communications facilities. This resulted in 1974 in the creation of a consultative process, in which the FCC and the American carriers would participate in joint consultation and planning sessions with the foreign telecommunications entities. However, with the release of the FCC's international facilites plan in December 1977, which postponed construction of a new undersea transatlantic cable—TAT-7—the foreign entities may discern that the problem does not simply lie with satellite-promotional policies. Rather, they may infer that the problem lies more deeply in the administrative process and its legislative framework.

Services, Common Carriers, and the Government—Participants in International Telecommunications

Services

Voice-grade telephone services. International voice-grade services are largely AT&T monopoly services that are virtually identical to domestic telephone services. AT&T makes arrangements for phone service with foreign correspondents regarding the facilities that are used and the division of revenues from international traffic.

Public message telegraph. Message telegraph services was the first form of overseas record communications offered by the IRCs, but has declined in relative importance because of the growth of telex and leased-channel services.

Telex. Telex service is a teletypewriter exchange service that has become a profitable offering for the IRCs. The IRCs compete in the form of service configurations, not rates. The IRCs provide Telex machines to users at a nominal $10 per month, and each IRC attempts to maintain and service its machines better than its competitors.

Leased channel. Services provided by leased channel (i.e., a circuit dedicated to a single customer's use) include data, alternate and simultaneous voice-data, facsimile, and teleprinter. The leased-channel services are competitive in terms of rates, since the United States government is the principal user (over 40 percent) and secures circuits under competitive bidding. Leased-channel services are offered at a monthly rate that depends on the transmission speed of the offering.

Television. International television service, on an occasional-use basis, is provided via INTELSAT facilities by the IRCs and AT&T on a rotational basis.

Common Carriers

Domestic carriers. Although its portion of the domestic communications market has been eroded by the authorization of competitors for its more profitable services, Western Union remains the dominant domestic record carrier. With respect to message telegraph service to international points, Western Union is principally responsible for pickup and delivery to an IRC for overseas transmission. However, the availability of a low-cost, nationwide telephone service has resulted in some diversion of traffic in placing messages with IRCs. With its acquisition of TWX from AT&T in 1971, Western Union also became the most significant entity—outside the gateway cities where IRC tielines are permitted—for the domestic leg of international Telex calls. However, Western Union's portion of the domestic leg of leased-channel communications is not significant. Most of these facilities are provided by AT&T, although the advent of specialized common carriers and domestic satellite systems poses some competition to both Western Union and AT&T in this field.

International record carriers. There are five major international record carriers: ITT Worldcom, RCA Globcom, WUI, TRT, and FTC Communications.

International voice carriers. AT&T's Overseas Department provides all message telephone service and private lines (leased circuits) for voice-only use from the United States mainland to international points. AT&T is also authorized to provide leased circuits for data, AVD, and simultaneous voice-data use, as well as telegraph-grade leased channels between the United States mainland and Hawaii. The FCC also authorized AT&T to extend its Dataphone Digital Service internationally, an authorization that could overwhelm the IRC's data communications offerings. This authorization, however, is being challenged in court by the IRCs. The Hawaiian Telephone Company provides international and interstate, as well as intrastate, services in Hawaii. As AT&T's correspondent, it provides all those services AT&T is authorized to provide between Hawaii and the mainland. All America Cables and Radio, Inc. and ITT Communications Inc.—Virgin Islands, two ITT subsidiaries, provide overseas circuits for telephone message service, leased circuits for voice-only use, and program transmission services from Puerto Rico and the U.S. Virgin Islands, respectively.

International satellite facilities: COMSAT/INTELSAT. As detailed in the previous section, COMSAT is a privately owned corporation with the unique legal authority to own and operate the United States portion of the commercial satellite communications system that the 1962 act anticipated would be established with other countries.

In addition to its designation as the United States participant in the International Telecommunications Satellite Corsortium (INTELSAT), COMSAT is a common carrier fully subject to the Commission's regulation under the Satellite and Communications Acts. As such, it must obtain Commission authorization to participate in the construction of INTELSAT satellites and establish lines of communication via these satellites between American earth stations and foreign earth stations.

At the initiative of the United States, COMSAT led the way to the creation of INTELSAT. In 1964, 19 governments signed an interim agreement for a global communications system, the system anticipated by the Communications Satellite Act. The interim agreement was the agreement among governments. It was matched by a special agreement which was the technical and operating document, signed, for the most part, by representatives of the designated telecommunications organizations for each nation. COMSAT was the United States representative. Each of the signatories to the special agreement owned shares in the INTELSAT space segment in proportion to their contributions to the system. Earth stations were to be owned separately from the global system.

At the outset, COMSAT and the United States had the dominant influence in INTELSAT. COMSAT owned 61 percent of INTELSAT and was the manager of the system.

The 1964 agreements led to the definitive agreements, which were completed in 1971 and became effective in 1973. Similar to the interim arrangements, the definitive agreements involve two separate agreements: one among governments and the other an operating agreement among the actual telecommunications participants in INTELSAT. Again, earth stations are not owned by INTELSAT. However, the structure of INTELSAT has changed and COMSAT'S role has been reduced sharply. There are now 90 signatories to the INTELSAT agreements. Ownership of INTELSAT relates more closely to actual use of the system and voting corresponds to ownership. COMSAT'S ownership interest has been reduced to approximately 25 percent. A board of governors controls INTELSAT and a director general is the chief operating officer. At present, COMSAT, as manager of the system, reports to the director general rather than to the board of governors. COMSAT's management services contract expires in 1979 and there is every indication that management and supply services for COMSAT will be more broadly distributed than they have been in the past.

United States Government Participants

Federal Communications Commission. The mandate given the FCC under the 1934 Communications Act applies broadly in the field of international communications. As part of its overall mission, the FCC is empowered to regulate "interstate and foreign commerce in communications by wire and radio." The FCC also is authorized to foster "nationwide and worldwide communications service with adequate facilities at reasonable charges." All international communications by wire or radio that originate or are received within the United States and all persons engaged in providing these communications are subject to the FCC's jurisdiction under the Communications Act. The obligations imposed by the Communications Act on international carriers are substantially the same as those imposed on domestic common carriers. The FCC essentially has full regulatory authority over them, their rates and charges, their construction of new facilities, and the terms and conditions under which they offer services to the public.

In addition to the 1934 Communications Act, the Communications Satellite Act of 1962 authorized the FCC to perform several other functions with respect to COMSAT, which the Commission does not have regarding other carriers. For example, the FCC has broader authority over COMSAT to assure that there is competition among COMSAT's suppliers and that there is greater supervision over COMSAT's accounting and rate-making practices.

In addition, the FCC also exercises some authority delegated by the executive branch, particularly the President's power to withhold or revoke cable landing licenses. However, license grants or revocations receive the perfunctory prior approval of the Department of State, preserving the President's primacy in the foreign affairs aspects of international telecommunications.

Under the FCC's statuatory authority, international carriers must come to the FCC on two separate occasions to receive approval for new facilities. First, carriers must obtain approval to construct facilities, and once the facilities are constructed, the carriers must return to the FCC for permission to activate and use the new facilities. Given the long planning time required, the large capital expense, and the multiparty nature of international facilities construction, the FCC has sought to make its new facility authorization procedures more responsive to the needs of international communications. Rather than reacting merely to a construction permit application submitted to it, the FCC now engages in facilities planning with carriers, other governmental agencies, and foreign organizations to provide guidelines for the facilities planning process.

Office of Telecommunications Policy/National Telecommunications and Information Administration (NTIA). The Office of Telecommunications Policy (OTP) was created in 1970 to fill the gap that was thought to exist in the government's ability to engage in long-range planning with respect to telecommunications policy issues. Although created by executive order, OTP had support in the Congress, since it was generally conceded that the FCC was mired in so many day-to-day activities that it had neither the time nor the inclination to engage in policy planning.

The executive order creating OTP paid more attention to international telecommunications than did the Communications Act of 1934. OTP was charged with coordinating interdepartmental and national activities to prepare for international communications conferences and negotiations and to advise the Secretary of State on conducting foreign affairs relating to telecommunications. Moreover, OTP was delegated the President's powers and authorities under the Communications Satellite Act of 1962, particularly the power to assist in the planning and development for the worldwide satellite communications system and the power to advise in governmental use of such a system.

During its lifetime, OTP was actively involved in the major international telecommunications issues relating to facilities planning and the planning process. OTP also advised COMSAT on its INTELSAT activities, coordinated and assisted in conferences, such as those leading to the INTELSAT definitive agreements, and commissioned several significant studies of international communications. OTP also developed proposed legislation to amend the 1962 Communications Satellite Act and various portions of the 1934 Communications Act relating to international communications. However, OTP was not successful in devising a comprehensive long-range policy for international communications.

OTP was phased out in 1977 and most of its functions were absorbed by the recently established NTIA in the Department of Commerce. It is too early to say whether the new division of of international telecommunications responsibilities between NTIA and the State Department will affect policymaking.

State Department. In addition to the ministerial functions allotted to the Department of State by the Cable Landing License Act, the 1962 Communications Satellite Act, and the new executive order creating NTIA, the State Department seems to have broad but vague powers in the field of international communications. It is clear, however, that the department is the principal repository of the President's foreign affairs power in the field and, as such, participates in international conferences and planning sessions as the "foreign office" element. The point of view represented by the State Department, however, has little to do with expertise in the substantive issues presented. Rather, it is one that stresses the impact of telecommunications policy decisions on the United States' relationships with foreign nations.

Problem Areas

A review of the background and development of the international common carrier industry discloses three major, interrelated problem areas: industry structure and regulation, the legal monopoly imposed on satellite technology, and lack of effective division of responsibilities organization among the United States government agencies involved in telecommunications.

Industry Structure

The international carrier industry functions in an environment of rapid technological development, increasing demand for service, large investments, and economies of scale. Because of this structure, competition in the market is difficult to sustain. If one were to analyze the history of the industry—removing the influences of regulation and government policy, particularly after the development of voice-grade cables—one would probably see a monopoly in which AT&T, as dominant supplier of the facilities and services, would be the single carrier of all voice and "record" communications between the United States and the rest of the world. As technology reduced the cost of providing services, the rates to the public would be lowered progressively, just as costs of domestic long distance communications have been reduced over the years. The pace of innovation in introducing new technology and new services perhaps would not have been very rapid, but one could imagine an industry structure that would be fairly responsive to demand for service at a generally decreasing cost.

It seems that the effect of legislative and regulatory policies has been to insulate the industry from the demands of the marketplace. The compartmentalization of the industry into domestic and international components, into service components (i.e., "voice" and "record"), and into technology components is the result of regulation. If the purpose of regulation is to serve the interests of the consumer of communications services, regulation of the international industry has not only failed but may disserve their interests.

Regulation of rates. One traditional element of the FCC's common carrier regulation, which is intended to protect the public from the abuses of monopoly, is rate-base regulation.[9] However, rate considerations do not seem to play a significant part in the regulation of international communications. For example, there has been no general review of rates and international communications since 1958. Moreover, other FCC policies, such as the "interim"—now permanent—formula for the distribution of nonrouted telegraph messages, actively undercut rate competition among the carriers.

The most significant action the FCC has taken with respect to rates is the investigation of the, and subsequent order regarding, COMSAT rates. COMSAT filed its initial tariff rates in 1965 when it began service. Those rates were under FCC review from the beginning. In November 1975, the FCC issued its rate reduction order and limited COMSAT's rate of return to between 10.8 and 11.8 percent and reduced the COMSAT rate base from more than $339 million to more than $123 million. To reflect these adjustments, COMSAT has proposed a one-eighth percent reduction in revenues.

In general, this type of remedy cannot be as easily implemented in the international sphere as a rate reduction order would be in the domestic sphere. The establishment of rates, just as the planning and construction of facilities, is a joint effort between American and foreign carriers. American carriers may be ordered to lower their rates, but, so far, the foreign correspondents' rates have not been considered under the FCC's regulatory jurisdiction.

Thus, the presence of foreign correspondents adds to the complexity of any rate supervision in the international communications area, including any substantial likelihood of rate competition among the carriers. Each rate for service in international communications is really two rates: a collection rate and a settlement rate. The collection rate is charged by the carrier to the user initiating the communications. The settlement rate is the basis for the division of tolls between American and foreign carriers, since the facilities of both are used to complete the communications. The settlement rate is a negotiated rate between the carriers and takes into account such factors as the rates of currency exchange and the differentials in traffic flow between the two carriers. With respect to rate competition, one of the American IRCs—TRT—attempted to gain a greater share of the market by lowering the rate for telex calls between the United States and Britain by some 21 percent. The collection rate was lowered from $2.55 to $2.00, with the agreement of the British Post Office. The settlement rate was lowered from $2.25 to $1.70. When the other IRCs objected, the FCC imposed its policy of "rate uniformity" and determined that the rate reductions were not in the public interest. Eventually, TRT was permitted to lower its collection rate but was not allowed to reduce the settlement rate. The FCC decided that its 1943 policy of settlement rate uniformity among the IRCs should govern. TRT, however, has little interest in absorbing the entire collection rate reduction alone, even if the three major IRCs had not filed collection rate reduction applications to match the TRT rate.

Of late, the FCC has shown slightly more interest in international communications rate questions, as it undertakes an audit of rates for international services and reviews AT&T rates for specific submarine cable circuits. These matter, however, have yet to be resolved. In the past, the FCC has not pursued rate regulation vigorously. If this action follows previous rate reductions, the rate reduction will be mandated by the FCC carriers' suggested rate reductions, which are based on the carriers' anticipated reductions in revenue requirements. The Commission does not seem to calculate independently the carrier's projected cost reductions. Thus, the rates are reduced by negotiating the amounts that the carriers have suggested would be reasonable reductions. In general, it does not appear that rate-base regulation has been an effective regulatory tool in the internationl sphere.

Regulation of facility construction. The other major function of the FCC in common carrier regulation is to oversee the construction of new facilities. This function—rather than rate regulation—has had the greatest impact on the international industry in the past 10 years and promises to have increasing importance in the coming decade. The FCC's regulation, in this regard, has led to substantial intermodal competition among the carriers, but it occurs at the Commission and not the marketplace.

The FCC assumes, with some justification except for television transmission, that both cable and satellite technology are interchangeable for international communications. As noted previously, the FCC developed the proportional fill rule to allocate traffic between cables and satellites so that each would reach its theoretical capacity at approximately the same time. The rule was enforced by requiring activation of already constructed facilities on a circuit-by-circuit basis. However, the rule had little effect on the carriers for market shares by proposing the construction of new facilities. As Professor Owen has explained, since investment in new capacity is required to meet growing demand, carriers can use the investment process as a device for allocating market share. Thus, unlike the usually routine Section 214 process in domestic carrier regulation, there is considerable controversy surrounding each new proposal for cable or satellite facility construction in the international field.

Indeed, given the incentives of the carriers in this regard, there may be substantial overinvestment (i.e., too much capacity) in international communications facilities. If one takes the public benefit as the touchstone of regulation, the policies that the FCC uses to guide facility planning and construction appear irrational. They become rational only if one returns to the hypothesis of the regulated oligopoly, with the FCC performing the cartel manager's role. In this role, the regulator is primarily interested in ensuring that each member of the industry receives and maintains a fair share of the business; that, overall, the industry is healthy and free of "destabilizing" competition; and that the structural framework created by previous decisions is maintained.

Effects of Imposing a Legal Monopoly on International Satellite Technology

Under the Communications Satellite Act of 1962, COMSAT was created as a private corporation charged with serving the governmental purpose of creating a global communications satellite system. The legislation infused COMSAT with a flavor of a quasi-public agency (e.g., with presidentially appointed directors, more government oversight over COMSAT's activities than over the activities of typical communications common carriers, etc.).

In the more than 15 years since COMSAT's creation, its role and function have changed dramatically, but the legislation creating it has not changed at all. For a variety of reasons and in numerous actions, the FCC has encouraged the carriers to sell their interests in COMSAT. The worldwide system of satellite communications has been created and is being administered increasingly by INTELSAT, without COMSAT leadership. Only the vestiges of presidentially appointed directors and additional governmental oversight of its contractual activities now distinguish COMSAT from any other private communications common carrier. There are, however, few who would argue that these legislative requirements serve any useful purpose.

The principal aspect of the 1962 legislation, that continues to have a substantial effect on international communications, is the legal monopoly the statute gives to COMSAT as a purveyor of satellite technology. As the industry structure discussion makes clear, this encourages competition for market shares through investment in new, sometimes unneeded, capacity with its resultant impact on the rate-paying public. The policy of the 1962 act also impedes the full use of and innovation in satellite technology as a means of international communications. Geosynchronous satellites are uniquely suited to global communications—a feature that was recognized by the proponents of the Communications Satellite Act of 1962. They did not realize in 1962 that the satellites evidenced few aspects of a natural monopoly, as later developments in the domestic use of satellites have shown. Indeed, it is not clear that the supporters of the 1962 act fully appreciated the difference in the implications of geosynchronous satellites and lower-altitude orbiting satellites. While creation of a legal monopoly may have been an appropriate legislative response for orbiting satellites, it has proven to be a straight-jacket for the more efficient, less costly geosynchronous satellite systems. Indeed, the 1962 act may well be one of the few instances in which the Congress undercut the flexibility of an administrative agency to deal with rapidly changing technology—one of the presumed benefits of an expert agency.

Thus, in international communications, the FCC is not free to deal with satellite technology, as it was with the high-frequency radio technology of the 1930s and 1940s. Satellite technology is encumbered by COMSAT and by INTELSAT. Any judgment, any decision, any action affecting satellite technology also affects their institutional structures and the relationships among them.

Moreover, one need only contrast the domestic use of satellite technology with the international use to see that the presence of a legal monopoly may have taken its toll on innovation in service offerings. Under the FCC's domestic satellite policies, any financially and technically qualified organization can engage in domestic satellite communications and can employ that technology through user-owned earth stations. The results have been cost reductions in services and facilities, competition with terrestrial carriers, and the development of new cable television and public broadcasting services for the public. While it is difficult to prove conclusively what could have been, one senses that, in the international sphere, there could have been greater cost reduction, and more innovative use of the inherent advantages of the technology. By conscious policy decision and technical design, INTELSAT employs relatively low-power satellites and high-cost earth stations to offer essentially point-to-point telecommunications services over wide areas of the earth. The more lucrative, high-traffic routes are served by undersea cables, which also are owned by INTELSAT's owners except COMSAT.

Given the inevitable politicization of procurement and services that arises in an organization comprised of more than 95 countries, an obligation to provide service along low-traffic routes, and the large members' ability to "cream skim" with cable on the heavy-traffic routes, INTELSAT cannot be relied on to innovate a tailored technology to provide new services. Indeed, present and proposed specialized uses of the technology occur outside of the INTELSAT structure. Unfortunately, in such instances, the INTELSAT model has led to the creation of new international entities to provide the specialized satellite service—for example, Aerosat for aeronautical satellite services and Inmarsat for maritime services. Thus, each new satellite service becomes further encumbered with new international legal entities, and the structure of relationships among governments, carriers, the public, and the entities become more and more complex.

United States Government Organizational Structure for International Communications

The two problem areas described above are exacerbated by the third—the lack of a coherent United States government organizational structure to deal with the issues and to develop and adapt policies in the field. Just as the industry is compartmentalized in services and technologies, the federal government agencies with international communications interests are fragmented and have overlapping, sometimes vague, responsibilities.

The FCC has the clearest and most extensive responsibilities. Recently, the Commission has used its authority in an enlightened, progressive way, often seeking to apply to the international sphere the regulatory innovations that have proven successful in domestic communications. Yet, despite its recognition of the anomalies and irrationalities of the present system, the Commission is more the prisoner of the regulatory structure than the master of it. As with any regulatory agency, the FCC cannot be expected to initiate an overhaul of its organic statute. The legislative framework constrains the FCC's freedom in international communications and, it is unlikely that the constraint will be removed in the near future.

Moreover, the regulatory structure created by the legislation and the weight of past decisions traps the FCC in the role of cartel manager. The natural inclination is to extend that role to the point at which the FCC is making basic business judgments for the carriers and their foreign counterparts. If present trends continue, the FCC will decide new facility questions using its own traffic forecasts, its own cost estimates, its own traffic allocation formulae : in effect, making the investment decisions directly, and making operational judgments as to facilities and services.

In addition, even if the FCC could somehow improvise a new legislative policy for international communications, the presence of the foreign governments' telecommunications organizations as one-half of any international communications project, would limit the Commission's freedom of action. Although the FCC might seek to remove entry barriers and to introduce new competition into the market, the foreign organizations can exercise their sovereign right to maintain the status quo.

In fact, a major deficiency of the United States government's organizational structure for international communications is the absence of a government agency that is equipped by interest, expertise, and authority to deal with the foreign governmental organizations. In other fields such a role is usually performed by an executive branch department, recognizing the President's traditional foreign relations powers. In international air transportation, for example, treaties govern the basic forms of reciprocal arrangements and the President has the veto power over the international route decisions of the regulatory agency. While this approach might not work well in international air transport—at least it reflects a fundamental recognition that an independent regulatory agency cannot have the first and last word.

The Department of State is intended by law to advise the FCC on certain foreign policy aspects of international satellite communications and informally is able to express other concerns to the FCC. However, the department's interest usually is limited to the foreign policy aspects of the matter and does not have a significant effect on the substance of the FCC's decision.

In addition to the foreign relations aspect of executive branch interest in international communications, the executive is the largest user of international communications services and facilities, a promoter of foreign commerce, and a policymaker. The United States government's views, as user, are usually presented to the FCC by the Department of Defense. But the executive branch influence as another pleader before the FCC is vastly overshadowed by the influence the executive exerts as a purchaser of services. Yet, the government can purchase only what is available for sale and it cannot, as a purchaser, completely restructure the market.

The executive's policymaking leadership role was assumed for a time by OTP. However, OTP never had the authority to do more than perform studies and make recommendations to the FCC and to the Congress. This is a significant function, and one not performed by any other executive agency with telecommunications expertise, but it is not likely to compensate for the lack of a coherent policymaking organization in the international communications field, able to deal on a government-to-government basis. It is not likely that NTIA will have more success at its international communications responsibilities than did OTP.

Conclusions

Realistic observers know that in international communications solutions are a long way off. There are at least three interrelated reasons why "reform" of the international communications industry is unlikely: (1) the gravitational pull of the status quo; (2) the presence of foreign governmental organizations as members of the industry; and (3) the sense that regulatory "reform" would not produce better results, given the nature of the market.

The forces supporting the status quo are strong in any regulated industry, but, in the international telecommunications field, these forces are intense. Maintenance of the status quo is not simply the desire of the industry, it seems to be the essence of the policy for international telecommunications. Moreover, it is not even a current status quo that is being preserved, but one of a marketplace and industry structure that reached equilibrium before 1943. Whenever some technological or other development upsets that equilibrium, the direction of governmental action—whether by the FCC, Congress, or the courts—is to restore the equilibrium.

When the legislative and regulatory policy pulls in the direction of the status quo, if there is to be any change, the Congress must step in. To accomplish this, it would take a concerted effort of strong interests to induce the Congress to adopt fundamental reforms. With the international industry resisting change and the FCC caught up in its regulatory structure, it is not likely that there are strong interest groups that are sufficiently dissatisfied with the current situation to launch a major campaign for reform.

The second factor working against substantial reform is the presence of foreign governmental organizations as coequal members of the international telecommunications industry. Having all the attributes of both government agencies and private monopolies, foreign PTTs would seriously inhibit any FCC attempt to achieve a more open, competitive international communications marketplace. Strongly identifying with the views of the established American carriers, the PTTs effectively can block significant departures from the status quo by refusing to enter into operating agreements with the new entrants, as has been the fate of Graphnet and Telenet (see, e.g., *Graphnet Systems, Inc.*, 63 FCC 2d 402 (1977). Since there can be no international communications without them, the FCC presently has no way around the PTTs.

The third reason why substantial reform is unlikely is the deep uncertainty that, given the nature of the marketplace, reform would produce greater benefits than the status quo. The feeling is strong that, if the international telecommunications industry were not a regulated oligopoly, it would be an unregulated oligopoly. As Professor Owen has stated it:[10]

> While there is little evidence that regulation produces results superior to those of unregulated oligopoly, there is also little evidence of the reverse proposition, and legislators may prefer to preserve the mechanism and appearance of fairness through regulation.

However, the writer's urge to offer a proposal to solve the conundrum has overcome the sense that adoption of any solution is unlikely. In its most radical form, the proposal would involve a complete overhaul of the legislation governing international telecommunications. Such legislation would give the Executive Branch a larger role to create an effective mechanism for accommodating the interests of foreign governments, would remove the constraining monopoly on satellite technology, would eliminate the communications service distinctions between international and domestic and voice and record communications, and would allow open entry to the international communications market for all providers of communications services.

Using an approach similar to that used in international air transportation, the Department of State—assisted by the FCC, the Department of Commerce, and industry representatives—would negotiate agreements with their foreign counterparts. The executive agreements would reflect the results of a planning process conducted free of the constraint of the Administrative Procedure Act. The agreement would authorize use of both cable and satellite circuits and deal generally with rates and other basic terms and conditions. The agreement also would make clear that national carriers of each of the signatories will be subject to the regulatory jurisdiction of all of the signatories.

Satellite circuits could be ordered from INTELSAT or could be provided through other satellite facilities, which would be owned and operated by any carrier, PTT, other service provider, or by a combination of these groups. Authorization of such satellite facilities could be covered in the bilateral or multilateral agreement.

The FCC would certify circuit activations if they were found to be consistent with the terms of the agreement. The foreign PTT's power to refuse to enter into operating agreements would be constrained in the government-to-government agreement. If necessary, the agreement could even specify the number of American carriers that would be allowed to interconnect with the PTT's, without actually designating them. If such limitations could be avoided, there would be open entry for all carriers and other organizations, including foreign PTTs. The removal of the satellite monopoly, the removal of service distinctions between international and domestic, voice and record, and the termination of rate regulation all could serve to break down the oligopolistic market structure.

This proposal does not have to be adopted *in toto* in order to make the structure of international communications more rational. There are, however, three features of the proposal that should be considered as the essence of any reform effort, even if the exact form of what is proposed here is not followed. At least one of these features will require congressional action. The others, and any additional embellishments, can be left to the FCC and the executive branch agencies.

The first essential requirement is to create a mechanism to allow government-to-government negotiations and agreements regarding international telecommunications facilities. This would probably not require legislation. The responsible government agencies could agree that the facilities planning and consultative process would end in a multilateral agreement rather than in an application to or policy ruling from the FCC.

Second, the foreign PTTs, who jointly own transoceanic cable with American carriers, should be made subject to some aspects of the FCC's regulatory jurisdiction, particularly its Section 214 jurisdiction. The FCC has never been able to adapt to the presence of foreign PTTs in cable facility ownership, which became necessary with the development of voice-grade cables. The present fiction that ownership shifts from American to foreign organizations at ocean mid-point deprives the FCC of direct regulatory oversight over the joint owners of cable facilities. Other U.S. regulatory agencies have authority over foreign governmental organizations that carry on commercial activities in the U.S. There seems to be little justification for not subjecting foreign telecommunications entities to some reasonable level of FCC oversight. While new legislation might facilitate matters, it may not be required. The ground-work for the FCC's assertion of limited jurisdiction could be established through State Department and foreign office channels and through negotiations at the operating company and FCC levels. There would be little need for the FCC to have direct regulatory authority over INTELSAT, since it would have such authority over the principal members of INTELSAT.

The third essential feature of the proposal for international communications is the termination of the COMSAT monopoly. This would require legislation, at least to repeal the 1962 Satellite Act, and could involve the modification of the "definitive arrangements" for INTELSAT or the withdrawal of the U.S. from those arrangements. At present, INTELSAT is an anachronism wrapped in an international bureaucracy that does little to exploit satellite technology for the public's benefit. To the extent that it serves as a useful mechanism to provide service on low-density traffic routes, it has a function that can be retained voluntarily by its members, without conferring a legal monopoly on INTELSAT. In fact, as satellite technology has developed, as specialized satellite services are offered outside of INTELSAT, as regional satellite systems are created, and as international facilities are used to serve some nations' domestic needs, INTELSAT's legal monopoly has become somewhat of a legal fiction. The fiction should not be allowed to constrain international satellite usage that may be in the best interests of the United States. Moreover, if the third essential change were to be made, the FCC would have more flexibility in authorizing the construction and use of international satellite facilities outside of INTELSAT. Therefore, intelligent use of satellite facilities would not depend completely on the FCC's direct or indirect control over INTELSAT.

In highlighting three essential features of a reform proposal, I do not mean to slight other significant aspects. I believe, however, that the FCC can be relied on to continue its efforts over the past five years to improve the international communications service market, even within the constraints of the 1943 amendments to the Communications Act. It would be preferable to repeal Section 222 of the act. Separation of the service market into international and domestic, based on fears of abuse arising from a domestic telegraph monopoly, does not make sense in the present domestic market for communications services. If, however, the policy continues, it ought to be applied to AT&T's telephone traffic. But rather than applying a bad policy equitably, it would be better to eliminate it. Similarly, the separation of voice and record services has been outmoded for decades and should be eliminated. The point is, however, that the FCC probably can muddle through these problem areas, although it would be better to have legislation to protect the FCC from court challenges to its efforts.

There is probably no answer to the question of whether any of these changes would make a difference to the American public. After all, as some are fond of pointing out, the American system works and even our international system works better than almost any other domestic telecommunications system. This, however, cannot be the criterion of the policymaker or legislator. Almost anything can be made to work, but at some cost and some level of efficiency. One cannot study the industrial, regulatory, and legislative aspects of international telecommunications without concluding that it could operate more efficiently and provide services that cost less. No one is claiming that more efficient, less costly international communications would lead to world peace or hasten the millenium. It is simply a goal worth pursuing in its own right.

Notes

1. The discussion in this chapter draws heavily from the FCC's Report on International Telecommunications Policies submitted to the Subcommittee on Communications of the Senate Committee on Commerce, Science and Transportation, July 13, 1977, and from "Government Regulation of International Telecommunications," a paper prepared for the Office of Telecommunications Policy by Columbia Law School's Legislative Drafting Research Fund. The views expressed in this chapter are solely those of the author and cannot be attributed to his law firm or its clients. Given the subject, it should be noted that one client is FTC Communications, Inc.—an international record carrier.

2. The Government Printing Office, Washington, D.C., March 1951, pp. 151-2.

3. *F.C.C. v. RCA Communications,* 346 U.S. 86 (1953).

4. *Id.* at 94, no. 5.

5. *Id.* at 97.

6. The economic analysis reflected in this paper is derived from and influenced by Bruce M. Owen, who served as chief economist of the Office of Telecommunications Policy during

much of the author's tenure as general counsel of the same agency. In particular, Owen's study, "The International Communications Industry, the TAT-6 Decision, and the Problem of Regulated Oligopoly," offered the most comprehensive explanation for the industry and regulatory conduct the author found to be prevalent in the recorded experience of the international communications industry.

7. Bruce Owen, "The International Communications Industry, the TAT-6 Decision, and the Problem of Regulated Oligopoly," Office of Telecommunications Policy, Washington, D.C.

8. A year after the TAT-4 decision, the FCC took a slightly different approach, but one motivated by a similar view as to industry structure. The issue involved the carriage of transoceanic television programming, a new service made possible only with the development of communications satellites. Unlike the AVD decision, the FCC did not attempt to categorize television transmissions as either voice or record communications, but simply allowed AT&T and the three major IRCs to provide this service on a rotating basis.

9. Rate-base regulation is, however, thought to have a particular impact upon the facilities issue. One reason that the carriers prefer cables to satellite circuits is that the cost of cables are included in the carrier's rate base. COMSAT/INTELSAT for satellite circuits are not included in the carrier's rate base, but are carried as expense items. Accordingly, a number of carriers and government entities, including the FCC, have urged that satellite lease circuits be capitalized and included in the carriers' rate base. While this second layer of capitalization may be an acceptable means of eliminating the carriers' bias against satellite circuits, therefore allowing more effective use of both technologies for international communications, it would probably mean increased rates for the public. Moreover, the artificiality of the concept, both in economic and regulatory terms, masks more fundamental problems with facilities planning and construction, which would not be solved.

10. Bruce Owen, *op. cit.*, p. 37.

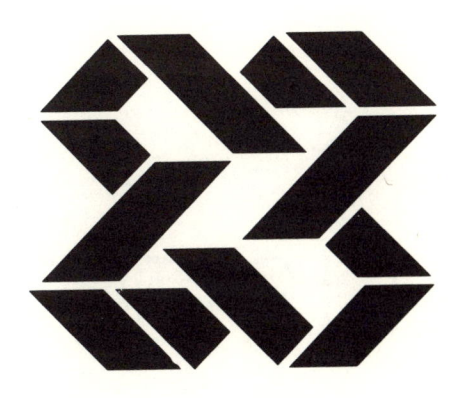

Part 3.
Applications of New Electronic Media

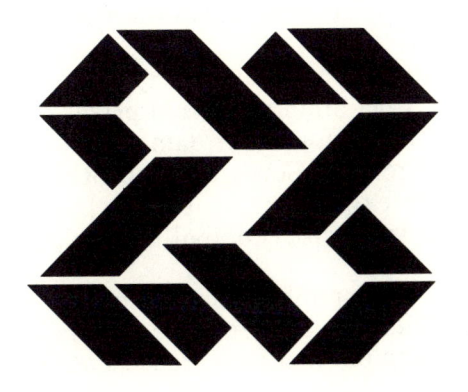

5.
Pluralistic Programming and Regulation of Mass Communications Media

Benno C. Schmidt, Jr.

If law echoes intuitions of public policy, as Justice Holmes taught us, American society's intuitions about mass communications media are strikingly ambivalent.[1] During the past half-century, when American law has had to deal with mass media, legal approaches to print publications and to broadcasting have rested on divergent, even opposite, basic premises. For print, the First Amendment has been the prime support for a legal theory that commands government not to interfere with the substance or process of publication. Anyone with a printing press and the means of distribution may publish just about anything without legal interference, although publishers are subject to laws governing labor relations, antitrust compliance, taxes, and other regulations generally applicable to business enterprises. This does not mean that everyone has the capability for mass communication. Among countless other differentiating factors, communications skills, organizational ability, hereditary privilege, luck, wealth, and popular acceptance determine who enjoys the power of effective expression.

Nor is freedom with respect to the content of expression completely unlimited. The laws of libel and invasion of privacy, for example, impose liability on publications that damage individual interests in reputation and repose. Copyright and other laws protect proprietary interests in expression. Still other laws punish expression thought to damage specific, compelling social interests.

But regulation of the content of print publications is the rare exception in our legal system. There is no recognized broad principle of legal control, under which publications are regulated to advance the general good. Even narrow content controls, designed to achieve specific social benefits, must overcome intense constitutional resistance. For print media, the utilitarian premise of the First Amendment has been that "right conclusions are more likely to be gathered out of a multitude of tongues, than through any kind of authoritative selection."[2] Holmes captured this idea in the laissez-faire imagery of a "free marketplace of ideas," and in his pragmatic view of "truth" as the shifting residue of competition in ideas. The notion that society can enjoy the benefits of freedom of the press if government simply leaves the press alone was a natural outgrowth of Enlightenment notions, reflected in our 18th century Constitution, that liberty would result from purely negative controls on government power. The First Amendment's injunction to government, like the other amendments in the Bill of Rights, was honored if people were left alone.

The legal status of radio and television is grounded in a different tradition. Created by statute in 1927 in response to the explosive growth of radio, the law of broadcast regulation contradicts the First Amendment's central historical principle—that government may not treat the right to publish as a privilege allowed only on condition of prior official approval. No one may broadcast without a license. Government licenses only broadcasters it believes will serve the public interest. And beyond licensing, the law of broadcast regulation tolerates controls and influences on programming that would fly in the face of constitutional tradition if applied to newspapers or books. A few of these controls bar or inhibit objectionable program content. More significant, and more dramatically at odds with the law of the First Amendment for print media, are regulations of broadcasting designed to enhance the amount or the quality of expression about public affairs. These require balance in public affairs programming and provide access to the airwaves for persons or points of view the broadcaster might not choose to cover as a matter of editorial discretion.

The divergent approach to mass communications media in the United States rests on the belief that print and broadcast media are fundamentally different. The broadcast spectrum is thought to be physically scarce in ways that the means of production of print media are not. Many have viewed the spectrum intrinsically as a public resource—like navigable waterways, not suitable for private exploitation—whereas printing presses, paper, and ink are deemed private goods. (The President's statutory power to take over all broadcast facilities in time of war or national emergency is a current manifestation of the view of the spectrum as a public resource.) Moreover, because rights to use portions of the broadcast spectrum turned out to be valuable, and are "given away" by the government, it has seemed fitting to many that some public duties should be imposed in return. Differences have also been discerned in the impact of the two types of media. Broadcasting is widely perceived to be a more intrusive medium. Its audience seems more captive, less selective in advance about what is taken in and less critical of messages received. Children have free access to it. Perceptions of special intrusiveness are matched by perceptions of special power. Many people believe that broadcasting has vastly more influence than has print media over political and cultural attitudes. These perceptions have prevented the non-interference principle, which is at the heart of First Amendment thinking about print media, from dominating the law of broadcast regulation.

Because the divergence in legal approaches to print and broadcasting is so striking, there is a tendency to overlook how much the two branches of law have in common. Both reject government ownership and operation of mass communications media, although this approach to broadcasting has commended itself to most countries with which the United States shares close kinship. Our law also turned away from treating broadcasters as common carriers, akin to point-to-point communicators such as the telegraph and telephone companies. The model that Congress followed in broadcast regulation was much closer to private print journalism than to the Post Office or Western Union. Despite rejection of the full tradition of private ownership and noninterference that characterizes our law of freedom of the press, the law of broadcast regulation also reflects grave doubts about government control over expression.

Thus, even in the face of the contradiction of licensing, First Amendment values cling to the electronic media with surprising persistence. Concern for freedom of expression and editorial autonomy have sharply limited official control over the content of broadcasting, and have become even more formidable forces in recent years. The First Amendment tradition for print and the law of broadcast regulation are not rigorously divergent. The differences in the two bodies of law are offset by some pronounced common threads. Viewed from the perspective of the constitutional tradition for print media, the law of broadcast regulation seems noteworthy for official controls and influences over expression. However, in terms of the way other nations treat electronic media, or even in terms of the stunted, inactive nature of First Amendment doctrine in the late 1920s when the law of broadcast regulation was established, this branch of mass communication law is remarkable for its commitment to private decisionmaking and its faith in the ultimate social benefits of a free marketplace of ideas.

Even when one focuses on the undeniably important differences between the two bodies of law, one must recognize that the differences are not frozen. Second thoughts about the wisdom of our law's divergent approach to mass communications are growing. Pressures seem to be pushing toward a unitary legal approach to mass media, or at least for a substantial shift within each separate legal tradition toward the other. As newspapers have tended to become local monopolies and have assumed the size, profit-orientation, and composite character of large corporations, doubts have arisen about the First Amendment's laissez faire assumptions. In a one newspaper town, does the absence of governmental interference really allow for a "multitude of tongues" that leads to diversity of expression? Questions such as this must eventually lead to pressures for legal efforts to regulate newspapers for fairness and balance, as electronic media are regulated.

On the other side, the emergence of cable television and other broadband methods of transmitting audio and visual signals challenges the assumption that broadcasting is an inherently scarce medium. The number of television or radio signals available to consumers is limited now by economic, not physical, constraints. The reasons for scarcity of broadcast outlets thus coincide with the reasons for scarcity of newspapers. Moreover, from the average consumer's viewpoint, there are likely to be many more community-based broadcasters than newspapers, magazines, or other publications. At least on a simple view of the matter, scarcity no longer seems a persuasive basis for regulating broadcasting along lines that differ fundamentally from the legal tradition for print. Broadcasting stations are more abundant than newspapers, where the assumption of diversity underlies the

Programming and Regulation of Mass Media

traditional view of the First Amendment as requiring government to keep hands off the press.

In sum, the distinction of scarcity versus abundance, traditionally offered to explain why broadcasting is regulated and print is free, has become rather timeworn in view of economic developments in the newspaper industry and technological developments in electronic media. Indeed, in terms of this distinction, the two branches of mass media may be not only merging, but even changing positions. Developments in electronic media seem to call for a shift from regulation toward First Amendment laissez faire principles. But the economic characteristics of print media seem to call for some of the patterns of official control developed for radio and television. Thus the challenge to the dual legal tradition for mass media is growing, but the direction of change is not clear.

Even if some notion of scarcity originally led to a legal approach to broadcasting quite different from the First Amendment tradition for print, one should not suppose that the end of scarcity necessarily will, or ought to, produce a radical revision in the fundamental tenets of broadcast regulation. As a matter of practical politics, a half-century of broadcast regulation has vested private and bureaucratic interests that stand in the way of fundamental reform of statutory or administrative law. Moreover, one must recognize that scarcity and abundance have not been, and certainly are not now, the only important differences between print and electronic media that might support different legal approaches. Theoretically, as well as practically, the divergent legal tradition will continue to attract strong support, even after the demise of the original basis for the different treatment.

Finally, one should not ignore the possibility that it is really the basic ambivalence about the idea of freedom for mass communications media that supports the dual legal tradition. The doubts that have emerged about the laissez faire First Amendment tradition for print, as monopolistic newspapers exert ever more obvious influence over political attitudes, tend to reinforce the tradition of regulation for broadcasting. Desire to preserve the half-loaf of legal control provided by the law of broadcast regulation may account for the legal dichotomy as much as a continuing sense that there are differences in print and electronic media justifying divergent legal approaches. Learned Hand said of the First Amendment, in words that have echoed through the years: "To many this is, and always will be folly; but we have staked upon it our all."[3] Our law's approach to mass communications media suggests that we have in fact hedged our stake on the First Amendment; if it is indeed folly, the regulation of electronic media perhaps will protect society from much of the damage that irresponsible mass media could inflict. And at the same time, the constitutional status of print media can provide society with the benefits of a free press. Perhaps we view the choice between freedom of mass communications media and governmental control for the greater good as one of those insoluble dilemmas for public policy to which our law should respond by looking firmly in opposite directions.

Among the many questions posed for policymakers by the rush of new communications technology is whether the mix of freedom and government control over mass electronic communications that was established in the early days of radio should govern our future. The question arises at a time when not only new technology but also economic developments affecting the structure of the mass communications industry challenge long-held legal principles governing print and electronic media. Directions for change, however, are obscured by uncertainties about future developments, by the tangled intricacy of policy considerations, and by the formidable importance of mass communications to the future of our society. Institutional capacities for reform, always puzzling, are more than usually complex; the First Amendment, among its many consequences, has the effect of thrusting the judiciary into the midst of policy formation about the place of mass media in our society.

Broadcast Regulation and the First Amendment

The statutory foundation for our system of broadcast regulation was established 50 years ago by the Radio Act of 1927 and has remained essentially unchanged.[4] The statutes were set in place before the existence of television, networks, and even before recognition that the commercial basis of broadcasting would be advertising. The passage of a half-century is not necessarily an overdue notice for revision, even in as volatile a field as electronic mass communications media. Many other basic regulatory statutes are of comparable vintage—the antitrust statutes, for example, are even older, and the Interstate Commerce Act, the Federal Trade Commission Act, and the basic statutes governing labor relations are about as old or older. To the extent that statutes establish processes for ongoing legal adjustment rather than ordain legal outcomes, the passage of years need not signify petrification.

The story of the genesis of broadcast regulation has been told often, and only the central elements contributing to the legal solution need highlighting here. Radio broadcasts over the same or adjacent frequencies of the limited spectrum available for broadcasting produced a confusion of interfering signals that led to pressure for government to allocate spectrum rights. The allocation system that emerged was a curious mix of preferences for the advantages of private enterprise and free expression over the airwaves and, on the other hand, a conception of the spectrum as a natural resource, like national parks or navigable waters, that should be utilized in the public interest. A Senate resolution stated in 1924: "the ether [the spectrum] and the use thereof for the transmission of signals ... is hereby reaffirmed to be the inalienable possession of the people of the United States and their Government."[5] The notion of full property rights in the spectrum was rejected. However, government operation of broadcasting (repeatedly urged by the Navy Department in the early days of radio) was likewise rejected. Instead, a compromise emerged of private broadcasting under official supervision through a system of licensing and renewals of individual broadcasters administered under a vague "public interest, convenience, and necessity" standard drawn from other early regulatory statutes. Private broadcast licensees were not designated as common carriers. Congress intended that private broadcasters should retain substantial editorial control over programming, though a government agency would determine who should be allowed to broadcast, would exercise some vague and limited power over what was broadcast, and would enforce a few specific programming controls.

Somewhat surprisingly, the Radio Act of 1927 included explicit protections for free expression. It must be remembered that when broadcast regulation was cast, free speech principles had yet to win a victory in the Supreme Court. Up to that time, the courts had shown virtually complete deference to legislative prerogatives in regulating expression. Moreover, early radio made no pretense of being a news medium. It appeared to have more in common with motion pictures than with the print media, and the Supreme Court had held in 1915 that motion pictures, which were in its unanimous view more like "circuses" and "spectacles" than like the press, had nothing to do with the guarantee of free expression.[6] The Radio Act, therefore, went well beyond what First Amendment notions of the day would have compelled when it barred censorship and included express statutory protection for broadcasters' right of free speech. However, the Act included a number of specific programming controls that had no parallels, then or now, in the laws applicable to print media. Moreover, the licensing function under the vague public interest standard has fostered a variety of techniques whereby the FCC influences the nature of programming without directly regulating its content. The next sections of this chapter discuss the most significant of these specific controls and general influences over program content.

Explicit Programming Controls

Equal opportunities. The most important of the relatively few explicit controls on programming is the "equal opportunities" provision in Section 315 of the Communications Act, which requires that broadcasters grant equal broadcast opportunities to opposing political candidates.[7] Section 315 does not require broadcasters to put political candidates on the air. Rather, its obligation is contingent: only if a broadcaster elects to give or sell a candidate airtime (other than in the course of a newscast or coverage of a news event) must that broadcaster provide a similar opportunity to other candidates for the same office. The equal opportunities provision was the legislative residue of proposals for much broader regulation of political programming in the debates on both the Radio Act of 1927 and the Communications Act of 1934. Proposals were advanced to make broadcasters virtual common carriers for discussion of all public questions, but these proposals were reduced to the existing, conditional rights for political candidates because Congress perceived that the broader obligation would expand to fill virtually all broadcast time.

The Congressional debates leading to Section 315 do not reveal any theory as to why broadcasters should be required to give evenhanded treatment to political candidates, when print publications traditionally had been free of any such obligation. Concerns were expressed about the potential power of a broadcaster to unbalance elections by inundating a community with a favored candidate, or by freezing out a disfavored one. But similar power could be exercised by a dominant newspaper. Perhaps the failure to advert to an underlying justification for Section 315 simply reflected a Congressional intuition in the 1920s that broadcasting had the potential for much greater influence over political contests than did the newspapers of the day. Or perhaps Congress viewed evenhanded treatment of candidates as a self-evident obligation of a public trustee. Probably most members of Congress did not think seriously about theoretical justifications for Section 315, and wished merely to protect themselves against broadcasters who might devote their radio stations to the advantage of opponents.

Section 315's relatively narrow right for candidates suggests one of the dangers of contingent access obligations. In theory such a requirement seeks to guarantee equality to promote fairness and public awareness of election issues and candidates. But in operation, contingent access obligations may have the effect of inhibiting so much expression of value that the gains in fairness and evenhandedness are not worth the cost. For example, Section 315 unquestionably has had a chilling effect on broadcasters' coverage of political candidates. If broadcasters give free time to major candidates in elections of popular interest, Section 315 gives candidates of minor and even frivolous parties a right of access. The effect of this is suggested in two Presidential elections. In 1960, Congress suspended Section 315 for the Presidential race, and three television networks donated over 39 hours of broadcast time to the Presidential candidates. In 1968, when Section 315 was in force, only 3 hours were donated.[8] The FCC recently ruled that candidate debates and press conferences not arranged by broadcasters should be viewed as within a statutory exemption to Section 315—reversing an earlier interpretation of the scope of the exemption and allowing the Ford-Carter debates to be broadcast without generating rights of access to Eugene McCarthy and other minor candidates. This ruling recognized that rigorous equality of treatment for candidates can have the effect of inhibiting much useful coverage of public affairs.[9]

The Fairness Doctrine. The most controversial regulation of programming is the Fairness Doctrine, developed by the FCC without explicit statutory authorization but generally approved by Congress in a 1959 amendment to Section 315.[10] The FCC has long regarded broadcasters as "public trustees," who are obliged to devote broadcast time to issues of importance to the public, rather than merely to reflect the view of the licensee. This concept gradually grew into the Fairness Doctrine, which imposes a two-part duty on broadcasters: to devote reasonable time to controversial issues of public importance, and, when an issue has been raised, to provide reasonable opportunity for contrasting viewpoints to be heard.

Although the Fairness Doctrine in principle would justify limitless official supervision of broadcasting, its actual impact has been cushioned by important constraints. Particular programs need not be balanced; the test of fairness is the broadcaster's overall programming. Thus, the Doctrine does not interfere with particular programs, no matter how one-sided. Even so, if the government were to insist on fairness in some mathematical sense, the Doctrine would substantially inhibit programming about controversial public issues, because further programming would almost always be required to balance the fairness ledger. This potentially suffocating official control has been avoided by a crucial principle based on the Communications Act's rejection of censorship and common carrier status for broadcasters: Fairness obligations are left initially to the editorial discretion of the broadcaster. Only for abuse of discretion will the FCC overturn a licensee's judgment on such questions as whether an issue is so important and controversial that some programming must be devoted to it, whether such an issue has in fact been addressed in programming, and, if so, what constitutes a "reasonable opportunity" for contrasting viewpoints. Recent court decisions, sensitive to First Amendment concerns to protect editorial autonomy and reduce official supervision of programming, have given wide scope to broadcaster discretion in Fairness Doctrine controversies.[11]

The result is substantial freedom for broadcasters under the Fairness Doctrine, though the threat of administrative sanctions remains in the background. For example, the first part of the Fairness Doctrine—the obligation to cover important issues—has generally been left entirely to the discretion of broadcasters. In response to one extreme situation, the FCC recently decided that a West Virginia radio station had failed its affirmative obligation by broadcasting nothing on environmental and economic concerns about stripmining during a period of intense legislative consideration of the matter.[12] Moreover, the FCC also recently refused to renew licenses of broadcasters that failed to schedule programming for the specific needs of significant minority populations in their service areas.[13] Thus, after years as little more than an exhortation, the affirmative side of the Fairness Doctrine may be cutting teeth, at least in instances of extreme neglect by broadcasters.

The second part of the Fairness Doctrine—the obligation, once an issue has been raised, to be balanced in presenting contrasting views—has led to much more significant official interference with programming. In the 1960s, the FCC issued increasingly specific rules about this contingent side of the Fairness Doctrine. Complaints were taken up on a case-by-case basis, rather than considered during review of a broadcaster's overall performance at license renewal time. This change in procedures brought pressures for immediate and particular remedies for fairness violations. Explicit access obligations were soon imposed. "Personal attacks" during discussion of controversial issues of public importance were judged to give the person attacked a right of reply. Likewise, the Commission ruled that editorials favoring one candidate gave competing candidates a right to respond.

In the well-known *Red Lion* decision of 1969, the Supreme Court unanimously sustained these new requirements.[14] Relying on the scarcity of the spectrum and the fact that broadcasters are given merely a temporary privilege of spectrum use, the Court upheld the FCC's power to force broadcasters to present views for the good of all: "It is the right of the viewers and listeners, not the right of the broadcasters, which is paramount."[15] The editorial autonomy of broadcasters, a prominent value in earlier statements of the Fairness Doctrine, disappeared in the Court's sweeping approval of the notion of the broadcaster as public proxy for the community at large.

The *Red Lion* decision appeared to go beyond affirming the Fairness Doctrine in the special context of broadcasting. The rhetoric seemed to embrace a broad power of government to promote diversity of expression in mass communications media of all kinds, including print. For example: "The right of free speech of a broadcaster, the user of a sound truck, or any other individual does not embrace a right to snuff out the free speech of others."[16] Even more sweeping were suggestions that the First Amendment might itself mandate government intervention against not only broadcasters, but also private publishers, to increase diversity of expression: "It is the purpose of the First Amendment to preserve an uninhibited marketplace of ideas in which truth will ultimately prevail, rather than to countenance monopolization of that market, whether it be by the Government itself or a private licensee."[17] And perhaps most supportive of government authority to regulate mass communications media to foster diversity: "It is the right of the public to receive suitable access to social, political, esthetic, moral, and other ideas and experiences which is crucial here."[18]

After the *Red Lion* decision, the FCC experimented with broader access rights in the area of political broadcasting and product commercials. When President Nixon delivered a series of television addresses on his Vietnam policy, the FCC judged that a Democratic spokesman should be given a right of reply.[19] But when a Democrat blasted Nixon's policies generally, the Commission found itself caught in a spiraling logic that compelled it to order a further right of reply for the Republicans. A blistering reversal by the Court of Appeals for the District of Columbia no doubt encouraged the Commission the following year to conclude that presidential addresses would not, in the future, give rise to rights of reply, and that any access rights in this area must come from Congress by statute, if at all.[20]

A similar situation occurred in the case of access rights in response to commercial advertisements. In 1967, the FCC held that cigarette commercials presented one side of a controversial issue of public importance, the desirability of smoking, and thus triggered a Fairness Doctrine obligation to broadcast messages opposing the use of cigarettes.[21] The Commission tried to prevent this ruling from becoming a Pandora's box that would trigger counter-commercial rights whenever an advertisement promoted products that might be unhealthy to individuals or society. Unable to come up with a limiting principle that would not encompass high-octane gasoline, phosphate-based detergents, and many other products routinely advertised on radio and television, the Commission finally retreated. It admitted its error regarding the cigarette ruling and announced that it would not consider commercial advertisements for products as subject to the Fairness Doctrine.[22]

Fairness as a rationale for regulating broadcast programming or promotion of commercial products is a ubiquitous concept that would support limitless official control over broadcasting. If fairness were pushed to its logical extreme, an official "fairness editor" would have the last word on every broadcast. Such totalitarian consequences would obliterate private editorial discretion. To mediate between these two rudimentary principles, fairness and editorial freedom, the Commission must steer a middle course that is almost certain to meander, to rest on vague standards, and to disappoint the advocates of each of the conflicting principles. Constrained as it is by the principle of deference to reasonable broadcaster judgments and administered under the eye of federal courts watchful of First Amendment values, the Fairness Doctrine is not nearly the censorial cleaver that it is sometimes depicted to be at broadcasters' conventions. The Doctrine does, however, impose costs on controversial public affairs programming. Broadcasters may prefer not to devote further airtime to discussion of an issue addressed in a previous broadcast. Whether this attitude is right or wrong is not important in assessing whether the Fairness Doctrine has an inhibiting effect. If forced to put on programming that they would not present as a matter of editorial discretion, broadcasters may conclude that controversial issues should not be raised in the first place. Fred W. Friendly has uncovered evidence that during the early 1960s the Democratic National Committee sought to use the Fairness Doctrine, especially the personal attack rules, to deter radio stations from putting on broadcasts by right-wing extremists who were vilifying the Kennedy Administration.[23] When a Fairness complaint is raised, a broadcaster may find itself in a process that takes the time of top personnel, requires consultation with lawyers, involves correspondence back and forth with the FCC, and threatens the station's reputation. Even if it is ultimately vindicated, a broadcaster may think twice about controversial programming that seems likely to stir up fairness disputes. As with Section 315's equal opportunities requirement, a major question for the future of telecommunications policy is whether in practice the Fairness Doctrine has chilling effects that outweigh whatever enhancement of expression the Doctrine provides. Another question, likewise applicable both to equal opportunities and the Fairness doctrine, is whether these rules produce benefits worth the administrative costs of enforcing them.

Although the primary justification for imposing the Fairness Doctrine has been the scarcity of broadcast outlets, as the Supreme Court emphasized in *Red Lion*, other interesting justifications for it have been advanced. Broadcasting is not the only productive way to use the electromagnetic spectrum, of course. Military and other government uses, land mobile communication, and countless private uses as described by William Lucas in Chapter 7 and by Raymond Bowers in Chapter 8, all compete for spectrum space. The FCC has often stated that it allocated as much of the spectrum to broadcasting as it did because of broadcasting's potential to contribute to an informed electorate. The Fairness Doctrine has been justified as a measure to ensure that broadcasting does, in fact, provide the service that resulted in so much spectrum space being given to it. Other justifications have looked directly to First Amendment values of diversity and balance in expression about public affairs.

Obscenity and indecency. Several regulations of broadcast programming rest on the elastic notion that broadcasters are public trustees obliged to program in the public interest. Scarcity of outlets is not the direct justification for these controls. Rather, other features of broadcasting that do not characterize print media are discussed in support of content controls not found in American law's treatment of newspapers, magazines, and books. For example, the Commission has sought to bar programming that is clearly not obscene in the constitutional sense that would justify penalties on print media. Congress invited this by a criminal provision in Title 18 of the United States Code, which prohibits the broadcast of "indecent" and "profane," as well as obscene material.[21] The Commission has taken the position that the "unique characteristics of broadcasting" justify greater governmental supervision of broadcasting than would be appropriate for other media. The public trustee notion often comes up—broadcasters are said to have a "fiduciary responsibility" to their communities, which especially includes the protection of children from indecent programming. More interestingly, the FCC and the courts have emphasized the special intrusiveness of radio and television. Unlike other media, broadcasting comes into the home without significant affirmative activity on the part of the viewer. After the television or radio is turned on, the theory goes, viewers or listeners are frequently passive; unconsenting listeners or viewers may be exposed to offensive material without warning. Thus, general privacy concerns may justify a greater range of content controls for broadcasting than for print. Moreover, because broadcasting is virtually a free good once the receiving device has been purchased, and because children often watch or listen without parental supervision, particularly before the evening hours, special rules to protect children from sexually explicit or other arguably harmful material, may be more sensible for broadcasting than for other media. The combination of intrusion into the home, exposure of passive and unconsenting adults, and susceptibility of juvenile audiences, suggests the analogue of nuisance regulation, whereby offensive but arguably legitimate activities are channeled to avoid unnecessary injury, but not totally suppressed.

A step by the FCC toward channeling offensive language away from children recently was struck down by the U.S. Court of Appeals for the District of Columbia in 1977. The FCC had barred the broadcast of certain vulgar words describing sexual and excretory functions or organs during times when a large number of children would be in the audience. To bar such words, regardless of context, was judged to be inconsistent with the no censorship provision of the Communications Act and, in addition, vague and unnecessarily broad because the Commission's ruling would prevent broadcasts of clear merit, such as some passages from the Bible. Moreover, the Court of Appeals felt that banning such expression, when children make up a sizable portion of the broadcast audience (over one million children watch television even after midnight), would limit adults to what is fit for children.[25] However, the Supreme Court has decided to review the case, and as of this writing, the extent of the FCC's power to prevent indecent broadcasts must be regarded as an open question.

Whether the intrusiveness of radio and television and children's access to it justifies more stringent controls on broadcast expression to prevent indecency and sexual explicitness than is appropriate for print media or motion pictures is a matter for congressional consideration. The courts will probably show greater deference to a clear congressional judgment than to FCC rulings under the existing rather amorphous statute.

Miscellaneous prohibitions. Other regulations not generally applicable to other media include bans on the broadcast of

Programming and Regulation of Mass Media

lottery or gambling information[26] (with some exceptions for information about state-run lotteries). A specific statutory ban applies to cigarette advertisements.[27] There is a ban on fraudulent contests and quiz shows, and the FCC holds broadcasters responsible for running false or deceptive advertising, whereas in the case of print media the advertiser, not the publisher, is held responsible for misleading ads. The rationales for these special regulations of broadcast expression have not been elaborated much beyond cursory references to the public trustee concept. It seems doubtful whether there are peculiar properties of radio and television that justify these regulations, which are not applicable to the print media.

Official Influences on Programming

Beyond the direct controls on programming described above, the FCC exercises an indirect influence over programming by controlling activities that in turn affect programming, by threats or persuasion, and by inducing industry self-regulation. Since these indirect influences are usually accompanied by denials from the Commission that it is influencing programming at all, justifications for these controls are rarely elaborated. Nevertheless, these indirect approaches, taken together, suggest perhaps the broadest basis for official control of radio and television programming. As such, they are an important part of the legal status of electronic media; they underline the extent of departure from the First Amendment tradition for print media; and they raise important questions about the underlying justifications for legal controls on programming.

Ascertainment and responsive programming. From the early days of radio regulation, the FCC has sought to make broadcasters present "balanced" programming that would be responsive to all segments of the particular local community served. The theory behind this requirement, in the words of the Commission, is that "the principal ingredient of a licensee's obligation to operate in the public interest" is a positive effort to "discover and fulfill the problems, needs and interests" of the public in its service area.[28] At the same time, the FCC has not been willing to dictate specific programs or program categories that broadcasters must carry; the Commission has been deterred no doubt by the practical difficulties in formulating and enforcing any such requirements and constrained by the First Amendment. Compromise positions have evolved to encourage responsive programming without official mandates; the current version is called ascertainment.

A broadcaster is required annually to ascertain the issues of concern to its community, list ten such issues, and indicate illustrative programming devoted to each. A formal process of interviews with "community leaders" for a "checklist" of typical "institutions and elements" must be supplemented with community surveys and demographic data. (Examples of these categories are agriculture, culture, environment, military, minority and ethnic groups, recreation.) In a city with a population of 200,000 to 500,000 persons, for example, 180 interviews with community leaders are required, at least half of which must be conducted by management-level employees of the broadcaster. The interview and survey work cannot be delegated to a professional research firm. There are dozens of highly specific requirements. Ascertainment requires extensive paperwork, elaborate filing systems, and endless hours of the time of broadcasters, agency personnel, and community leaders.

In addition to ascertainment, several other Commission practices seek to foster broadcaster responsiveness in nonentertainment programming to all significant elements of the community. For example, when applying for licenses and renewals, broadcasters are required to specify the percentages of programming in a typical week to be devoted to news, public affairs, and "all other programs, excluding entertainment and sports." Although the FCC has never set minimum program-category requirements, there are rules of thumb governing which program proposals will be approved routinely by the Commission staff, and which will be delayed and perhaps require additional hearings. In view of the great expense of hearings, broadcasters, not surprisingly, seek to abide by those rules of thumb that lead to routine approval. For example, approval is routine when total nonentertainment programming exceeds 6 percent for commercial television stations, 8 percent for FM, and 10 percent for AM. Commercial VHF television stations also must show, as a condition for routine approval, at least 5 percent news and public affairs, and 5 percent local nonentertainment programming between 6:00 a.m. and midnight. Again, the rationale for these official influences on nonentertainment programming is the expansive regulatory concept that broadcasters must undertake public service programming that is responsive to the needs and interests of the local community.

The radio format controversy. Whether the FCC should regulate entertainment program formats of radio stations in the interests of diversity has given rise to an unusual controversy between the Commission and the Court of Appeals for the District of Columbia. The Commission has taken the position that radio entertainment formats should be left to market forces and licensee discretion. The Court of Appeals, however, ordered the Commission to prevent changes in formats from, say, classical music to rock when the existing format is unique in the particular community and financially viable.[29] Although the precise legal issue posed in cases regarding format is whether the FCC must hold a hearing when a license sale includes a plan to change a unique format, or instead can simply approve a license assignment without a hearing, the underlying dispute between the Court and the FCC is whether the agency should maintain diversity of entertainment formats, at least to the extent of preventing abandonment of unique formats. The Court of Appeals has taken the view that market forces will not produce format diversity because listeners do not pay for broadcasts. Because broadcasters sell audiences to advertisers, the Court reasons, the tendency is to aim for a mass audience, or for a particular audience which would have the greatest appeal to advertisers, such as young adults who tend to have larger discretionary incomes. This may lead to programming that ignores substantial segments of the community. The Commission, by contrast, has objected that the Court's position rests on inherently vague differentiations between format categories, that the logic of the court's position calls for pervasive regulation of formats to ensure diversity, that the FCC cannot provide a better measure of the social utility of various entertainment formats than the advertisers' marketplace, and that format regulation might inhibit experimentation by locking broadcasters into formats regarded as unique.

The radio format controversy is a minor issue in practical impact; it does not concern television (because all stations have general formats), but rather only those radio stations with unique formats. The issue does, however, raise questions that go to the core of government regulation of broadcasting. To what extent does advertiser-supported broadcasting generate programming that satisfies the preferences of various audiences, particularly minority preferences? Is mass-audience, homogenous programming the result of outlet scarcity or the lack of direct payment for programming by listeners and viewers? Is a government agency better equipped than an imperfect radio market to determine what formats are in the public interest? Is it consistent with the First Amendment for the FCC to require a given format, and thereby prevent another from being presented?

Wholesome programming: the "raised eyebrow" technique and induced self-regulation. The least direct method of inhibiting expression without banning it is known to the trade as the "raised eyebrow" technique. In practice the technique has been devoted to concerns about the wholesomeness of programming, particularly for children. Some examples will illustrate how the technique is used and its effectiveness.

In the early 1970s, certain popular songs considered to promote the "drug-culture" came to the attention of the FCC. The Commission issued a notice "reminding" broadcasters of their obligation to make reasonable efforts to determine in advance the meaning of what is broadcast, and to judge whether it is in the public interest. The Commission was careful to state that it was not banning songs that promoted drugs, but merely noting an instance of the general requirement that broadcasters know the content of their programs. This was disingenuous. It was fairly clear that the notice was designed to indicate Commission displeasure and to induce cautious, or perhaps prudent, broadcasters to ban songs thought to promote illegal drugs.[30] Given the threat of nonrenewal of licenses on account of broadcasting objectionable material and the expense of a contested renewal proceeding, it is not difficult to see that the "raised eyebrow" technique is a potent instrument for suppression.

Officially induced self-regulation by broadcasters is another controversial recent method of discouraging objectionable programs. The "Family Viewing Policy" adopted by the three television networks and the National Association of Broadcasters after discussions with the Chairman of the FCC was recently set aside by a federal district court as a violation of the First Amendment. In essence, the policy provided that only programs appropriate for viewing by children would be broadcast in the first hour of prime time (8:00 p.m. to 9:00 p.m., eastern time zone) and in the preceding hour. In case of exceptions, warnings were to be displayed. The court found that "the uncertainty of the relicensing process and the vagueness of standards that govern it" allowed the Chairman to put pressure on the networks and the National Association of Broadcasters to adopt a programming policy that violated the First Amendment. The Court found that "family viewing" involved essentially vague standards for judging the appropriateness of programming. Moreover, the policy was held to contravene the First Amendment requirement of "diversity in decision making" with respect to the content of programming.[31] As of this writing, the decision is before the U.S. Court of Appeals for the Ninth Circuit for review.

The "raised eyebrow" technique suggests the broadest basis for official regulation of broadcast programming: that broadcasters must program "in the public interest," and that this justifies the exercise of an unfettered paternalism by the government to control what is good for audiences to see and to hear. Such a concept of official responsibility for the content of broadcast programming submerges the tradition of private editorial discretion. Not suprisingly, it tends to appear only in the appealing posture of an effort to protect children. However, in view of the large numbers of children that watch and listen to television and radio until far into the night, concern for children would justify regulation of the vast preponderance of broadcast programming.

The legality and wisdom of the various controls and influences on broadcast programming described above have been matters of intense controversy. Some of the controls have been attacked as inconsistent, even with the regulatory assumptions adduced to support them. The equal opportunities requirement of Section 315 and the Fairness Doctrine, as has been pointed out, were justified on the theory that they would increase diverse and robust expression about public affairs. Critics have charged that these regulations inhibit more expression than they enhance. Moreover, vagueness and inconsistency of application is a common complaint about some of the Commission's content control rules. Another objection frequently voiced is that content regulations cost broadcasters, the Commission, and the courts much more to administer than any benefits the public receives. And, of course, objections based on the First Amendment are common.

The Advent of Pluralistic Programming

The central question for policymakers now is whether new technologies of broadband distribution promising many more channels for viewers and greater diversity of programming render obsolete the premises of traditional broadcast regulation. The assumptions of spectrum limits, frequency interference, and scarcity of outlets are vanishing in the advance of new telecommunications technologies. In approaching this question, it is helpful both to review the response of the FCC to the advent of cable television, and to speculate about future regulatory directions.

Cable television—augmented or eventually supplanted by optical fibers—is the most significant new mass communications technology. It presages a new communications environment, though at present cable is still in a state of uncertain adolescence. Now available in about 17 percent of all households with televisions, its likely growth over the next two decades is a matter of dispute.

Until 1972, the FCC's response to cable reflected chiefly a concern to protect local broadcasters. The Commission believed that cable might so disperse the audience for television that advertisers would support only network programming. The result would be that the audience for non-network broadcasting, particularly that of small rural stations or UHF stations, would either lose programming options or be obliged to pay cable installment charges and monthly rates for programming previously supported by advertisers. Thus, some amount of "free television" (a somewhat misleading term in view of added product costs, induced consumer spending, and viewer time spent watching commercials) would be replaced by television that is not free. Some people undoubtedly prefer "free television" with limited program selection to more abundant viewing options that cost money. And if cable growth created strong competition for free television, the FCC expressed the concern that cable might "siphon" certain very popular events, such as the World Series, away from network broadcasting. Some alarmists go farther, and fear that unrestricted cable might destroy broadcasting altogether.

Unrestricted cable growth undoubtedly poses a threat to television for rural, out-of-town areas. Cable television requires sizable capital costs in laying trunk lines and providing branch and feeder lines to individual households. Ironically, towns in rural areas have been attractive prospects for cable because many of them have had minimal service from over-the-air broadcasting. However, some rural areas with low population density do not have enough potential subscribers to justify the initial capital costs, and servicing costs may also tend to be higher in such areas. The accepted rule of thumb in the industry is that a cable system needs at least 30 to 40 households per cable mile to generate a sufficient return on investment. Aside from signal interference problems, broadcasters can provide rural areas with a few signals quite cheaply, but if cable drives broadcasters out of business by decimating its audience in population centers, service might be denied to rural areas that were previously reached by electromagnetic waves but are uneconomical for cable.

Many critics of the FCC believe that the Commission's initial policy toward cable was not aimed so much at protecting rural areas as at maintaining the monopoly profits most broadcasters receive in major markets. The Commission restricted cable systems from importing "distant signals" (programs from stations outside the broadcast service area) into the top 100 markets (which serve about 86 percent of the viewing public), except when such importation was proven not harmful to local UHF stations. This policy resulted in a virtual freeze on cable growth. At the same time, this policy also served to maintain the scarcity of outlets and programs which has been the primary justification for the traditional approach to broadcast regulation.[32]

After 1968, however, an unusual movement in support of cable gathered momentum. Cable became a symbol of pluralism and localism that many hoped would end the bland uniformity of broadcast television and counteract the homogenization of American society. Egalitarians saw in cable the promise of abundant channels with open access for all to television audiences; new mechanisms for participatory democracy were envisioned. The video avant-garde hoped that cable could provide an outlet for radical programming concepts. Futurists saw cable systems as the first step toward elaborate home consoles that would deliver the morning paper, provide access to computer information banks, let people vote from their homes in vast town meetings, as well as provide access to richly varying entertainment possibilities.

Cable enthusiasts found a receptive audience among policy planners who tended increasingly to believe that market forces rather than regulation should determine the future of mass communications technologies. The result was a cautious move away from the virtual freeze on cable, following an FCC-initiated agreement among the broadcasters, program producers, and cable interests. This consensus was reflected in the issuance of comprehensive rules to govern cable in 1972.[33] Cable systems were allowed to import at least two distant signals, and were obliged to carry local broadcasters' signals without simultaneous duplication. Moreover, the FCC required new cable systems in the top 100 markets to be capable of two-way circuitry and a minimum 20-channel output. For each broadcast signal retransmitted, a cable system was required to make a channel available for cablecasting. In addition, "access" channels had to be set aside for the general public, educational institutions, and local government. Remaining channels reserved for cablecasting were to be available for lease to any program producer, with the cable system acting essentially as a common carrier. In 1976, the FCC relaxed most of these requirements, other than two-way circuitry, because they were found to be unduly burdensome for most existing cable systems. New systems were subject to these requirements, however.

In 1975, the FCC also relaxed certain restrictions on "pay cable" (whereby subscribers pay on a per-channel or per-program basis), although restrictions remained applicable to certain feature films and sports programs. The intention of these rules was to avoid siphoning these popular programs away from free television. In

March 1977, the U.S. Court of Appeals for the District of Columbia struck down these pay-cable restrictions.[34] The Court held that the FCC did not have sufficient evidence of siphoning to justify the rules, that the rules were designed to protect broadcasters from cable competition, while the proper regulatory goal should be to let cable grow to become equal to broadcasting, and that the scarcity justification for regulating broadcasting, upheld in Red Lion, had no application to cable because scarcity of cable channels was solely a result of economic conditions rather than spectrum limitations. As the Court stated: "There is nothing in the record before us to suggest a constitutional distinction between cable television and newspapers..." The Court emphasized that cable could be regulated; but regulation would have to satisfy general First Amendment standards applicable to nonbroadcast expression, as in newspapers and magazines.

The Commission has struggled with the advent of cable television without any legislative direction from Congress, although there is no reason to suppose that the FCC's approach is not in accord with what Congress would wish. Indeed, informal reflections of Congressional sentiment have been even more protective of broadcasting than the Commission has been. The Commission has sought to protect traditional broadcasting without killing cable in its infancy. In the process, the Commission has managed to maintain the scarcity of telecasting that is the foundation of its regulatory power over electronic mass media.

To be fair to the Commission, one needs to bear in mind the absence of legislative guidance (except on copyright questions); the difficulty of predicting the pace of cable growth; its economic consequences for broadcasters; its impact on "free television" consumers in terms of programming quality and variety, cost of viewing, and service to rural areas. Although our legal system has failed to grasp either the potential or the problems of cable television, the blame cannot be placed solely on the Commission. That Congress has remained on the sidelines during the advent of what promises to be a telecommunications revolution spawned by cable and its successor technologies, such as fiber optics, is the most compelling reason for thorough rethinking of the Communications Act of 1934 in relation to radio and television.

Pluralistic Programming and Traditional Broadcast Regulation

It is widely accepted that cable calls into question traditional principles of broadcast regulation. Even without legislative guidance, the Commission has partially recognized the inappropriateness of imposing public trustee obligations of programming responsibility on certain cablecasts. The most obvious case for different treatment is the access channels. Because they are open to all on a first-come, first serve basis the FCC has exempted them from the equal opportunities rule, the Fairness Doctrine, and other public trustee constraints. On the other hand, the FCC has maintained that programming originated by cable operators should be subject to the same public trustee requirements as traditional broadcasting. The basis for this is not clear. Public trustee obligations deriving from the scarcity of frequencies in the spectrum available for broadcasting do not fit cable television, and the capacity for channel abundance will be even more dramatic with the eventual advent of fiber optic distribution systems. Scarcity of cable or fiber video channels is the result of economic forces. And the Supreme Court has recently and unanimously rejected the thesis that economic scarcity justifies public trustee regulation of newspapers to ensure fairness, balance, and the public's "right to know." As systems of broadband distribution make television resemble print in moving toward market determination of the number and diversity of programming options, what justification remains for a fundamentally different legal approach to the two types of media?

The Commission grappled with this problem in one fascinating ruling. Although cable origination programming has been generally subject to the same public trustee constraints as over-the-air broadcasting, the Commission has held that these regulations would not apply if a newspaper were "delivered" via facsimile reproduction over a cable system.[35] Intuitively, this ruling seems sound. That a newspaper is delivered electronically rather than tossed on the front porch does not seem grounds for a radical change in constitutional status. Presumably, the FCC would also exempt from programming regulations the use of cable or other broadband channels to teleprint material from the wire services or books. What about a newsperson who reads material from the wire services over the air? Almost all television programming emanates from printed scripts. The exception that broadband transmission of printed matter should not be subject to the traditional regime of broadcast regulation calls into question the general rule; it suggests that all broadband programming should be treated as a print medium for regulatory purposes.

In Chapter 6, Bruce Owen warns that the technological coalescing of electronic and print media conceivably might mean that print could be brought within broadcasting's regulatory net. The danger is a serious one. Broadband telecommunications, like the introduction of electronic transmission technology into newspaper production, challenges our law's divergent regulatory approach to print and broadcasting.

New Directions for Broadband Regulation

Whether the elimination of scarcity as a technological necessity in telecommunications justifies a move away from the public-trustee regulatory regime that has governed broadcasting since 1927 is a major policy question for Congress. Congress should not uncritically extend to pluralistic telecommunications the regulatory approach for broadcasting that evolved in the era of spectrum scarcity. The FCC should not be left in the dark about congressional intentions as it attempts to grapple with a revolution in communications technology. There are many possibilities for a fresh legal approach, from the "hands-off" approach that exists for the print media, to common carrier status that would open broadband channels to all program producers, to a public-trustee status similar to that of broadcasters, with numerous possible combinations. The various regulatory options cannot be assessed in detail here. Rather, this chapter in its remaining pages identifies the underlying policy questions, the answers to which will determine the directions, if not necessarily the detail, of an appropriate constitutional response.

Scarcity as a rationale for regulation. The first step should be to reject frequency scarcity as a rationale for broadband regulation. This does not mean, however, that concerns about scarcity of other sorts may not underlie a new legislative response. Scarcity of programming options will persist as an economic reality, especially in rural areas; economic scarcity may call for special subsidies to encourage a multiplicity of outlets, following the pattern of special postal subsidies for delivery of mail to rural areas that developed during the 1930s. Where scarcity of outlets persists for economic reasons, regulation of programming to require balance and access to various points of view may seem desirable. The existence of scarcity, and the resulting absence of diversity, may seem a more important regulatory consideration than its causes. Such an approach, however, must contend with the fact that our tradition precludes regulating newspapers to enhance balance and access, even where economic scarcity has reached the extreme point of monopoly. Regulation of broadband program content solely because of outlet scarcity could only be imposed by Congress and sustained by the courts on grounds that, in theory, would threaten the editorial autonomy of the print media.

In addition to spectrum scarcity and scarcity deriving from economic causes, there is a third type of scarcity which may call for attention as Congress considers its response to broadband telecommunications. This might be termed, hypothetically, a scarcity of viewer preferences and habits. It may turn out that even when viewers can choose from an abundant array of programming options, a few networks will continue to capture most viewers' preferences. This seems an unlikely prospect. Even if viewers concentrate their preferences only on a few channels, the growth of broadband is almost certain to provide the economic base for a fourth and fifth network. And if pay television enables viewers to pay directly for what they watch, considerable diversity of actual viewer preferences would be assured. But if a handful of national networks dominate viewers' preferences even after the advent of abundant programming options, what are the regulatory implications?

There would seem to be little reason to impose public trustee pressures for responsiveness and diversity of the sort that have been applied to traditional broadcasting. If viewers can turn elsewhere, why bother with ascertainment or other rules designed to make programming "relevant" to all significant elements of the community? If viewers have abundant options on other channels, including presumably locally originated programs addressed to various minority tastes, but prefer the national, mass-audience fare, regulations that encourage that locally oriented fare on each channel simply impose on viewers something they do not want, and in any event can get elsewhere. Likewise, there is little reason to encourage minimum requirements for such programming as news, public affairs, educational, religious, and agricultural program categories, and so forth. The more programming options that are available, the more one can assume that market forces will give viewers what they want. To regulate the content of popular programming to enhance balance, diversity, and access could find its justification only in a candid paternalism, not an uncommon feature in American law, not even unknown in First Amendment law (as our law of obscenity attests), but a legislative outlook that runs counter to strong values of personal autonomy reflected in the First Amendment.

What about the requirements of equal opportunities for political candidates and the Fairness Doctrine in a system of scarcity as measured by viewer preferences? Here also, pluralistic programming options will almost certainly offer coverage of political candidates and controversial issues of public importance to those viewers who want such fare. To force such issues onto particular channels thus accords better with our traditions respecting castor oil than individual taste in exposure to political ideas. Aside from paternalistic notions, affirmative requirements to put on candidates and public issues might be viewed as a subsidy to politicians exacted from program producers, broadband distributors, and viewers. Questions of efficiency and fairness must be added to the questions of First Amendment principle that one would hope policymakers would address before imposing such obligations on broadband telecommunications. To these considerations must be added the problem of inhibiting effects that is such an unfortunate feature of contingent obligations under the equal opportunities requirements and the Fairness Doctrine. Finally, policymakers must consider how the Fairness Doctrine and equal opportunities requirements would be administered in a system of pluralistic programming, assuming that these rules are thought to be sensible as a matter of principle and policy. As mentioned above, these rules do not mandate even-handedness in particular broadcast programs, but rather apply to a broadcaster's overall programming. The FCC has fallen back on the fiction of a perpetual viewer or listener who never changes the channel, and thus refuses to consider broadcasters' claims that a station's one-sided programming has been balanced by another station's presentation of the other side. Would this approach make sense in an era of pluralistic programming? If abundant options lead viewers to choose their programs the way readers now choose books, it is hard to see how even-handedness obligations make any sense unless they are applied on a program-by-program basis. Yet this is a method of enforcement the FCC has rejected because it would intrude the influence of Government too deeply into the content of broadcast and would exacerbate the problem of inhibiting effects. In short, policymakers must consider not only whether statutory and administrative law doctrines developed for traditional broadcasting make sense in principle as applied to broadband telecommunications. Basic changes in the extent, the causes, and the characteristics of program scarcity also would affect the ways in which traditional broadcast regulations would be administered.

Intrusiveness and privacy. Although spectrum scarcity is the most frequently articulated basis of traditional broadcast regulation, and the only one embraced by the Supreme Court to date, at least two other general rationales for regulation have been advanced. One is the intrusiveness of radio and television, a quality that is not changed whether the signal is received by electromagnetic waves or over shielded transmission paths, and whether there are many program choices or few. These media come into the home without the active decision for consumption embodied in the choice to read, to go to a motion picture, or converse with a friend. When a consumer subscribes to a newspaper or magazine, the precise content is not known in advance. But editorial continuity provides far more knowledge, and therefore "consent" to exposure, than can be extrapolated from the decision to buy a television set and turn on a channel. Recognition of privacy interests is well-established in our First Amendment law. To protect persons from unwanted intrusion, particularly in the home, expression in the form of trucks equipped with loudspeakers or door-to-door solicitation can be regulated to a substantially greater extent than forms of expression that do not penetrate uninvited into the home.

Thus, policymakers should consider whether the intrusiveness of electronic media consumed in the home by choice calls for programming regulation. In particular, control is needed to protect children from pornography, and perhaps from violence and other vulgarities. Parents cannot control what their children watch and hear, particularly when parents are away from home. Certain types of programming might be restricted to times when parents are usually at home and have the opportunity to exercise supervision. Alternately, relatively simple technologies should be encouraged that allow certain channels to be set aside for adult programming with access under parental control, through channel-locking or similar devices.

A second type of programming regulation sensitive to intrusiveness concerns might aim to prevent certain types of programs from entering specific homes. An analogy from postal regulation is suggestive. If an addressee of mail so instructs the Post Office, mail from a designated sender may not be sent to the complaining party. The purpose of this regulation is to protect nonconsenting persons from mail advertising considered obscene by the recipient. To avoid problems of official definition, a complaining recipient can designate anything as obscene and the sender is ordered to strike that recipient from all mailing lists. Comparable consumer control of programming seems appropriate in a system of pluralistic programming. To respond that an unwilling viewer or listener may change channels to avoid obnoxious programming seems no more the answer than that an unwilling recipient of obnoxious mail should toss it in the trash. In both cases there is exposure, however brief, and the possibility of anger and embarrassment.

Whether comparable consumer control of programming is feasible at reasonable cost is the critical question. If the effort entangles government in the inhibition of certain types of programs by relegating them to secondary channels on the basis of vague standards, the effort raises First Amendment problems not worth the gains in protecting privacy. But in principle, there is no reason to deny consumers of electronic mass media protection in the home similar to that afforded to recipients of unwanted mail. If public-trustee constraints are removed from radio and television programming, nonpaternalistic consumer-based controls to protect privacy will assume greater importance because freer programming will be offensive to some consumers.

The converse is also true. A great incentive for policymakers to devise methods to allow parents to control their children's viewing habits in the home, and to protect all persons from unwanted exposure to offensive programming, is that there will then be much less pressure for across-the-board content controls. If the problems of parental control and nuisance intrusions can be solved, there will be no basis for government suppression of offensive programming other than a paternalism that is odious to First Amendment traditions. The analogy is to the right to choose freely what books one includes in a private library at home.

Political and social impact. A third rationale sometimes offered to justify regulation of electronic media is the enormous but mysterious impact that television exerts over politics and culture. Television is the symbol of the media's huge power in the post-industrial age; it is a prime focus of anxiety about the future of our society.

Television has such an enormous and far-reaching impact on our society that, like the automobile, it is impossible to untangle its various influences. A particular concern is that television is widely thought to shape our politics, not only in the familiar sense of promoting certain candidates, but also in more systematic ways, such as emphasizing news at the national rather than the local level. Although many other factors account for the increasingly national focus of American politics, it is suggestive of the impact of television that some observers believe it is largely responsible for the fact that only 7 percent of the nominees for President and Vice President were from Congress between 1900 and 1956, but since 1956 the vast majority has been from Congress.

Others believe that television is partly responsible for greatly expanding the powers of the President at the expense of Congress. Recent Presidents have dramatically increased their use of free, prime-time, simultaneous broadcasts over the networks, leading one observer to liken television to "an electronic throne."[36] Television has taken from Congress the advantage of contact with localities, which Madison asserted to be its chief power in competing with the President.[37]

The television consumption statistics do not suggest a modest assessment of its impact. Television is present in over 97 percent of American households. The average television is on for seven hours each day. Two-thirds of the adult population look to television as their primary news source; more than one-half find it the most reliable news medium. Annual figures show a constant increase in television's apparent dominance over other media.[38]

The advent of pluralistic telecommunications provides an occasion for policymakers to think about the regulatory implications of television's large but amorphous impact on American society, politics, and culture. The idea that electronic mass media should be regulated because it is so powerful was not a justification for the regulatory regime developed for broadcasting a half-century ago because the power of radio was not apparent then and television did not exist. Moreover, in later years the FCC has not attempted to justify particular regulations because of the social or political impact of radio and television, and for reasons that are understandable. For one thing, assessments of impact are necessarily intuitive and debatable. It is difficult, perhaps impossible, to measure radio and television's impact on opinion formation, or on human behavior.

Even if we knew more about electronic media's impact than we do, there would remain a serious constitutional problem with justifying official control of expression because of the expression's impact on society. The First Amendment does not recognize a justification for official control of expression because the power of expression to affect political and cultural attitudes. Even when expression may lead to concrete harmful acts, as when an incendiary speaker urges a crowd to riot or threatens minority with injury, the First Amendment prevents interference with speech or press unless the specific harm threatened is almost certain to come about. First Amendment principles reflect a deep skepticism about official justifications for controlling expression because of its antisocial effects. Historically, repressive and politically motivated controls over expression have been defended by officials because of supposed bad effects.

Another important strand of First Amentment principle also stands as a barrier to regulating a communications medium because of its power. One of the functions of the First Amendment is to put the press beyond the reach of government control so it can check government abuses of power, reveal incompetence and corruption, and shape public opinion free from political controls. In terms of the function of a free press to check governmental abuses, the social and political power of a communications medium is not an argument for regulation; on the contrary, the more powerful a medium the better it can serve to check government abuses and the greater the risks of government control.

Even though it may conflict with First Amendment principles, and even though it may rest on untested assumptions, the impetus to regulate telecommunications because of its social and political impact is probably an unexpressed motivation of many policymakers. Those who feel this way should consider the implications of pluralistic telecommunications. Concerns about the vast political power of television news may be the dominance of three networks rather than the power of television as such. If so, the pluralism promised by broadband transmission should allay these concerns, and undermine this particular argument for legal regulation. Even if the total political and social impact of telecommunications is very great, division of the cumulative impact among a great number of programmers may alleviate the most serious concerns.

This is not to say that assumptions about the power of electronic media should have no effect on government policy. But the concern should be with concentration of market control, not cumulative media power, and the appropriate regulatory response would appear to be structural, antitrust approaches that will multiply the number of independent media outlets rather than regulate the programming of existing outlets. Thus, concerns about the social and political impact of television seems to point in the same direction as the traditional concern about scarcity of outlets: toward encouraging many outlets for diverse expression, essentially unregulated as to program content.

Other Mass Communications Media

Technologies that allow consumers to select programs for home viewing from a very broad range of options—for example, by purchasing videodiscs for the home film library or by leasing a program from a computer-controlled electronic warehouse—are essentially like books or long-playing records in terms of legal regulation. Where wide diversity of programs and extensive consumer choice is offered by mass media technology, First Amendment values of free personal choice would appear to submerge any general governmental interest in content control. Certainly, there is no basis for extending traditional broadcast regulation to videodiscs or computer-library selection systems. Scarcity will not exist because of spectrum limits, and intrusiveness and the passive/captive audience rationale for regulation have no application to a system of active consumer choice among a wide range of programming options.

It should be noted that current broadband systems do not include the capacity to transmit a separate signal to a single receiver that would be required for data retrieval or program ordering from an electronic library. Instead, existing systems send a number of messages to all users who may choose among them. Thus, program selection from electronic libraries is a possibility for the future.

Return-Path Communications

Many cable systems currently have the capacity for simple return-path communications from the consumer to the transmitting source. A yes-no switch to record consumer responses to programming questions is now feasible. More elaborate modes of return-path communication are currently under development. The conversion of television from a one-way, mass-audience, passive consumption system to one that allows for feedback communication in connection with broadly pluralistic programming raises a number of critical policy questions that have not yet been addressed in traditional broadcasting. The policy issues fall into six categories: privacy, political participation, mandatory features, the common carrier approach and the separation principle, pay television, and public television.

Privacy. The growing commitment to privacy—both freedom from official surveillance in the home and from revealing certain types of information about personal preferences and habits—must be given substantial weight in regulating return-path communications from consumers to programming sources. Most Americans would be outraged if the government or uninvited private entities gathered information on the books they read, television programs or motion pictures they watched, or even how many hours were spent in front of the television in a given period. Our privacy in such matters is protected now by tradition and also by the absence of any technology for gathering the information. Return-path communications technology could fill this void, and policymakers should therefore consider limits on the information that can be gathered without the consent of telecommunications consumers.

Political participation. One major potential for return-path communications is political. If a school board meeting is one of the programming options available on a two-way broadband communications system, viewers might be asked to offer opinions or even to determine outcomes by their "votes" transmitted to the meeting via fiber or cable. The examples could be multiplied. Two-way broadband systems could become a major focus of the perennial demand for participatory democracy and decision-making. As with the privacy values mentioned above, the absence of a convenient means of registering popular choices except on infrequent occasions has played a large part in fashioning the representative democracy we enjoy. One of the reasons our system does not operate by taking continuous nose counts on daily government decisions is that mechanisms for doing that have not been available.

A two-way broadband system is such a mechanism. Its potential impact on politics suggest that policymakers should make explicit what is usually left unstated, or treated as a matter of convenience, that is, the values of representative rather than rigorous democracy. Our system rests on periodic elections, followed by decisions taken by representatives that may or may not accord with fluctuating popular majorities, and followed in turn by the electoral opportunity to hold representatives to account for their past decisions and future prospects. This pattern serves a number of important values. It allows representatives occasionally to lead rather than follow; it muffles somewhat the power of interest groups to affect the outcome of specific political decisions by dominating single issue referenda; it allows the wisdom of many governmental actions to be judged at large by the people in the fullness of time rather than singly and beforehand. Moreover, given the impossibility of submitting more than a handful of political questions for referendum even with the most efficient nose counting technology, a plebiscitary system can be manipulated by controlling the issues presented, by the timing of votes, and by orchestrating the information made public. To articulate the dangers as well as the attractions of plebiscitary democracy will be a major task of sound policymaking with respect to return-path broadband communcations.

Mandatory features. In Chapter 7, William Lucas points out several useful features of return-path communications in addition to serving as a program-ordering device for consumers. Cable or optical systems can provide many services, such as fire and burglar alarms, marketing channels, the opportunity to extract information from computer information banks, and facsimile delivery of newspapers. Some of these services are similar to those offered by the telephone or the computer. Congress should consider the feasibility of mandating certain return-path capabilities in cable and comparable broadband communications systems. This will require a careful weighing of costs and benefits. The possibility of public subsidies for public interest features such as early-warning fire alarms should likewise be explored.

Common carrier and the separation principle. In general, return-path communications suggest the wisdom of the common carrier approach to the informational services provided. Several studies have gone farther and urged that the programming functions of cable systems should also be subject to common carrier regulation. The difficulty, perhaps impossibility, of distinguishing between information services and programming would seem to support regulation based on what has come to be termed the separation principle: that idea that cable or fiber systems should provide channel facilities but have neither control nor ownership interest in the programs and information distributed through the system.

The major justification for this separation principle is that broadband systems exhibit vast economies of scale; hence, it is likely that a broadband distribution system would be a natural monopoly in each locality. If broadband technology is not to exert undue monopoly power over the content of telecommunications, some policy (similar to the separation principle) is indicated. Delivery without control of content is the appropriate policy for the postal service, the telephone, and even public libraries. A similar approach to broadband systems would eliminate the pressure to regulate the content of broadband messages because distribution systems are local monopolies. However, distribution aspects of the system might be regulated. Whether government should regulate rates for delivery of programs, free-access channels, and channels reserved for special uses along the lines of the FCC's 1972 cable ruling, which set aside channels for local government and educational institutions, raises questions that Congress should address.

Pay television. Major policy issues are raised by pay television. Because traditional broadcasting does not sell programs to viewers, but rather sells audiences to advertisers, and because there have been relatively few broadcast channels, television programming has naturally focused on attracting the largest possible mass audience. Minority audiences with intense preferences for special types of programming tend to be ignored. But with many channels available, payments from viewers could offer economic incentives for a great variety of special interest programming. The diversity that such a development would bring to mass communications deserves strong support from Congress. The FCC has adopted a generally restrictive policy toward pay programming because of fear that pay television will take the most popular programs away from "free television." Congress should consider whether the potential diversity in largely unrestricted pay programming should be dampened because of this fear. Probably a few extremely popular sporting events should be protected from siphoning.

Although pay television should be encouraged, Congress may wish to consider some balanced approach that would allow for the diversity of direct-pay programming, while taking care not to undermine advertiser-supported television as the source of most mass audience programming. Advertiser-supported television is worth saving in our society. It puts the rich and poor on virtually the same footing with respect to the most important leisure activity.

Public television. A vexing question of policy is whether a system of pluralistic telecommunications leaves room for government-funded public television. One is tempted to say that broadband services, including pay television, could satisfy all audiences that public television serves. On the other hand, we are surely far from the day when we can say that public television serves no useful purpose. And in view of the great financial problems facing the performing arts, and the difficulty of devising workable means of public support, perhaps our system of public television is worth keeping even in a future of pluralistic programming. In theory, however, there would be no need for a public television network, and government contributions should probably take the form of program-production subsidies administered on a decentralized basis by agencies such as the National Endowment for the Humanities or the National Science Foundation.

Conclusion

"It cannot be helped, it is as it should be, that the law is behind the times," said Justice Holmes in a famous epigram.[39] But Holmes viewed law as typically foreclosing choice, and he approved the idea that we should be cautious before displacing freedom and private ordering with legal solutions. Our law of broadcast regulation is about to be behind the times, but we should not take heart from Holmes. For the nature of our backward law is to foreclose rather than enhance choice, inhibit rather than allow for expansion of news technologies, and maintain scarcity of expression where there might be abundance. The pluralism promised by modern mass telecommunications should move policymakers in the direction of traditional First Amendment values of diversity of expression, dispersal of media power, greater freedom of choice for consumers, and government withdrawal from regulation of the content of radio and television programming.

Fears are occasionally voiced that pluralistic programming will erode social and political cohesion, confuse consumers by an embarrassment of choices, and fractionalize audiences beyond the point necessary to support expensive program production. These fears seem exaggerated as a matter of prediction. They can be tested by looking at the effects of communications media where pluralism has been accepted without much question. Records and books are media that offer consumers a bewildering abundance of choice. Yet each medium generates very substantial social cohesion. To some extent, this may be attributable to the influences of other, less pluralistic media—radio, for example, in the case of records. But if radio is viewed as a force for cultural cohesion, we are dealing with as pluralistic a medium as emerging telecommunications systems are likely to offer in the foreseeable future. Moreover, even if some loss of social cohesion results from pluralistic telecommunications, this will be compensated for by the stimulus of diversity, by the opportunity for consumers to move from a passive to an active role, and by reducing the fears of manipulation of opinion that inhere in the present system of scarcity and concentration.

Notes

1. O.W. Holmes, *The Common Law* (Howe ed.), 1963, p. 5.
2. Judge Learned Hand, writing in *United States v. Associated Press*, 52 Fed. Supp. 362, 372 (Southern District New York), 1943.
3. *Id.*
4. The Radio Act of 1927 can be found at 44 Stat. 1162, 1927.
5. S. 2930, 65 Cong. Rec. 5735, 1924.
6. *Mutual Film Corp. v. Industrial Comm.*, 236 U.S. 230, 1915.
7. 47 U.S.C. Section 315(b), 1970.
8. B. Schmidt, Jr., *Freedom of the Press vs. Public Access*, Praeger Publishers, New York, 1976, p. 145.
9. Aspen Institute, 35 P. & F. Radio Reg. 2d 49, 1975.
10. 73 Stat. 557, 1959, 47 U.S.C. Section 315, 1970.
11. *See*, for example, *Straus Communications, Inc. v. F.C.C.*, 530 F. 2d 1001 (District of Columbia Cir.), 1976.
12. *Representative Patsy Mink*, 37 P. & F. Radio Reg. 2d 744, 1976.
13. *Alabama Educational Television Comm.*, 32 P. & F. Radio Reg. 2d 539, 1974.
14. *Red Lion Broadcasting Co. v. F.C.C.*, 395 U.S. 367, 1969.
15. *Id.* at 390.
16. *Id.* at 387.
17. *Id.* at 390.
18. *Id.*
19. *Committee for the Fair Broadcasting of Controversial Issues*, 19 P. & F. Radio Reg. 2d 1103, 1970.
20. 24 P. & F. Radio Reg. 2d 1917, 1972.
21. *Application of the Fairness Doctrine to Cigarette Advertising*, 9 F.C.C. 2d 921, 1967.
22. 30 P. & F. 2d 1261, 1974.
23. F.W. Friendly, *The Good Guys, The Bad Guys, and The First Amendment* Random House, New York, et seq., 1975, p. 32.
24. 18 U.S.C. Section 1464, 1970.
25. *Pacifica Foundation v. F.C.C.*, 40 P. & F. Radio Reg. 2d 99 (District of Columbia Cir.), 1977.
26. 18 U.S.C. Section 1304, 1970.
27. 15 U.S.C. Section 1335, 1970.
28. *Primer on Ascertainment of Community Problems By Broadcast Renewal Applicants*, 35 P. & F. Radio Reg. 2d 1555, 1975.

Programming and Regulation of Mass Media

29. *Citizens Committee to Save WEFM v. F.C.C.*, 506 F.2d 246 (District of Columbia Cir.), 1974.
30. *Yale Broadcasting Company v. F.C.C.*, 478 F.2d 594 (District of Columbia Cir.), 1973.
31. *Writers Guild of America West, Inc. v. F.C.C.*, 423 F. Supp. 1064 (Central District Calif.), 1976, appeal pending, No. 77-1602 (9th Cir.).
32. See generally, B. Schmidt, Jr., *supra*, ch. 13.
33. *Cable Television Report*, 36 F.C.C. 2d 143, 1972.
34. *Home Box Office, Inc. v. F.C.C.*, 40 P. & F. Radio Reg 2d 283 (District of Columbia Cir.), 1977.
35. *Memorandum Order and Opinion (Cablecasting)*, 23 F.C.C. 2d 825, 1970.
36. See generally, N. Minow, J. Martin, L. Mitchell, *Presidential Television*, Basic Books, New York, 1973.
37. *The Federalist No. 49* (Cooke ed. 1961).
38. See generally, Robinson, "American Political Legitimacy in an Era of Electronic Journalism," in *Television As a Social Force*, Praeger Publishers, New York, 1975, p. 105.
39. O. W. Holmes, *Collected Legal Papers*, Little Brown, Boston, 1921, p. 231.

Selected Readings

On Freedom of the Press

Near v. Minnesota, 283 U.S. 687, 1931.

Associated Press v. United States, 326 U.S. 1, 1945.

New York Times Co. v. Sullivan, 376 U.S. 254, 1964.

Miami Herald Publishing Co. v. Tornillo, 418 U.S. 241, 1974.

T. Emerson, *The System of Freedom of Expression*, Random House, New York, 1970.

B. Schmidt, Jr., *Freedom of the Press v. Public Access*, Praeger Publishers, New York, 1976.

V. Blasi, *The Checking Value in First Amendment Theory*, 1977 American Bar Foundation Research Journal, p. 521.

On Broadcast Regulation

National Broadcasting Co. v. United States, 319 U.S. 190, 1943.

Red Lion Broadcasting Co. v. F.C.C., 395 U.S. 367, 1969.

Columbia Broadcasting System, Inc. v. Democratic National Committee, 412 U.S. 94, 1973.

Citizens Committee to Save WEFM v. F.C.C., 506 F.2d 246, 1974.

F. Friendly, *The Good Guys, The Bad Guys and the First Amendment,* Random House, New York, 1975.

G. Robinson, "The FCC and the First Amendment: Observations on Forty Years of Radio and Television Regulation," *Minnesota Law Review,* Vol. 52, 1967, p. 67.

H. Kalven, Jr., "Broadcasting, Public Policy and the First Amendment," *Journal of Law and Economics,* Vol. 10, 1967, p. 15.

L. Bollinger, Jr., "Freedom of the Press and Public Access. Toward a Theory of Parital Regualtion of the Mass Media," *Michigan Law Review,* Vol. 75, 1976, p. 1.

L. Powe, Jr., "Or of the (Broadcast) Press," *Texas Law Review,* Vol. 55, 1976, p. 39.

D. Bazelon, "FCC Regulation of the Telecommunications Press," *Duke Law Journal,* 1975, p. 213.

On Cable Television

On the Cable, Report of the Sloan Commission on Cable Television, McGraw Hill, New York, 1971.

D. LeDuc, *Cable Television and the FCC,* Temple University Press, Philadelphia, Pennsylvania, 1973.

Cable: Cabinet Committee Report on Cable Communications, U.S. Government Printing Office, Washington, D.C., 1974.

6.
The Role of Print in an Electronic Society

Bruce M. Owen

This chapter contains some speculation about the policy implications of the present trend toward the use of electronic and computer technologies for the production of print media. There is also a warning here: the "electrification" (injection of computer and telecommunication technologies) of print technology may lead to government regulation of the press, with attendant dangers to our civil liberties. Some observations are made concerning economic trends in the traditional print industries to shed light on the structure of these media and their role in an electronic society. This trend may result in a shift away from local values and toward new communities of interest. Attempts to thwart this shift to preserve local values may lead to unwarranted regulation of the press. Our regulatory institutions are insufficiently flexible to deal with the problem of regulating a new technology without also regulating the uses that are made of it. In this case, the implication is that if we regulate the medium, we seem hardly able to avoid regulating the message.

Communications for Tomorrow

Functions of the Print Media

Implicit in the title of this chapter is the picture of yellowing pages of print antiquated by dazzling new electronic devices. In the science fiction world of the future electronic society, people would watch "news fax" machines rather than read newspapers and slip compact tape capsules into electronic "reading machines." Books would become museum artifacts. Libraries would be computer memories. In one view, everyone would have instant access to vast storehouses of data as we enter the Information Age. In another view, television and its technological progeny would create a massive illiteracy in which many or even most people would not know how to read.

We have a long way to go, however. The printed page is merely a repository for symbols embodying information. As such, it has certain advantages: durability, compactness, portability, cheapness. We use it because it provides the best combination of these characteristics. It is not as durable as stone, not as compact as microfilm, not as portable as memory, and not as cheap as talk. For the electronic society to overcome the economic advantages of the printed page, it must produce a substitute that has these qualities in a superior combination. It may well do so for some individual uses.

A second feature of the print media is that they typically supply a high level of editorial service. That is, people are willing to pay something to avoid the task of sifting data for themselves, and editors compete for the readership market by compiling packages that suit the tastes of individuals. Indeed, in the Age of Information, editors assume an even greater importance; people will pay *not* to be deluged with unedited data.

The printed page can be viewed in various ways. Functionally, the printed page is a transmission medium as well as a storage medium for information. The two functions are not necessarily joint. *The Wall Street Journal*, for instance, uses satellites and related electronic devices as a medium of transmission to its regional printing plants. The newspaper itself is then a communication medium to readers and a temporary storage device. Eventually, libraries maintain back issues on microfilm. But neither electronic nor microform transmission is available to readers in their homes because these media require expensive and cumbersome interpretive equipment. Even if microform reading machines or electrofax receivers were inexpensive and portable, they would have a disadvantage, depending on consumer demand; most are capable only of linear presentation of the material, so that overview scanning and random access are more difficult than they are for print. These difficulties can be overcome only by much greater effort at indexing, if at all. Indexes are an imperfect substitute for editors. Even scholars might prefer a good survey article to relatively inexpensive access to a good library.

This last point highlights the general importance of consumption technology, which refers to the decisions that consumers make and the processes they use to organize their consumption activities.[1] In a sense, people do not consume "automobiles"; rather, they consume "transportation services," which involve the use of various inputs, such as cars, gasoline, highways, and time. People do not consume books or newspapers; they consume information or entertainment, together with numerous other consumption activities, many of them simultaneous. A book may be one input into a leisure activity that also requires a hammock, good weather, iced tea, and peace and quiet. A newspaper may often be used as an input into a complex activity that also requires the presence of family members, breakfast, and so on. New media of communications must be fit into these consumption activities in appropriate ways if they are to be accepted, or if they are to replace printed communications. At the same time, new communications technologies can create entirely new opportunities for consumption activities, with important social and cultural effects. Unfortunately, these effects are most apparent to historians.

Television, obviously, has had an effect of this kind. Television did not replace print. It does not perform the same functions, or at least does so only poorly. In fact, the decline in motion picture attendance, commonly attributed to television, appears to have begun *before* television diffusion was significant.

The Environment of the Print Media

Newspapers, periodicals and books are the major printed mass media. Each medium is composed of competing firms (information sources) and the media themselves compete for the custom of consumers and advertisers. Motion pictures, radio and television—technologies new to the present century—substitute to some degree for printed media. The appearance of substitutes for print has not, however, prevented considerable growth in the size and revenues of the print industry. Real per household expenditures on print media grew 75 percent between 1930 and 1968; real disposable income, however, more than doubled. More than half of the print media expenditure has traditionally come from advertisers. The year 1968 is significant because it was in that year that per household expenditure on nonprint media caught up with and surpassed expenditure on print media. It is interesting to note that of the $185 per household in 1958 dollars spent on nonprint media in 1968, only a third was from advertising revenues. Consumer expenditures on receiver equipment and repair constitute an important and often neglected cost of the electronic media system. The costs and technologies facing the household, quite different now than 50 years ago, have not led to drastic declines in expenditures on printed matter.

We can turn to the evidence, at this point, from household "time budget" estimates. For instance, the average adult male spent 22 minutes per day reading books in 1934 and 9 minutes in 1966, while time spent watching television increased from 0 to 90 minutes per day.[2] Time-budget studies, unfortunately, are difficult to do well or to interpret because even slightly obtrusive measuring methods create distortions in behavior. Nominally, if we believe Nielsen,[3] people spend a staggering amount of time consuming television, and this time must at first blush be spent at the expense of other leisure activities. Hours spent working have not changed much in this century, despite the shorter work day and work week, because of increases in commuting time.[4] The data on motion picture attendance in the post-war period indicate clearly where some of the time was spent. In 1946, the average citizen over five years of age spent about 10 out of every 1,000 hours at a motion picture theater. Today, we spend less than one-tenth of that. But television viewing and reading are not mutually exclusive activities. People can and do engage in them simultaneously, and Nielsen measures only the sets that are on, not who is watching them attentively.

Trends and Economic Forces

The present century has witnessed a steady decline in the number of daily and weekly newspapers, an increase in the number of cities with daily newspapers, and therefore a decline in the number of competing newspapers in individual cities.[5] In 1900, there were about 2,100 daily newspapers and about 12,000 weekly newspapers. By 1973, there were about 1,566 daily newspapers and 9,300 weeklies. But the average daily circulation of the surviving newspapers has increased—for dailies, from about 6,800 copies in 1900 to 39,000 in 1973; total circulation has also increased, from 15 million copies per day in 1900 to 62 million in 1970. When these increases in circulation are corrected for population growth, however, we find that the number of newspaper copies per household grew from .94 in 1900 to 1.37

The Role of Print in an Electronic Society

in 1930, and then declined to .99 in 1970. The changing structure of the newspaper industry is explained partly by the introduction of new competing advertising and consumption technologies—motion pictures, radio, and television—and partly by the adjustment of the industry to economic forces that have made the one-newspaper town the typical equilibrium situation. At present, the forces leading to decline in the number of newspapers per city appear to have diminished, there now seems to be an offsetting increase in the number of cities large enough to support a daily newspaper. We should therefore begin to see an increase in the number of daily newspapers during the rest of this century, one to a city. Within this overall trend we are also likely to see a continued diminution in the relative importance of large newspapers with regional circulations and an increase in the importance of newspapers with local circulation. The reasons for this are that electronic media have greater relative competitive impact on the largest newspapers and that the smaller newspapers have some advantages in adopting the new electronic production technologies.

The policy concern about concentration in the newspaper industry, i.e., "monopoly,"—a concern that may serve as a rationale for regulation—is easily exaggerated. Although one-newspaper cities are clearly more concentrated than multi-newspaper cities, the former are not pure monopolies; they compete with neighboring firms and with other media. They do not earn monopoly profits.[6] There seems to be an increase in the extent of newspaper chains; although, by any national measure, the newspaper industry is not highly concentrated, and it is far from clear that the increase in chain ownership is a worrisome trend at this stage. It is against this rather nebulous background that one must ask whether electronic technology will reduce or increase concentration, a point which we develop below.

Turning now to the book and periodical trade, we find that economic conditions are quite good. Statistics on these industries that would allow careful examination of long-term trends are not available. Although there is no sign that television or the other electronic media have as yet made a significant impact in this print market. There is little doubt that television was a major factor leading to the demise of several mass periodicals, such as *Look* and *Life*, which were close substitutes for television, but these magazines have been replaced by other, more specialized ones.

As Marc Porat tells us in Chapter 1, information production and consumption is a large and growing portion of the economy. Despite the salience of the effects of electronic media on a very few mass periodicals, the magazine and book trades have continued to prosper as part of this overall trend. With the advent of television, periodicals had a comparative advantage in serving relatively specialized interests and audiences but it cannot be said that they have become less important or less influential.[7]

It should be noted that substitution effects exist in advertising markets as well as in readership-viewership markets. Indeed, it is in advertising markets that we find the most plausible explanation for the effects of television on print media. Mass-circulation general-interest magazines depend heavily on advertising revenues, as do newspapers. More specialized periodicals depend to a greater degree on subscription revenues and on specialized advertising, for which television advertising is a poor substitute. Thus, it is not surprising to find that television has had the greatest impact on mass circulation periodicals and on newspapers in the largest cities.

These considerations suggest an interesting hypothesis: that the FCC's restrictive policies toward spectrum allocation and cable may have served to protect the print media. There seems to be little doubt that more channels and more specialization and, especially pay television, would result in television programming that was a much closer substitute for a much larger segment of the print industry.[8] By resisting such policies, the FCC may have prevented a substantial portion of the print media from becoming extinct, simply because it has not allowed advertisers access to the full range of opportunities made possible by television technology.

If cable television is eventually deregulated and attains the form that so many of us have hoped (or at least predicted), the development of specialized audiences, including those specialized by location, may result in a substantial shift in advertiser demand away from print, particularly periodicals. If cable eventually becomes more sophisticated, with direct consumer access to information bands that fascinate so many futurists, further substantial inroads on the local newspaper advertising market are likely. It seems that the major impact of this is that print media (or their electronic analogs) will be increasingly dependent on direct subscriber support. The convergence of print and electronic technologies, however, makes such an observation uninteresting; that which is lost is not lost to consumers; rather it is lost to the owners of a particular set of capital goods embodying obsolete technology.

Somewhat more interesting is the question of whether decreasing transmission costs will result in a trend away from local and toward national interest content. Other things being equal, a fall in transmission costs accomplishes just that. However, this effect will probably be small compared to the influences from the demand side that will result from the (presumed) ability of consumers to pay directly for content. The result of this is to increase specialization in the market; local content is one form of specialized content. The major threat for newspapers like *The Washington Star* is not so much from *The New York Times* or the *Daily News* as from the presently nonexistent *Georgetown Journal*, a new "neighborhood" electronic newspaper. It has always seemed to me desirable and rather likely that the electronic technologies will enable us to achieve a media structure that is essentially the same as the present periodical or journal industry.

It is government policy toward the electronic media, as much as their intrinsic characteristics, that has resulted in a relative absence of geographically local content. While it may be true that an electronic press would tend not to emphasize local interests and values, this might be attributable to the fact that an electronic press provides opportunities for the formation of new communities of interest, which are not specialized by geographical location. Local community values arise in part because of economic costs of communication and transportation. As these costs fall, and as the quality of communication and transportation rises, consumers naturally tend to substitute community-of-interest contacts for local contacts. It is true, however, that our political system is founded partly on representation of geographical interests. Perhaps this will be changed as well. The substitution of communities-of-interest for geographical communities does not pose any obvious threat to democratic or humanistic values. On the contrary, policy intervention designed to conserve the values of geographical localism may pose significant dangers to our civil liberties. If the decreasing price of communication does in fact result in a substitution of communities-of-interest for localism, such intervention can be successful only by placing constraints on the freedom of expression that might otherwise be attained.

Increased freedom of choice for individuals does not necessarily make society better off. The best example of this is the freedom, which some people used to enjoy, to discriminate against others on the basis of race or sex. It is no doubt possible that the increased freedom of choice which results from declining communication prices can have such bad effects. But it seems clear that we do not know enough about this problem even to state it in a way that would allow relevant research to be undertaken. It follows that our policy-making machinery should remain flexible. Flexibility in policy making requires policy institutions that are flexible, and in this respect it is apparent that the creation of regulatory agencies is not the proper way to solve the problem.

Subsidies

A major policy issue for both books and magazine is the postal subsidy. For many years postal rates for magazines, newspapers, and books were supposedly subsidized, but the postal reform efforts in the late 1960s set a new policy, under which each class of service was to carry its fair share of postal costs. Because these costs are largely shared, there has been endless controversy over metaphysical cost allocation plans, not unlike those confronting the FCC with respect to telephone prices. It is by no means clear that it is a bad policy to subsidize postage for the print media. Postage costs are only about 10 percent of total expenses for major magazines but they are nevertheless important, and even more important for small, specialized periodicals. Rising postal costs and other factors have led to a situation where today, for the first time in history, more than half of all copies of major consumer magazines are sold on the newsstand.

Postal rates represent an interesting issue for another reason. The distance-insensitive rates are structured to discourage localism. Indeed, as Daniel Boorstin has pointed out,[9] since the enactment of the rural free delivery (RFD) statutes, the postal service has acted to destroy local, and especially rural, community orientation. Moreover, by providing long-distance and especially rural service at subsidized rates, the postal price structure has discouraged the development of technological substitutes for these services, while perhaps artificially encouraging the development of technologies that can compete on heavy-traffic and shorter routes.

There are of course numerous subsidies and taxes, explicit and implicit, which affect the media. While the FCC's grant of "free" spectrum to broadcasters is a subsidy in kind, the Commission places a "tax" on broadcasters when it constrains the quantity of spectrum available. The postal system taxes localism; the FCC tries to subsidize it. These taxes and subsidies, taken as a whole, do not serve a consistent set of policy objectives, and this is symptomatic of the historical failure to coordinate communications policy in this country.

Proponents of subsidization of the electronic press offer different reasons. That we presently subsidize some communication activities at the expense of others or at the expense of taxpayers generally is no valid reason to subsidize electronic transmission, or aspects of it. Nor is income redistribution a valid reason for subsidization. Poor people are not going to be politically or economically disenfranchised by (for example) cable television. Cable requires that people pay, just as television does. It was noted above that only a third of the total societal expenditure for nonprint media comes from advertising. Most of the remaining two-thirds is consumer expenditure on television sets and repairs. Consumers bear a greater proportion of the total cost of television than they assume for newspapers, where the shares are roughly reversed. Surely no one would argue that television disenfranchised the poor. If anything, television has redistributed income downward. Nor is the electronic press likely to be different. There is no evidence to support the view that the new technologies would, on balance, make poor people worse off. Indeed, to the extent that the price of access to organized information and education is reduced, freedom of opportunity in society would be increased. The distributional effect of this would be in favor of those who stand to gain the most from cheaper information.[10]

Technological Change

Many people believe that there have been two revolutions in communications—one sponsored by Gutenberg about 1450 and the electronic revolution in this century. In fact, there have been three. In the nineteenth century, cumulative technical improvements in printing and favorable economic factors led to a massive increase in the circulation and number of daily newspapers. Indeed, it was during the late nineteenth century that the first truly "mass" medium (the daily newspaper) was developed. The technological inputs to this second revolution were the application of steam power to presses, the invention of the rotary press and linotype machine, and falling newsprint prices. The economic factors were the significant increase in demand for advertising space by manufacturers, brought about by the development of improved transportation for both people and goods, and an apparent shift in consumer tastes. The revolution was not limited to newspapers. The same forces led to the rise of the first mass circulation periodicals at the end of the nineteenth century.

In the past 10 or 15 years, the electronic revolution has finally begun to be applied to the technical process by which newspaper and magazine copy is printed.[11] Attention has focused on the use of computers, at first in composition, then in editing. The development of ink-jet, or even laser beam-electrostatic presses may well result before long in a newspaper production process completely controlled by computer—from the written to the printed page. Although dramatic, these developments do not have profound implications; they simply make it less expensive to produce newspapers, and they have some effects on the economic structure of the newspaper industry. We have already mentioned the case of using satellites as communication channels to transmit copy to the press. There also may be an increase in "remote-printing"—a substitution of communication for physical transportation. At one extreme, such an innovation could produce the equivalent of a newspaper press in every home, although it seems unlikely that this would be feasible. Another electronic device is in use in Britain, where several "teletext" service innovations use spare capacity in existing television channels to transmit potentially large quantities of textual material. Because transmission costs are low, the primary cost concern is the expense of the device that must be attached to each television set to receive teletext. There is a pricing problem because users cannot readily be charged in accordance with usage. These and other electronic devices are greeted with great enthusiasm by the communications fraternity, although it is too early to know whether they will be acceptable to consumers.

One way in which we can predict the direction of electronic technology's impact on print is to follow the investment strategies of corporations like Xerox and IBM. Both are investing heavily in office-information processing devices that can replace "hard copy" with electronic displays. In addition, IBM seems to be committed to the notion of communication links among computers and information devices. Various firms are investing in the development and marketing of consumer video-storage devices that eliminate the durability disadvantage of video-displayed information and entertainment. It is not clear how the portability problem will be solved, but it is not inconceivable that small-screen cassette players will be the answer to this difficulty. Corporate investment strategies seem consistent with the hypothesis that these alternatives to current print technology will eventually be economically viable. The hypothesis is further supported by the increasing relative cost of postal service, which is labor-intensive, and of paper itself.

The gradual elimination of printing as an information technology and its replacement by electronic technologies will occur only if consumers (and advertisers) find that the latter offers an attractive combination of quality and price.

Undoubtedly some pleasant aspects that were worthwhile in the world of printing presses and newsboys will disappear but other features will accompany the new technology. Gutenberg's inelegant output, after all, replaced beautiful illuminated vellum manuscripts, which we do not mourn excessively. Similarly, electronic technology will create the potential for new freedoms. The danger in all this lies not in the technology but in our policies toward it. Gutenberg's invention destroyed a technology of information that was conducive to authoritarian control by the church. Electronic technology, by contrast, destroys a technology characterized by a tradition of freedom, and replaces it with one more readily subject to authoritarian control by the state.

The problem is that communication channels such as satellites are regulated, and there is a serious prospect that computers will be regulated as well.[12] And when this happens, the dependence of print technology on electronics will mean that print is regulated. We will then be faced with such issues as whether people ought to be allowed to pay for printed communication, whether there is too much violence in print, and how many people should be allowed the privilege of printing. By all the rules and precedents that are available to us, the computer-communications utilities are likely to be regulated, and we have ample historical evidence of the form and effects of that regulation. Unnecessary monopolies are created; necessary ones preserved beyond their useful span; decentralized decision-making, a crucial element of freedom, is subordinated to social engineering in pursuit of ill-formed and generally obsolescent social goals. Of course, one can always hope for the best. It is not impossible that regulation can make us better off than we would be in its absence. It is just, on the record, most unlikely.

The Impact of Future Technology

The future depends on exogenous technological developments and on economic conditions, shifting consumer tastes, and government policies. There is uncertainty about how these future developments will combine to determine the shape of communications for the remainder of this century and beyond. It would be risky even to seek to limit the range of possibilities. It is possible that difficult policy choices will not arise because print media, even if altered by electronic technology, will continue to be dominated by decentralized producers and by technologies that do not lend themselves to regulation. Possibly all that will happen is that we will receive cassettes, and floppy disks, and similar media through the postal system to replace, or largely supplement, printed pages. Then the only interesting questions are those dealing with the demise of the soon-outmoded libraries and the development of new copyright standards. Library policy, copyright laws, and privacy issues constitute an important set of closely interrelated problems that fall under the general heading of information production policy. Information has some peculiar economic properties, and it is important that these be understood in order to design policies that create efficient incentives for information production and dissemination. These issues have recently been receiving much attention in the economics literature, and deserve closer examination than that provided.

It is possible that some accidental conjunction of technology and policy will result in concentration at some production stage of media messages or will produce the use of some existing regulated transmission medium. Such situations could arise from cross-ownership policies. It is this sort of eventuality that would constitute the most interesting policy problems, and the ones most likely to show institutional inadequacy in our policy machinery.

There are several future scenarios that we might consider to examine these problems further. One example is found in cable television, which is the primary conveyor of mass media messages, including the electronic successors to print. If present trends continue, the sources of these messages will be regulated by the FCC in essentially the same way that broadcasters are regulated. It is highly significant that the FCC has chosen to regulate cable in the broadcast framework rather than the common carrier (telephone) framework. Enforcement of the FCC's pay-television rules, for instance, required the Commission to extend its jurisdiction to cover cable channel leasees. From the point of view of freedom of expression, this is the worst of all possible worlds.

A second scenario is one in which either the telephone company or its post-divestiture successors would supply this service, in conjunction with regulated computer utilities. (More or less equivalently, these services might be provided by cable, but under common carrier access principles.) It should be noted that there is no strong tradition of direct regulation of message content in common carrier communication systems. Nevertheless, regulation of these media has an effect on message content—by controlling the prices charged, the quality of service, and the degree of technological competition. The elaborate tariff structures of the utilities are designed to discriminate in price, and in policing this process, the regulatory generally make rules and decisions that are message-content and message-source specific. That is, the tariffs and the regulations discriminate among messages according to their content and their source. It is difficult to imagine that a regulatory agency of public utility could ignore the existence of socially unpopular messages, whether political or moral, if those messages seem to be qualifying for one or the other of the "subsidized" transmission rates. Thus, this second scenario seems only a little less likely than the first to lead to a diminution in freedom of expression, even in "normal" times, and seems much more susceptible to control than present media in extraordinary times (e.g., war years). The important point is that even if content regulation is not initially intended, regulation of the medium is very likely to imply regulation of the message and regulation of message sources.

The Present Danger

Print is a romantic medium to those who live by it, a romanticism to which commercialism contributes rather than detracts. Television is distasteful partly because it is new, and partly because the medium dwarfs the individual artist and the individual viewer, a consequence caused not by the technology itself, but by the policies of benevolent regulators with limited vision. The structure of the television industry is not naturally so monolithic as our public servants have made it. The present danger is that these public servants will do the same to print as printing and electronic technologies converge.

One can imagine a process by which that material which is most controversial, violent, subversive, and prurient is gradually concentrated in old-fashioned unregulated mechanical print media; electronic technology, regulated as it seems likely to be, would then increasingly assume a greater role in the information process. At first it would be intermediate transmission, then computer storage, then other stages would follow. As this happens, what is eventually transmitted will have been subject to electronic regulation at one or more stages of the process in increasing degree, with the effect of sanitizing the material involved. Controversial material that could not gain access to this process either because of direct regulation or because licensees exert caution so as not to endanger profitable licenses would be relegated to old-fashioned mechanical means—obsolescent presses and physical transportation modes. Eventually it will be possible to stamp out such material entirely.

Only two assumptions are required to generate this nightmare. One is that the regulation of electronic communication will continue on its present course.[13] The other is that electronic technology will continue gradually to replace the mechanical technology that produces printed pages. Of all the future possibilities, this is the one that deserves the most attention from policy makers.

Electronic technology need not imply a more monopolistic press and, indeed, offers the opportunity for increased competition and decentralization.[14] The important point is to separate the medium from the message and to treat the electronic transmission technologies in the same manner as we have treated the post office—as a neutral conveyor of material selected and controlled by a highly competitive and decentralized market of buyers and sellers.[15] Neither the owners of the electronic technology nor their regulators have any interest in the content of messages. It is for this reason, despite grave reservations on one level about efficiency and progressivity, that the idea is entertained of allowing the post office and the telephone company to control the new transmission technologies, rather than the cable companies. The regulatory seem much more likely to treat the latter as they have treated broadcasters, and somewhat less likely to do this with the traditional common carriers. Even then, however, as was pointed out above, the danger of less direct content and source regulation is not negligible. Whether or not we need to regulate this electronic transmission service (in the traditional public utility sense) is the next question. On the whole, given the performance of public utility regulators over the years, we would likely be far better off in the area of communication technology, in terms of both economic efficiency and freedom of expression, if there were no economic regulation except antitrust controls. This regulation could perhaps be supplemented with a statutory common carrier access obligation enforced through nonadministrative judicial forums.

The fundamental question, then, is whether the electronification of print will result in greater or lesser freedom of expression, e.g., the ease of access to the media by individual speakers. The answer appears to be that the technology itself allows as much freedom of expression as we now have, and possibly more. How much freedom will be possible depends on our communication policies. As Benno Schmidt so forcefully demonstrates in Chapter 5, content regulation is a widening gyre with an internal logic that seems impervious to fact or consequence. The lesson is that in order to avoid regulating the content of the print media we may have to start now to deregulate the electronic media.

Notes

1. See K. Lancaster, "A New Approach to Consumer Theory," *Journal of Political Economy*, Vol. 74, pp. 132-157, April 1966; G. Becker, "A Theory of the Allocation of Time," *Economic Journal*, Vol. LXXV, September 1965, pp. 493-515.

2. T. Scitovsky, *The Joyless Economy*, Oxford University Press, London, 1976, p. 163. The data include leisure and non-leisure reading.

3. Data on television viewing from the A.C. Nielsen Company are not publically available. Summaries of such data can be found in various issues of *Broadcasting* magazine or in *Broadcasting Yearbook* (Broadcasting Publications Inc. Annual).

4. See S. de Grazia, *Of Time, Work and Leisure*, Twentieth Century Fund, New York, 1962.

5. For a general discussion of newspaper economics see J. Rosse, "Daily Newspapers, Monopolistic Competition, and Economies of Scale," *American Economic Review*, Vol. LVII, May 1967, pp. 522-533. A number of working papers have been published on the topic in *Studies in Industry* by the Department of Economics, Stanford University: "Economic Limits of Press Responsibility," Working Paper No. 45, February 1975; "Trends in the Daily Newspaper Industry, 1923-1973," Working Paper No. 57, May 1975; "The Daily Newspaper Firm: A Twenty-four Equation Reduced Form Model," Working Paper No. 76, January 1977. See, in addition, Bruce M. Owen, *Economics and Freedom of Expression: Media Structure and the First Amendment*, Ballinger, Cambridge, Massachusetts, 1975, pp. 33-85.

6. J. Rosse and J. Panzar, "Chamberlin v. Robinson: An Empirical Test for Monopoly Rents," *Studies in Industry Economics*, Working Paper No. 77, Department of Economics, Stanford University, February 1977. However, we cannot exclude the possibility that potential monopoly profits are in fact earned in other ways (editorial discretion) or are appropriated by other groups, such as unions.

7. Trends in scholarly publications merit examination because microform and electronic substitutes for print might first appear here. Professor Fritz Machlup of Princeton has a major research project under way in this field, and one of his tentative conclusions is that the market for scholarly books and journals is subject to "economic strangulation." Circulation per item has been falling; library budgets have been unable to keep up with rapid increases in the volume and prices of scholarly output; and the industry itself is increasingly supported by government, foundation, and industrial sponsors of research. F. Machlup,

"Publishing Scholarly Books and Journals: Is It Economically Viable?" *Journal of Political Economy*, Vol. 85, No. 1, February 1977, pp. 217-226.

8. See M. Spence and B. Owen, "Television Programming, Monopolistic Competition and Welfare," *Quarterly Journal of Economics*, Vol. XCI, No. 1, February 1977, pp. 103-126; and J. Beebe, "Institutional Structure and Program Choices in Television Markets," *Quarterly Journal of Economics*, Vol. XCI, No. 1, February 1977, pp. 15-38.

9. Daniel J. Boorstin, *The Americans: The Democratic Experience*, Random House (Anchor edition), New York, 1973, pp. 131-136.

10. The economic rationale behind the desirability of subsidizing the electronic press is that the markets for information in particular, and the operations of monopolistically competitive markets in general, have imperfections that may otherwise lead to underproduction of certain kinds of goods and services.

11. For an excellent detailed survey of the new technology used in the American press, see Rex Winsbury, *New Technology and the Press*, Working Paper No. 1, the Acton Society Press Group and the Royal Commission on the Press, H.M.S.O., London, 1975. Winsbury points out that 84 percent of all U.S. daily newspapers used cold type in 1974, but that these were the smaller firms, accounting for only about one-half of total daily circulation. Similarly, in that year, offset replaced letter press in 63 percent of the firms (with 27 percent of total daily circulation). The use of optical character readers and other computerized newsroom technology is growing rapidly as well.

12. For a discussion of FCC regulation of computers, see Roger Noll, "Regulation and Computer Services," in M. L. Dertouzos and J. Moses (eds.), *The Future Study on the Impact of Computers and Information Processing* (tentative title, MIT Press, Cambridge, Massachusetts, forthcoming).

13. The greatest danger, of course, lies in the prospect that regulation will be like the laws governing broadcasters rather than those controlling common carriers. As discussed above, however, common carrier regulation itself is by no means "safe."

14. A more detailed treatment of this point is presented in B. Owen, *Economics and Freedom of Expression: Media Structure and the First Amendment*, Ballinger, Cambridge, Massachusetts, 1975, Chapter 1.

15. It is worth noting that the postal service has not always been a neutral or "transparent" conveyor of information. Messages containing controversial moral or political content have frequently been censored.

7.
Telecommunications Technologies and Services

William A. Lucas

The United States has long enjoyed a multitude of telecommunications opportunities. With the coming of satellites, sophisticated cable television systems, and new videodisc and optical fiber, the flexibility and number of potential applications will be as limitless as the human imagination. Even now, combinations of new services and advanced technologies are being field-tested. For example, the ATS-6 satellite provided a communications link between a physician in Washington state and village health aides in remote Alaskan villages, and at the same time was used to conduct in-service training of school teachers at 15 sites in Appalachia. The same satellite was then moved, and it directly broadcast agricultural extension, health, and basic educational television to hundreds of villages in India. The laser beam has been used by a senior anesthesiologist to supervise nurses in another hospital during operating room procedures. "Slow-scan," the transmission of live voice and intermittent pictures over telephone lines, enabled nurses employed by prisons in Dade County, Florida, to consult with inmates on health matters without visiting the prison. In a separate system, slow-scan technology is being used to create a consortium of five community colleges, ranging from North Carolina to California, which can share educational programs despite immense geographic distances.

Combined with the many uses of microwave, the telephone and other more conventional technologies, these experimental projects provide a rich base of experience. Indeed, no short review can capture the diversity of this field; it has become a constantly changing kaleidoscope of technologies, services, and organizations. The purpose of this chapter is to provide examples of promising applications to depict both the potential of and the barriers to the development of new communications services. Two short histories—local services and public programming—illustrate that as new services inevitably emerge, choices among telecommunications technologies will necessarily be made that could affect the welfare of society.

Technologies for Delivering Local Services

The range of technologies, organizational approaches, and possible service combinations that are logically possible are almost limitless. Obviously one should first define the social problem or need, and then choose the most appropriate technology. Unfortunately, advocates of any one technology tend to pick problems to solve. Perhaps this is unavoidable; service providers cannot be expected to study the full array of potential telecommunications alternatives. Nonetheless, the result is not always the most appropriate match between service and technology.

One major problem is that the choice of technology may involve some unintended choices among competing social values. Telecommunications technologies create new service opportunities by spanning great distances or by reaching large numbers of people, and some groups will benefit more than others. The technological choice that must be made between cable and telephone systems for the delivery of local services serves to illustrate the point.

The economics of the cable industry have been determined from its inception by the principle of providing improved television reception to a large number of homes at a low unit cost. Construction costs vary, and they can be extraordinarily high when underground construction is required. Still, many cable systems approach economic viability when they have an average of 40 or more homes on each mile of constructed cable, and approximately one-half of those homes are willing to pay a monthly fee of $6.50 to $7.50 a month for that service. As a result, major cable systems tend to be located in urban areas where there is a greater density of potential subscribers. In rural areas, the density is not sufficient to support a major cable system; and, if there is a cable, it takes the form of many small systems of a few hundred or less in isolated pockets. The number and quality of over-the-air television stations and restrictive regulatory policies have limited cable penetration in the past; nevertheless, 18 percent of American households subscribed to cable television service in 1976. Recently the regulatory environment has become more favorable, and the forecast is for continued growth—over one-quarter of the nation's homes on the cable by 1980.[1]

Once cable systems are in place and the costs of the cable system are paid through subscription fees, the technology can be used for a vast array of services—assuming the economic and organizational problems can be solved. With conventional technology, a single cable can easily carry 30 channels of programming. Even after the cable operator carries the signals from all the area's television stations, and perhaps imports signals from one or two stations in distant cities, there may be 20 unused channels. This is equivalent to all educational, cultural, and informational programming at an extraordinarily low cost.

A further dimension is added by "two-way" or interactive cable. Just as a cable is capable of carrying signals "downstream" from a central antenna or studio to homes and institutions, it is able to carry "return" or "upstream" signals from homes to a central facility or elsewhere in the cable system. Thus, one could show a live city council meeting by placing a camera in the city hall, and with data terminals in the home, citizens could express their evaluations of city services. The same home terminals could be used to educate students, to provide vocational training for handicapped adults, to order groceries from a catalog, send alarms to fire stations or to the police in emergencies.

These and hundreds of other ideas appeared in lists of new services which would be available in a "wired city" that would use two-way cable technology in the nation's cities. Particularly in 1970 and 1971, a rash of books and studies were published that were confident that a wired city was at hand.[2] Their common assumption was that the costs of maintaining the two-way cable system and the terminals would be shared by both the subscribers and by state and local agencies providing services. Any one service would be inexpensive because the cost of the two-way would be shared across hundreds of services. As more and more wired cities appeared, home terminal costs could be expected to decline, and an increasing number of cable subscribers would buy their own terminals to gain access to these services. The result, which was the central assumption of the wired city vision, was that the total cost of individual services would thus become negligible, clearing the way for an infinite array of them.

No cable system has come close to offering the range of services first envisioned. Few of the early projections recognized the enormous costs involved in providing high-quality programs for all channels. Cable operators were for a time expected to encourage and even to subsidize these program services because new programs were anticipated to attract new subscribers, but this hope proved to be unrealistic. Public agencies came forward in a few cities, but then encountered regulatory and organizational barriers that slowed or prevented implementation of new services.

There is sufficient experience with a few services, however, to make it clear that although cable technology does indeed lend itself to the provision of varied and innovative services, it also raises serious questions of equity. A more creative use of conventional telephone technology lacks the rich opportunities for programming possible through cable, but it might serve the disadvantaged and redress rather than exacerbate the inequitable access to services between those who can afford it and those who cannot.

One problem is that market forces and the structure of cable systems place a steady pressure on the development of services wanted by those best able to pay. The experience to date suggests that education and entertainment programs oriented toward the middle class are economically attractive, while new cable services designed specifically for the disadvantaged may not be. Services for the low-income population involve many serious operational problems, require a substantial subsidy, and may still be under-subscribed.

The experience of a Spartanburg, South Carolina cable program illustrates how the cable market affects the services offered. The Spartanburg system has been used to provide, among other services, adult education. The 1970 Census revealed that 62 percent of Spartanburg's adults over 25 did not have a high school education. Starting in 1975, a program was offered via cable by a local technical college to prepare adults for a state examination that certifies students as having the equivalence of a high school degree. The students watch 15 weeks of instruction in mathematics, English, science, and social science. When a teacher asks a question, the students use a simple, eight-button terminal to respond by selecting from among multiple-choice alternatives. The teacher then sees on a display screen the answer given by each student and can respond to students by name to correct or reinforce their learning. The students can also initiate requests for the teacher to slow down, repeat a point, or give an example. The system has proven to be as educationally effective as conventional classroom education, and it is particularly valuable for adults who are barred from completing their high school education because of lack of transportation or a responsibility to care for children or invalids at home.[3]

Unfortunately, the number of students who enrolled in the course has not been sufficient to continue the program. One explanation is that, despite the census data, there is no large-scale demand for this service. This view should be qualified somewhat because the cable system does not serve the poorest areas of the city where those needing adult education are most likely to be found. In addition, some low- and moderate-income housing units were subject to regulations that prohibited the installation of cable. The response to the course suggests, however, that one should be cautious about assuming that there is a large market among the disadvantaged for new programs.

Another reason for low enrollments is that educationally disadvantaged adults living within reach of the cable are less likely to subscribe to it, and if they do, they are less likely to hear about and be attracted by a cable course. To remove the possibility that the monthly subscriber fee might be a barrier, free cable service was provided. It became evident that installing cable service and two-way terminals and—after the course—removing the equipment brought headaches as well as expense. Under the best conditions, it is time-consuming to arrange for installations when there is someone at home, and setting up the two-way equipment involved several visits. The students wanted to retain their free cable service and did not go out of their way to keep appointments to remove equipment. While these problems seem to be trivial, it adds an average of $30 to $40 to the cost of a student's participation. Because the cost of tuition and books of a current conventional adult education class is $31, it more than doubles the cost of the service, as well as adds a nuisance that most cable operators would rather avoid. If cable is to be used for the home delivery of services, it will be most efficient if the terminal equipment can be left in place because the family has a continuing use for the terminal.

For many, this view means that new public service will follow on the heels of pay cable television. Already there are perhaps one million homes in the nation paying from $4 to $10 a month (in addition to about $6 for a regular service) for an extra channel carrying motion pictures, sports, and other entertainment, but pay cable has not yet gone beyond entertainment programming. One problem has been that pay cable is generally offered on a *per channel* basis, i.e., subscribers pay a flat monthly fee to obtain access to a special channel. Without the installation of extra equipment in the home, there is no way to document who is viewing. A public-spirited agency might be pleased about large numbers of citizens watching its program on, for example, consumer economics and home management, but unless it can demonstrate that welfare of other qualified recipients actually watch, it often cannot receive state or federal reimbursement. Consequently, significant new public services are more likely to come with the growth of sophisticated *per program* pay television. The same computer hardware that is needed for billing purposes to charge homes according to what commercial programs they are watching can also be used to monitor what educational programs are being viewed. Service agencies can then provide evidence about usage of their programs, and third-party vendors can be charged for the time that they actually use.[4]

Many in the cable field are closely following the development of Warner Cable's Columbus, Ohio system. In addition to being one of the first major per program, pay cable systems, the technology allows one to limit access to programs. Thus, an agency could offer training for parents of emotionally disturbed children, and only authorized homes with enrolled parents could view that channel at that particular hour. This technical flexibility gives educational agencies the ability to limit access to enrolled students, allowing them to recapture costs through standard student fees. It also allows agencies and groups to use existing program materials that are free if they are addressed to special audiences, but subject to copyright fees if they are made available to the general public. In light of the system's capacity both to control access in this way and to conduct per program billing, events in Columbus will reveal much about the willingness and capacity of local agencies and educational institutions to respond to this kind of opportunity.

As services begin to emerge in pay cable systems, market factors will lead to services, both public and commercial, which are oriented toward middle and upper-income consumers that can afford to pay more than the average $72 a year for basic cable service. Thus, the Columbus, Ohio cable system lists many potential services, but its major market is the middle income consumer. For entertainment and other commercial services, this

Telecommunications Technologies and Services

only reflects pre-existing differences in our society and must be accepted as such. The serious ethical concern over these developments comes, however, from the fact that it will have concomitant effect of exacerbating the gap between the urban middle class and the disadvantaged in their relative access to essential public services. By law, many individuals are entitled to certain minimum services in, for example, basic health care. The nation has made it mandatory for state and local agencies to offer a wide variety of minimum services for the handicapped and other client groups without providing funds sufficient to accomplish that goal. By default the mandated services are offered where it is cheaper to provide them—in many metropolitan areas—and then only to those who seek out and have transportation to the special services. As a consequence, many have looked to telecommunications, and particularly cable, as a means to offer these minimum services, but the distribution of cable systems will be driven by the economic forces behind providing entertainment for those who can pay. Thus, as public services become available on the cable, it will have the ironic consequence of providing greater access to precisely those urban and advantaged groups who already have the greatest access now.

Equity and Outreach

Telecommunications is not the only way of reaching out to citizens, however, so one could argue that equality of access to public services could be served by providing other forms of the same service. Take, for example, the effort of the civic involvement project in San Jose, California. In the Willow Glen school area, a one-way cable system has been used to increase citizen involvement in the governance of schools. When the project began, some 4,600 citizens responded by mail and telephone to register as "televoters." Then on a biweekly basis, the public access cable channel would carry a program reviewing the background of a set of some educational policy issues. Representatives of alternative points of view would discuss the pros and cons of different actions, and structured alternatives would be presented. Recognizing the equity problem, the project staff also spent considerable time and effort to reach citizens who did not have ready access to cable. These same policy choices were printed and mailed to every televoter, as well as printed on the front page of the local newspaper. Outreach in this sense was total—directed to everyone who had shown interest. But despite these efforts to equalize access to the televote system, the participation was higher among residents of the cable area. Thus, the system increased civic involvement, but the relative influence of the same suburban, advantaged citizens who already have the stronger voice in school affairs was increased relative to the central city residents.[5]

These problems of differential access can be avoided in new communities like Woodlands, Texas. They, too, have considered the use of two-way cable to provide citizen feedback, but there every home is on the cable. This has come about because Woodlands is a new town, and cable is installed as part of the original construction. The installation includes sensors that can detect fire and automatically send signals via a central computer to the nearest fire station. The computer would print out the address of the home or business and any characteristic of that structure that would affect the progress of a fire. Thus, the fire department can be on its way well before most citizens would know there was a fire.

This cable system highlights the advantages that telecommunications can offer when community development plans include communications. By introducing this sensor system into new structures, costs are lower, and the equipment is compatible with an overall communications system of services. The fact that all homes in Woodlands have access to these systems, however, only means that equity has been achieved within the narrow context of a specific political jurisdiction. Cable systems still do little to reach those who most need public services. Citizens in the poor areas of the city and scattered across the rural countryside where large cable systems are not economically viable are not now or in the future likely to gain from advances in new cable services unless major steps are taken.

One simple approach is for a city to use its franchise authority to require cable operators to build cable throughout a city. Cable operators would have to recover the higher construction costs with higher monthly subscriber charges, but at least all residents would have access to the cable system. Perhaps a lower subscription rate could then be arranged for hardship cases, but, again, the added costs would be subsidized by the average user. This logic prevails with the telephone, for the higher costs of rural telephone service are in fact partially offset by urban charges (see Chapter 3), and the telephone company offers a basic Lifeline service in some areas for hardship cases with a compelling need to have a telephone. In the past, service agencies have seen the telephone as a luxury and have not wanted to support home telecommunications. If local agencies found they could offer more cost-effective services over cable, they might pay to connect their clients to a communications system. As the telephone experience has demonstrated, subsidy schemes are complex, difficult to regulate, and in the case of cable, would help with only the within-city equity problem. The pressing needs of rural America for service would still remain unaddressed.

The Narrowband Alternative

For that reason, many now advocate telecommunications technologies that are based on the telephone system. Unlike cable, which is "broadband" technology capable of passing full motion video and voice, the telephone line is "narrowband," which is sufficient to carry data and voice signals. Copy machines use this narrowband capacity to transmit facsimiles of papers one piece at a time, so that a copy can be reconstructed at the receiving end. Now "slow-scan" technology can do the same with pictures. As you speak on a slow-scan terminal, your image is trickled through a telephone line and compressed at the other end so that, at regular intervals, a new picture appears. The practical consequence is a moving slide show, where the still picture on your screen periodically changes.

When cable advocates have spoken of the "wired-city," representatives of the telephone industry have grown weary of noting that the city is already wired. So is the countryside. The telephone system cannot be used to reach large numbers of students or clients at the same time and lacks the incredible richness of broadband programming opportunities that cable offers. But the conventional telephone system in some rural areas can now be used to offer some valuable services, as examples that follow will illustrate. Although it is important to note that most rural telephone exchange systems would require substantial upgrading before they could provide many innovative services, the telephone is a far more ubiquitous, and, therefore, equitable, technology.

Currently, one of the major debates in the field of telecommunications services is this choice between technologies for rural areas. A report compiled by staff of the Office of Technology Assessment in 1976 proposed that cable be explored for two-way broadband services in rural areas.[6] In a subsequent OTA conference on rural telecommunications in November of that year, objections were raised. All agreed that many cable systems in the East and Midwest could be constructed to aggregate 10,000 to 20,000 homes by connecting dense pockets of rural towns and villages. These systems could raise substantial revenue by charging consumers for the distribution of over-the-air commercial television signals. This disagreement begins with the recognition that such rural cable systems would require significant support from public funds to operate; some object to an approach that must rely on government support, while excluding homes on farms, along the back roads, and in small hamlets.

In the Western Plains and Rocky Mountain States where it is difficult to aggregate a sufficient number of homes to approach economic viability, the objections are that the concept has no validity, and cable is simply the wrong technology. The argument continues that translators or direct satellite broadcast could be used instead to retransmit over-the-air television signals, and microwave technologies could be used to interconnect agencies in the small towns and villages. Telephone service would reach into almost all homes, and narrowband technologies using the telephone system or radio frequencies are proposed to be much more cost effective.[7]

The rejoinder is that such systems lack the rich programming potential available in the multiple channels of coaxial cable, and that the cabling of the small towns is only an intermediate step. Once the first system has repaid its investment, the systems would extend their wires gradually down the side roads much the way rural electrification and telephone service did. The Rural Electrification Agency is a particularly vocal advocate for this concept and one of the political forces behind the drive for launching a major demonstration of rural broadband cable.

The point to be made is that in choosing technologies, we are often determining which consumers will have access to services. The conclusion is not that the more equitable approach will be selected. Almost every technology, including the telephone, benefited a privileged few before it reached the disadvantaged. When a valuable service can be offered more efficiently or effectively on existing communications systems, it would be unreasonable not to do so on equity grounds alone. But when the technology will require major investment of federal funds to be implemented, or when the system will require continuing support from public agencies, equity should be a guiding concern. In most circumstances, that will mean a preference for narrowband technologies either alone or combined in hybrid systems with more equitable broadband technologies. Since narrowband systems could substantially contribute to new services for the city as well as for the countryside, equity considerations, cost-effectiveness criteria, and common sense all support a more concentrated effort in developing narrowband services.

Public Programming

Just as there are choices to be made about which technologies will be used to deliver services, there are also decisions to be made for many types of public services about how to produce the content they will carry. The current movement is overwhelmingly toward national control of production—broadcast television, satellite, and video player technologies all put a premium on the aggregation of national audiences. This trend in telecommunications contributes, however, to the decline of state, local, and regional institutions. As we shall see, some of these same technologies can also be used to strengthen a sense of community, but not unless the technology is organized to provide interaction among geographically related groups.

The national bias of telecommunications stems from its ability to reach large numbers of people. When first radio and then television became technical realities, many predicted that society would be transformed by new cultural and educational programs. The transformation brought about by broadcasting and a new economics of programming has taken a somewhat different direction. A single VHF television station can reach an average of over two million homes in the top-ten media markets. When stations are tied together in a network, virtually the entire nation can share a single event. Extraordinary amounts of money can be used to purchase events, talent, and production skill, and yet, per viewer, the cost is measured in pennies. When its 1976 advertising revenue is close to a billion dollars, a network like NBC can afford to pay $85 million for the rights to the 1980 Olympics. These advertising economics, tied directly to the size of the audience, lead to programs that appeal—or at least keep the television sets on—in the largest number of homes.

Once it was evident that commercial television would be dominated by light entertainment programs, educational and noncommercial broadcast stations were established in an effort to harness the power of radio and television to more enlightened purposes. In most locations, the television space for the superior VHF frequencies was already taken, and public television was left to the UHF frequencies. This allocation put public television at a permanent disadvantage because its signal transmission cannot consequently be as clear or as distantly received as commercial signals. But even after accounting for this handicap, public television is not eagerly viewed. With the exception of children's programs, shows aired by local public television stations are viewed as much as once a week by less than one adult in five. WNET, the New York educational television station that has the unusual advantage of having a VHF frequency, found that 68 percent of its potential audience never watched the channel; another 26 percent of the homes could name only one or two programs that had been watched by members of the family in the past week; and only 8.7 percent of the adults were regular viewers.[8] These figures do not mean that public television has failed. A major purpose of public television is to create program alternatives to commercial television, so in that sense public television was an early, although limited, success.

Through the establishment of the Corporation for Public Broadcasting, Congress sought to strengthen noncommercial television so that new funds would be insulated from federal interference. Its 1967 Declaration of Policy emphasized the need for programming to be responsive to specific localities as well as to the nation as a whole.[9] Local public stations have each been allocated funds to support a diversity of programs, but the quality of the programs has not been competitive with commercial network productions. Following the logic of national networks, the Public Broadcasting System (PBS) evolved into a cooperative for the joint purchase of programs. Each local public television station that wished could offer a program, and a station could choose through a complex set of preference votes which offerings it would also like to air. In this way, programming reflects the choices of the local station, but most of the economic advantages of network production can be realized.

The evolution of noncommercial television seems likely to continue toward pooling resources and networking so as to aggregate larger national audiences. And there are major social benefits to be obtained from using broadcast television in this way. The most salient example has been provided by "Sesame Street," which has now developed 1,000 original programs reaching millions of children. Not only has it met the need of entertaining children, but it has achieved a most difficult goal—using television to educate them. It has been demonstrated that pre-school children who watch Big Bird and his friends make measurable and significant gains in educational skills.[10] And the program does this for one dollar per viewing child per year.

National Diversity and New Telecommunications Technologies

As alternative means are found to transmit and store electronic signals, diversity of programming will continue to grow. It will be national diversity, aiming at specialized audiences scattered across the nation.

One new means of transmission that has the same power to reach wide audiences and can foster diversity is electronic signals that are ancillary to broadcast. A promising possibility is the use of the broadcast signal as a vehicle for carrying public information and data services. A television signal is a series of

Telecommunications Technologies and Services

frames, and data can be encoded on each frame so that it does not affect the regular television pictures. Television sets can then be equipped with special decoding devices so that messages can be pulled off the air and shown either on the television screen or on some auxiliary device.

The first service to use this technology will likely be captioning for the deaf. The Federal Communications Commission has approved the use of commercial broadcast signals for this purpose. WGBH, the public television station in Boston, now prepares captions for the 11 p.m. ABC television news. The captions appear on the bottom of the screen much in the same way as English subtitles are used in foreign films, so that a deaf adult can both watch the news clips and read the content. In addition to broadcasting in Boston, the program is sent over PBS network to aggregate a specialized national audience of deaf adults who might not exist in sufficient numbers in any one area to justify the program. But all viewers see these captions, both the hearing and the hearing-impaired, and the system is unlikely to grow because the hearing viewer finds the captions distracting. The PBS is now exploring alternative designs of decoder devices so that the homes of the hearing-impaired can be equipped to receive encoded broadcast signals. The day will soon come when, with the flick of a switch, an adult with a hearing impairment can have captions for the same program a hearing viewer sees without the captions.

The same technology could be used for a wide variety of information services ranging from sport scores to descriptions of eligibility requirements for federal service programs. The British Broadcasting Corporation is testing such systems, and other nations have comparable systems under development. In this case, a terminal attached to the television set is used to select the information that is desired from a stream of data encoded in the broadcast signal. When the viewer sets the dial, information such as the local train schedule is shown on the television screen.

Other new technologies, such as satellites and video recorders, wil also support this trend toward a diversification of national programs. Satellites are already in use, and PBS will use one for its network interconnection. The Home Box Office, a programming source of sports and movies for pay cable television, is now using satellite transmission to provide programming to cable systems. Thus the aggregation of specialized, national audiences becomes yet easier. Video recorders will allow television programs to be transmitted like long-play records. One day it will be possible to buy a video recording of a production of Hamlet, a home instruction course in interior decorating, or a report on a new tax law. Electronics thus has the potential of serving highly specialized audiences the way journals and magazines do today. Producers respond to market economics by developing programs that reach the largest audiences. As advanced telecommunications make distance increasingly irrelevant, specialized diversity will prevail.

While few would object to the continued development of these telecommunications services, it should be noted that the net effect of new technologies is to emphasize national institutions and national perspective. It has long been the nation's goal to strengthen regional and local diversity as well. The current system of broadcast allocations was established so that each major city would have its own radio and television stations, as a conscious choice to strengthen local communities. Local origination is still a major requirement for commercial stations. Stations must broadcast programs not provided by the national networks, including "local origination" during prime-time viewing hours. But rather than producing programs locally to meet this requirement, many stations buy program packages from national studios or syndicators. National syndication of programming has undercut the local origination policy as nationally produced—albeit locally transmitted—and game shows and reruns of "I Love Lucy" now characterize the bulk of local television programs.

Developments in noncommercial television also have given the national audience priority. The Corporation for Public Broadcasting has been fairly successful in minimizing federal interference in noncommercial broadcasting, but it has not done as well in strengthening local programming.[11] Perhaps we should ask if there is not a contradiction in purpose in trying to serve both local and national goals. It creates diversity in programming and an important national resource to have quality production in local PBS studios outside New York and Los Angeles, but the current and perhaps irresistible trend at those stations is toward a concern for audiences defined more in national than local terms.

If localism is in fact a national purpose, then more support is needed for interactive technologies. Recent experience suggests that there may be a useful tradeoff between interaction and expensive production. The continuing problem in creating robust local program service is that the American viewer expects expensive, high-quality television. A commonly held view is that the public will not accept or watch mediocre productions, regardless of their content. There is some evidence, however, that viewers are quite attentive when technology allows the viewer to be engaged as a participant.

Telecommunications Technologies and Services

A two-way cable project in Reading, Pennsylvania illustrates this point quite clearly. Cable is used to interconnect three senior-citizen sites so that each can serve as a mini-studio. Each day, senior citizens gather and participate in a program based on local interaction among these citizens. Using cameras at these and other sites, including city hall and similar public agencies, participants see and are seen by each other as well as by the community on the cable system. All programs are produced by senior citizens. Videotapes are prepared to bring information into the programs, but basically all the programming is live. Many programs consider the arts and culture of the area, followed by questions and discussions. In other sessions, the talk among citizens at these centers is the program, e.g., when they question a city councilman or reminisce about the "roaring twenties." The senior citizens are the content of the program; the viewers are sharing in a communication process, not watching television.[12]

A similar two-way cable system in South Carolina used two-way video communications to provide community-based training for the staff of day care centers. Professionals led workshops that switched from center to center so that the staff could be seen as they asked questions and occasionally offered alternative views about child care. In addition to those in centers with cameras, another group of centers on the cable could only watch. Even though they were never seen, they too felt actively involved, for the staff they saw in other centers raised many of the same questions they would have asked.

For both projects, viewers in other cities would not gain as much from the programs. Like everyday human communications, time was occasionally consumed with pleasantries. Friends were recognized who had not been seen in years. They discussed procedures and problems at local agencies. But while these characteristics may have made only fair television, it was good community communications. The legacy of both projects has strengthened ties to the community, both for participants and viewers.

Barriers to New Services

Current experience with social service applications of telecommunications has led to a better understanding of what can be expected from telecommunications. Several projects have proven to be more expensive and less useful than what had once been hoped, producing a healthy reluctance to make sweeping claims about the revolutionary potential of new services. Ironically, even the successes have given rise to a more cautious attitude.

Proponents of applied telecommunications have recognized that projects of proven benefit are often discontinued and rarely replicated. Despite the expenditure of millions of dollars on new telecommunications services, successful demonstrations are often ignored.

One common explanation for this is that organizations that could be gaining from these new services are simply resisting change. Bureaucratic resistance is, however, at best only a partial explanation for the failure of new service uses to spread. Federal, state, and local service agencies err in the direction of avoiding risk, especially the sort of risks created by reorganization and technological innovation. But experience has shown that innovation is difficult for the managers of a service agency, even when they wish to introduce new telecommunications applications.

Barriers in Service Regulations

The barriers are the greatest in areas where a new telecommunications service would supplement, enhance, or substitute for some more conventional service. A preoccupation with potential abuses has far exceeded the desire to encourage innovation in the nation's service system, making it particularly difficult to implement new approaches. Tests of eligibility of clients are strengthened and narrowly focused to minimize risk that an ineligible applicant will receive public funds. Specialized professional credentials are required of service providers of each specific service as a means of ensuring that service quality is maintained. Agencies would rather risk not doing something of value for a potential recipient than to risk providing a service to a recipient that might open them to criticism. The U.S. Congress has put this attitude into legislation by concentrating on the special needs of many different categories of clients, creating at the federal level a system of categorical programs that tie appropriations to very specific services. Former Secretary of the Department of Health, Education and Welfare (DHEW), Elliot Richardson, noted that HEW administered some 300 such categorical programs, and other federal agencies administered perhaps another 300. HEW programs involved approximately 1,200 pages of regulations in the *Federal Register*, and he estimated that there were 12,000 pages of federal guidelines explaining the regulations.[13] Then each state receiving funds had its own interpretative guidelines. Some services are offered by state and local personnel, and these are shaped by a different set of political and professional concerns that lead to further constraints. The net result is that almost any innovation will be hampered. The personnel must have the proper certificates; the service must be as described in guidelines; and the clients must fit the definitions of acceptable recipients.

Perhaps the most interesting experiences with the programmatic limits placed on innovative services is in the field of health—for example, the Boston nursing home project. It has long been evident that health care in many nursing homes is a disgrace. Few homes can afford to hire physicians; visits by private physicians are often infrequent; and it becomes impossible to provide the continuity of health care essential to preventive medicine. The Boston City Hospital sought to meet this need by establishing a system of nurse practitioners who, after special training, worked in the nursing homes as part of a medical team. These nurse practitioners used a telephone to consult with physicians.

The results and the sequel are both impressive. An evaluation of the quality of the health care showed that the telephone was completely satisfactory for this purpose. The health care of patients in the nursing home was improved, and the cost of the system was less than the savings generated through reduced usage of hospital facilities. The savings are heavily affected by variation in hospital costs per day; but using typical Boston City Hospital costs based on an assumed population of 500 patients in the program, the yearly savings would be $365,000.[14]

The sequel is that it looked for a period as though the program might be discontinued. Medicare as a matter of policy will not reimburse nurse practitioners unless a physician is called into an encounter between a nurse practitioner and a patient (so the service can then be seen as an extension of physician services). Thus, the project could not obtain Medicare reimbursement. And it is worth noting that the savings would be realized by the federal Medicare program, not by the hospital or nursing homes. Had the project been located in other states, that may well have been the end of it, but fortunately Massachusetts state agencies could step in. Exceptions to Massachusetts rules were established so that the Massachusetts Medicaid Program and the Rate-Setting Commission agreed to fund the program for their clients. The program has continued for those patients in the program who are qualified for welfare under Massachusetts law.

Ironically, Congress is now considering new legislation that will remove this restriction on reimbursement for "physician extenders." The reasons for the revision have little to do with telecommunications, however. This illustrates that many of opportunities and limits on the uses of telecommunications will result from factors completely outside the communications field. It is fortuitous that the physician-extender amendment will make it easier to move forward in telemedicine, although barriers will continue to appear.

The potential use of telecommunications has also been demonstrated in a system in Maine. Many rural areas are unable to attract physicians, and patients must travel great distances or do without health care. Rural Health Associates, centered in Farmington, is serving Franklin County with three clinics, two staffed with nurse practitioners. Using a microwave link to carry picture and voice, the telecommunications system allows a supervising physician to see as well as talk with the patients and nurse practitioners. This capacity strengthens the physician confidence in his or her diagnostic judgments and the results of the prescribed treatments. Sometimes a patient must come in to the central medical facility, but the remote clinic with its telecommunications link has demonstrated that it can provide quality health care at an acceptable cost for a significant proportion of medical problems. The system as a whole has also had the important consequence of attracting and keeping well-qualified physicians in the area.[15]

The difference between these two health projects is instructive. The Maine project has not had to face the financing crisis; the Rural Health Associates in Maine is a group health arrangement supported initially by funds from the Office of Economic Opportunity and now from HEW. Patients with income below the poverty level are enrolled in a pre-paid medical plan. When Rural Health Associates can obtain Medicare reimbursement for services, it does so, but HEW pays the difference in service costs. Because of this mixed system and its ability to go outside the Medicare system, it has much better control over the use of funds for telemedicine. This is not to say that there were not severe problems in meeting federal requirements, for there were. But without the legislation that gives latitude to prepaid health care organizations and the new organization established by Rural Health Associates, the telemedicine link might never have been established.

Telecommunications as a Threat to Service Providers

Service providers can also be a source of resistance to enhancing or replacing conventional services. Even if agency managers are able to put a new form of telecommunications in place, it does not necessarily follow that their staff will make the system work. In some cases, professionals resist the increased scrutiny of

Telecommunications Technologies and Services

their day-to-day activities that is often a concomitant circumstance of telecommunications systems. In other situations, professionals resent a loss of control over schedule or administrative procedures that comes with a communications system. The greatest problems arise when new systems threaten in some way to replace the service provider.

The experience of educational television illustrates some of these points. The rhetoric of instructional television refers frequently to the fact that televised teaching is not as effective as a good teacher in the classroom. But it is said to be better than a poor or mediocre teacher. Thus, the power of educational television is that it makes the exceptional teacher available to vast numbers of students. Since television can reach these students, state educational agencies can afford to support exceptional teachers in the preparation of detailed lesson plans and scripts for each class. The programs portray these teachers at their best.

State legislatures were persuaded by this argument, and state educational television systems sprang up around the country. To win political acceptance, it was necessary to eschew language suggesting that teachers would be replaced in order to gain support of the educational community. The result was a supplemental system of educational television where the teacher became the user. The teacher continued to make the final decision on the classroom curriculum, whether or not available television programs were part of it. After a decade and the expenditure of millions of dollars, one-quarter of the schools in the nation did not have a single television set to view instructional programs. Based on a national usage survey, perhaps one school in twenty had a sufficient number of sets in classrooms to be able to take reasonable advantage of instructional television.[16]

The reasons for this failure are in large measure attributable to the refusal of teachers to surrender their professional autonomy and responsibility in the classroom. In some ways the resistance is warranted, for teachers know the instructional needs of their classes better than the producer at the educational television (ETV) studio knows them. To surrender control of the pacing and schedule of instruction is to abrogate responsibility to the students. But teachers are not usually prepared to admit whether or even when ETV instruction is better than standard classroom instruction.

South Carolina has one of the most advanced and extensive systems of instructional television in the nation.[17] The studies are centrally located in Columbia, and programs are transmitted over a network from ETV stations in all corners of the state. Already 80 percent of the state's elementary and secondary schools can receive instructional programs every hour of the school day. The instructional programs are generally of high quality and are used by several other states. Yet in 1974, at the high school in one of the most innovative school systems of the state—with an administration that actively encouraged teacher use of ETV programs—the teachers almost never used the programming in their classes.

Two years later, educational television had become a mainstay of the educational curriculum in that school system because the technology became responsive to teacher concerns. At considerable expense, the single broadcast channel available was supplemented by an additional five channels of programming to the secondary schools. A recording studio in the high school could tape and replay a program on the day and time of a teacher's chosing. Advanced French language instruction, for example, could be carried at several different times, making class scheduling much simpler. The total system returned the control of timing and the structure of the class curriculum to the teacher, and the results were dramatic. In the high school, ETV usage went from a situation in which a television set was rarely turned on to one where 40 television sets were given more or less continuous use.

The important point is that telecommunications will not be used in existing service areas unless it is designed to be responsive to teachers, caseworkers, physicians, or whatever the relevant professional might be, as well as to the students, clients, or patients. When the technology fails to provide some control for the service provider as a "user," it will be considered a threat and be resisted. The first designs of new service applications frequently fail to provide that flexibility of control because it is usually both less efficient and more expensive to do so. As a consequence, the introduction of new systems is frequently associated with a painful learning period and, often, unanticipated additional costs.

Economies of Scale and Increasing Problems

Many new telecommunications services become attractive only when they serve functions carried out by several agencies. By sharing the function's costs, telecommunications applications become both economically viable and add in turn new capabilities to the total service system. But the problems are then multiplied as each agency sees the function differently, has different regulatory latitude, and different professional concerns.

Telecommunications Technologies and Services

Most observers of the American system of service delivery agree that the client is the loser. The fragmentation resulting from categorical programs and a federal system of services separately provided in the same locale by federal, state, and local authorities has resulted in considerable confusion. One HEW document reported that generally there are several hundred public and private service organizations in an average city. Each agency has detailed guidelines that dictate who they can and cannot serve and what services are to be provided. Under such circumstances there is a continuing need for an effective information and referral system so that wherever the client enters the system, accurate information will be available about agencies and the specific services they provide. Then there is the further concern of whether, once referred, the client ever appears. Clients with more than one service need who must combine services from more than one agency can get lost in the system. Effective follow-up can insure that the client is receiving a meaningful package of services. Data management systems using any of several forms of telecommunications seem ideally suited to these tasks.

These problems were recognized by city, county, and service agency officials in Chattanooga, Tennessee. A survey found that there were over 100 agencies who sought to provide human resource services in the metropolitan areas, and that as a general rule the providers of services at one agency had only the vaguest notion of the location, hours, and eligibility requirements of most other agencies. In the 1960s a telephone-based Human Services Management Information System, using terminals and display devices, was established to link 98 public and private social agencies. The system used one standard application form, reducing administrative overhead and permitted standardized procedures for matching client needs with agency locations and services in an area containing 14 counties.

In addition to greater efficiency, more effective follow-up, and increased continuity of service delivery, the system also yielded a substantial cost savings. Fourteen federal agencies in and around Chattanooga had operated vans to provide transportation for their clients. The average cost had been $2.93 for each passenger mile. The telephone-linked system permitted the agencies to pool their vans. Another conventional technology, mobile radio, was used to dispatch vans to pick up clients and deliver them to agencies. Instead of an agency covering a wide area with a few vans, the vans could now collect neighboring clients even though they were going to different agencies. The cost per passenger mile dropped to less than 60 cents.[18]

Clearly one lesson of the Chattanooga system is that existing, ordinary telecommunications technologies permit substantial improvements in services. The promise of the future is for a telecommunications environment that is richer in new service opportunities, and the Chattanooga case could be a beginning. But another lesson is derived from events that followed the widespread discussion of Chattanooga's experience. The technology was inexpensive and available; the service problems that it could solve were evident; and the solution was clearly demonstrated. But no other metropolitan areas have successfully copied the Chattanooga system.

An attempt to set up a similar system for agencies in and around Greenville, South Carolina is illustrative. The County Health Department there did a design and feasibility study of an automated data file that would allow personnel at one agency to set appointments for their clients at a second. Each agency would have a terminal and display that would allow, for example, personnel at the Society Security office to check to see if there was an opening at the nearby food stamp office schedule for a client to save an additional trip into town. The computer would automatically print client-reminder letters to reduce the high rate of "no-shows," which creates major inefficiencies. A follow-up routine would allow service staff to determine whether a client had received services from other agencies to insure that the client was receiving consistent service. A transportation system was forecast as a second step, and the savings and other advantages of the Chattanooga system were frequently referenced.

A grant was obtained to support most development of the program, and arrangement was made to lease the terminals with an option to buy. After a year of negotiations and demonstrations, only ten agencies were prepared to go forward. Some objected to the necessary changes in the forms and procedures and doubted that the federal regulations permitted the local agencies to make those changes. Others denied the existence of the problems or doubted that the system would be worth the cost. Key agencies that did most of the referring decided that they would not use their operating funds to pay the necessary $332 a month for the terminals. For lack of support from agencies (particularly two central agencies), the remainder of the system was too expensive and not sufficiently comprehensive and has been abandoned. When it is necessary to aggregate functions across service agencies, the regulatory, professional, and organizational barriers are in turn multiplied.

The conclusions based on these and other project experiences suggest a variety of lessons. First, telecommunications service innovations will come neither easily nor swiftly. Innovations that seek to improve existing services may require new federal legislation, or, at a minimum, waivers must be obtained on a service-by-service basis from federal and state regulations. Telecommunications services may require new organizational structures to be used effectively. They will always require substantial periods of planning and education of the relevant personnel. The corollary that follows from these observations is that progress will be glacial without forceful leadership at the national, state, and local levels.

Another conclusion is that new telecommunications services have been proven to be effective at acceptable cost. If continued innovation is fostered, many groups in our society would benefit substantially. The error of the past has been undue optimism, which led to an inadequate consideration of professional, organizational, and regulatory problems and to an exaggeration of the immediacy and magnitude of the expected benefits. The current vogue is pessimism and doubt that any change will ever come, and many technologists who felt that it should be much more simple and rational to introduce telecommunications have retreated from the field. The truth lies between. Just as continued effort to establish new telecommunications services must be informed by a realistic understanding of the barriers, it should also focus on the real benefits that can be achieved.

The Federal Role in Developing Telecommunications Services

Some recommendations flow easily from this discussion. In general, of course, as a nation we must recognize the social consequences of different technologies and guide our policies accordingly. Several specific areas would seem to warrant federal attention. At a minimum, the federal government should remove the existing barriers it has created to the development of telecommunications services. Within the realm of communications, attention should be concentrated on removing barriers to the growth of new systems. The promise of two-way cable suggested in this paper, and the absence of compelling legal or economic grounds for the current restraints on cable discussed by others in this volume leads to the view that the current trend toward cable deregulation should be encouraged. The constraints on the translator industry have also been relaxed, and here the result will be to increase the variety of technical opportunities for reaching rural populations. The prevailing trend in Washington, D.C. is apparently toward deregulation in general, and there is little more to be said than that these actions will help new services emerge.

In the absence of positive action to stimulate new telecommunications services, the future will bring a diversity of national programming. Diverse commercial programs aimed at increasing specialized audiences seems inevitable. The question now is only which of several technologies will convey the programs, and the market appears quite able to make that determination. Public service applications will appear, but progress will be slow because of limitations in service agencies outside the communications industries, and as they come into use, these new applications will in general worsen rather than remediate inequities to access to services. And if the nation wishes to strengthen local perspectives and institutions, then action is indeed required. Federal initiatives should thus be considered on two fronts: (1) to foster the capacity of federal agencies concerned with service delivery to plan and to implement equitable and efficient uses of telecommunications, and (2) to strengthen the capacity of local governments and service agencies to use telecommunications both for the enhancement of community awareness and information and for the delivery of local services.

Strengthening Service Agency Capacity for Telecommunications Services

The barriers to developing public telecommunications services and local programs are formidable. Moreover, any recommendation for government action immediately encounters a dilemma in dealing with these problems. Federal support of telecommunications services is fragmented throughout the federal establishment to an extent that inhibits, or perhaps prohibits, the development of an aggregated systems view of services. But any effort to create an agency with cross-cutting responsibilities for telecommunications services is neither politically feasible nor particularly practical.

The fragmentation is evident. The agencies that might do the most to support telecommunications systems are the Department of Agriculture and the Department of Housing and Urban Development. Their components charged with regional and community development could make a major contribution to both localism and public services by recognizing the essential role a communications infrastructure plays in economic and social development. But they have been reluctant to view telecommunications in this light and clearly have not been charged with supporting such programming. HEW has an Office of Telecommunications that concentrates more on frequency space questions. It did help start the ATS-6 satellite project, but that responsibility has moved to the National Institute of Education. HEW also has an Educational Television Facilities program, which now has difficulty exercising its relatively broad services mandate from within the Office of Education and may be moved. Each mission agency in the federal government has one or more administrative units that have investigated some aspects of telecommunications services, but very few have continuing programs in telecommunications that learn from both success and failure. The telemedicine work at the National Center for Health Care Research and Development and the Lister Hill Medical Library have been exceptions, but their work alone is unlikely to influence issues outside the health field.

Neither these nor other organizations have the capacity to deal with the fundamental barriers to new telecommunications services. Currently there is no organization charged with or capable of following, much less guiding, the developments in more than a single-service field. That is not by definition bad, because federal direction is not always beneficial. There are many directions that can and should be tried with new telecommunications services, and it would probably be both impossible as well as foolish to pursue a single coherent plan of development. But it remains that many of the gains to be achieved with telecommunications will come only through serious revision and/or consolidation of existing services, and that will not occur without federal leadership.

Simply put, many service agencies feel they do quite well without telecommunications. A health official concerned with Medicare fraud and the problems of maintaining the integrity of the entire health system is not going to be enthusiastic about telemedicine. In the early 1970s, the education component of HEW artfully avoided participating in a program that only suggested it would want to do research on combining services. The organizational, professional, and regulatory barriers to integrated telecommunications are supported, not opposed, by the structure and inertia of the federal government. Without an effective counterweight in the federal government, little is likely to change.

But what form would that counterweight take? At a minimum, somewhere in the federal government there must be an office or division charged with looking at communications services across agency functions. It could take on investigations for other agencies so that they would have access to expertise in telecommunications. But the organization should be structured so that there is a continued focus on the content of the services, service regulations, service staffing, and opportunities to aggregate services. If nothing else, a stable unit that merely informed one agency about the past experiences of others would be a useful addition to the federal scene.

The various service programs themselves must continue in the appropriate federal agencies. While it could be valuable for a telecommunications services group to have grant authority to fund the integrative functions of projects that traditional federal agencies would or could not, it is important that it cooperate rather than compete with established agencies in supporting telecommunications services. Whatever programs and projects are established to offer new services will be far more useful if they endure when successful. And success comes from using existing agency channels and drawing funds from ongoing programs.

Another initiative at the federal level that could help deal with service barriers would be a systematic assessment of regulatory barriers and the establishment of a waiver system. These activities should be supported by the major federal agencies. Work has already begun in this area by an inter-agency task force to consider a rural telecommunications demonstration. Inter-agency committees have considered communications services in the past, however, and these efforts usually become a rather useless, symbolic gesture. Washington D.C. has countless coordinating committees that have no effect on the agencies that are represented by them.

This brings us to the need for the rural telecommunications demonstration. Whether such a demonstration is launched, and whether it succeeds or fails, it may be a valuable vehicle for forcing agencies to cooperate in communications system planning. It seems appropriate to create an inter-agency group with its own staff (or supported by the new office concerned with services) that can study the feasibility of implementing communications services in several specific sites. The barriers can thus be cataloged and solutions found for real problems at existing sites. Inter-agency agreements in general are rarely as compelling or as informed as agreements on shared services for actual recipients.

Of course, this purpose could be equally served by an urban demonstration, but the rural area should be given priority because of the equity consideration. If the nation is to invest substantial funds in telecommunications activities, it should do so in areas that are in the greatest need of assistance. Such a demonstration could entail a survey of the adequacy of rural telephone systems, spur the development and application of narrowband terminals using telephone and other carriers, and encourage the use of translator and microwave to provide services. These steps, coupled with setting precedents in waiving service regulatory barriers, would make it far easier for rural uses of telecommunications to develop.

The Local Public Telecommunications Organization

Thus far, these conclusions have focused on the national level. There will be little progress in the public service sector, however, until there is a widespread organizational capacity at the local level to deal with telecommunications services. For years, those believing in the opportunities of advanced cable systems have placed the burden of prompting new services on the cable operator. Some have responded, but for the most part, local cable personnel lack the expertise to deal with local agencies and have little incentive to become embroiled in their problems. Service agency personnel tend to be suspicious of the cable operator and the profit motive generally. There is such virtual unanimity among those who have worked at the local level with new telecommunications services that, with rare exceptions, a third party must act as a catalyst.

Localism is another factor that suggests a broad role for this local organization. As networking and national programming grow, the local perspective will receive less attention. The local broadcasters may continue to offer local news, weather, and sports, but there will not likely ever be a return to the days of a local television broadcaster who produces substantial hours of programming each week in local studios. If we are to realize the goal of widely available local programming in depth on the culture and public affairs of the state and community, it will again be provided by a local organization broadly interested in public telecommunications. Like the services provided by local agencies, this programming would use interaction heavily as a means of involvement.

There are many existing organizations that could evolve into this local public telecommunications resource. A local college or school district, or a consortium of schools and colleges, could quite reasonably assume the role where the tradition of continuing education is strong. A group working in the office of a major in a metropolitan area might be a good location for this activity. The public broadcaster is another.

Public broadcasting as we know it will come to a severe challenge. The diverse national programming for specialized audiences that will accompany the 1980s will offer intense competition for audiences that public television now considers its own. While some of the public television stations clearly have the production capability to meet this pressure, many may not. And even if they could, it is not evident that funds from the public treasury, much less a special-use tax, should be used to support a decentralized system of noncommercial programming if a commercial system can provide many of the same cultural, educational, and public affairs programs supported only by direct user fees. Satellite interconnection of public television stations with multiple-channel capacity will lead naturally to a robust public television network, and its governance and programming decisions about the selection of national programming would have "local" guidance from the various stations. But without a goal of localism, most public broadcast stations will have very little purpose beyond serving as transmitters.

There is no barrier to local public television stations taking a broader view of their role except their own self-definition as broadcasters. Each community can use a local organization sensitive to its needs, capable of producing programs, and able to guide and encourage the use of telecommunications by local public agencies. The new Carnegie Commission could envision a broader purpose for a redefined system of public telecommunications. Such a system could include the current public broadcasting by perhaps 10 or even 20 production centers and also provide state and regional centers that would demonstrate the new services made possible by the full array of telecommunications technologies.

For the most part, there is no way to avoid new telecommunications services. The future will see rich opportunities being realized as our society eventually learns to exploit more fully the power of communications technologies. In light of the serious economic, organization, and regulatory problems that need to be resolved, the introduction of new telecommunications services will come far slower than many would hope. The question here is thus whether society should seek to influence the direction and pace of change. For those concerned with existing inequities in access to public services, or with the strength of local institutions, or who simply feel that the government should actively promote earlier realization of this potential, new federal initiatives and new organizations are clearly called for. And it seems only prudent to establish the capacity to assess, if not guide, the way new telecommunications services will shape our society and its institutions.

Notes

1. James D. Livingston, *CATV Networks and Pay-TV, Feasibility and Prospects,* Knowledge Industry Publications, Inc., New York, 1975.
2. See, for example, National Academy of Engineering, Committee on Telecommunications, *Communications Technology for Urban Improvement,* Report to the Department of Housing and Urban Development under Contract No. H-1221, Washington D.C., June 1971; Sloan Commission on Cable Communications, *On the Cable: The Television of Abundance,* McGraw Hill, New York; and Ralph Lee Smith, *The Wired Nation—Cable TV: The Electronic Communications Highway,* Harper and Row, New York, 1972.
3. William A. Lucas, "Moving from Two-Way Cable Technology to Educational Interaction," *Proceedings of the National Telecommunications Conference,* Dallas, November 1976.
4. This capacity also raises serious problems concerning individual privacy. See the discussion by Benno Schmidt in Chapter 5 on privacy and return path communications.
5. Vincent M. Campbell, *The Televote System for Civic Communication: First Demonstration and Evaluation,* AIR-38700-9/74-FR, American Institutes for Research, Palo Alto, California, September 1974.
6. U.S. Congress, Office of Technology Assessment, *The Feasibility and Value of Broadband Communications in Rural Areas, A Preliminary Evaluation,* OTA-T-33, Washington, D.C., April 1976.
7. *Evaluation of Two-Way Telecommunications Systems for Interaction Between Professionals for Delivery of Social Services to Rural Populations,* Practical Concepts Incorporated, Washington, D.C., undated.
8. Jack Lyle, and Donna Ellis, *A General Analysis of the Audiences of Public Broadcasting in the New York Area,* CPB/OCR 211, Corporation for Public Broadcasting, Office of Communications Research, Washington, D.C., 1973.
9. Public law 90-129, 90th Cong., S. 1160, November 7, 1967.
10. Gerry Ann Bogatz, and Samuel Ball, *The Second Year of Sesame Street: A Continuing Evaluation,* Vols. 1 and 2, Educational Testing Service, Princeton, New Jersey, November 1971.
11. R.J. Blakely, "Rethinking the Dream: A Copernican View of Public Television," *Public Telecommunications Review,* Vol. 3, No. 5, September/October 1975, pp. 30-39.
12. Mitchell Moss, Jacqueline Park, and Red Burns, "Experiment I, Public Service Uses of Interactive Television, New York University/Reading, Pennsylvania," *The Access Workbook,* Al-

ternate Media Center at New York University School of the Arts, January 1976, pp. 1-17.

13. Department of Health, Education and Welfare, *Interim Report of the FY 1973 Services Integration R & D Task Force*, internal report, Washington, D.C., 1973.

14. Roger C. Mark, et al., *Final Report of the Nursing Home Telemedicine Project*, Boston City Hospital, Boston, Massachusetts, July 15, 1976.

15. R. Bashur, *Rural Health and Telemedicine: A Study of a Rural Health Care System and Interactive Television*, Department of Medical Care Organization, University of Michigan, 1976.

16. Dave Berkman, "Instructional Television: The Medium Whose Future Has Passed?" *Educational Technology*, May 1976, pp. 39-44.

17. Ralph Lee Smith, "A State ETV Network: What's in It for Higher Education?" *Planning for Higher Education*, Vol. 5, No. 1: 7/7, February 1976, pp. 1-8.

18. "New Federalism V/Coordinating Services," *The National Journal*, Vol. 5, No. 9, March 1973, pp. 305-311.

8.
Communications for a Mobile Society

Raymond Bowers[1]

This chapter concerns land-based communications systems involving mobile units in vehicles or carried by individuals.

Land mobile communications deserve a special place in any assessment of the future of electronic communications in the United States. During the past 35 years, many important but specialized applications of mobile communications have been developed for police, safety services, and commercial purposes. But these uses represent only a fraction of the potential. A major expansion appears near, not only within these specialized areas, but also in a broader range of uses for commercial, public, and private purposes. Taken together, these uses of land mobile communications could have an extensive and significant impact on our society.

Our highly mobile society is characterized by long trips between home and place of employment, a large amount of intercity travel on an advanced highway system, a high degree of dependence on trucks to transport goods, an extensive public service delivery network that uses cars and other vehicles, and an elaborate network for providing commercial services to homes and offices. Electronic communications already are involved in many of these areas.

In 1974 there were approximately 105 million automobiles and 25 million trucks and buses in this country, most of which had standard radios. The recent acceleration of Citizens Band (CB) radio sales and other forms of land mobile communication raises the question whether deployment of two-way devices could become as widespread. Seen in a broader perspective, eliminating the wire leash from the telephone—the most common two-way communication device—adds new dimensions to the usefulness of electrical communications. The consequences could be profound.

The attempt to make electrical communications mobile is more than 100 years old. Once the telegraph became important in the evolution of the railroad, the need to communicate between the fixed stations and the moving train was recognized. Wagon trains carried the first portable field telegraphs during the Civil War, but substantial freedom of movement was not possible until Marconi invented a wireless method of communication.

By 1895, when the first wireless telegraph signal was transmitted and received, the limitation of wired systems was apparent and the need for wireless communications widely understood; there was none of the skepticism that had greeted the telephone's invention. Early work in radio communication focused primarily on the need to communicate with remote mobile units, such as ships, and with less accessible locations, such as islands, where wires were not practical. Some of the earliest applications of the wireless demonstrated its importance in calling for emergency assistance, and the sinking of the Titanic in 1912 brought this use to the attention of the population at large. This event led the Detroit police commissioner to suggest that his department develop a radio communication system. Thereafter, the police and the military played a significant role in the evolution of the new technology. By the late 1920s, police confirmed the utility of two-way radio systems in apprehending burglars, car thieves, and other criminals, and the use of such systems expanded rapidly during the 1930s. Other sectors adopted mobile communication technology during the late 1930s and 1940s; rapid growth in industrial and commercial uses began in the mid-1940s. Expansion after World War II was attributable partly to technological improvements introduced by the military during the war and to the return to the civilian sector of veterans with considerable experience and contact with the technology.

Land mobile systems represent a relatively recent development in electrical communications. Although extensive use of the telegraph can be traced back more than 125 years, and of the telephone more than 75 years, two-way land mobile communications began to spread extensively throughout our society only about 30 years ago. In this respect, land mobile systems can be compared with television, though their dissemination is far less. Despite rapid growth in recent years, land mobile devices still are relatively rare compared to the telephone, radio, and television. The potential for massive growth is real, however, and precursors are already apparent. This is why any survey of the evolution of telecommunications in the United States must give special attention to the land-mobile sector.

The Technology: Its Services and Uses

The Technology

The distinction between one-way and two-way mobile communication systems is important. In a one-way system, signals can be transmitted either from a central base station to mobile units, or in the reverse direction. This technology is used most commonly for paging systems in which a signal from the base station can activate a tone or light, or convey a short verbal message to the mobile unit. But the user cannot respond via the receiving unit.

Two-way systems permit conversations between the base station and the unit. The most common type is known as a conventional system. In its simplest form, the base station and the mobile unit operate on the same frequency, which often is allocated for exclusive use by a single organization, and frequently those conversing must take turns using only one voice path. More than one channel may be available, but in the simplest systems a free channel is selected manually. Conventional systems of varying complexity are used widely in taxis and in commercial delivery and public service vehicles. Because of increased demand for these systems, much effort has been devoted to improving their efficiency and convenience. Approaches used are characterized by 1) sharing blocks of channels; 2) frequency reuse, or the repeated use of the same frequency in separated areas; and 3) sophisticated message-transmission techniques, often involving digital format.

Multichannel trunked systems (MCTS) are a good example of the first approach. Users of these systems share a block of channels and, when they wish to communicate, search among this "bank" of channels for a free one. This arrangement leads to steadier and more efficient use of the channels than would be possible if each were dedicated to a single organization or user. As a result, either waiting time can be reduced or the number of users increased.

The more advanced MCTS systems select and assign channels automatically. Although they can provide all of the services associated with simpler conventional systems, at the moment the more advanced systems are being used only for mobile telephone service. Such systems allow relative privacy because each simultaneous communication occurs on a different frequency, but the number of users is limited by the number of channels assigned to the system, and capital costs are greater than those for the simple conventional systems.

One of the most advanced systems exploiting frequency reuse is the cellular system,[2] in which a large service area is broken up into cells. Within each cell, a subsystem operates similar to that of the multichannel trunked system. The cells may be from one to twenty or so miles in diameter, and the MCTS transmitters in adjacent cells operate on different sets of frequencies. The major purpose of the cellular system is to increase mobile communication capacity within a given spectrum allocation. Short-range transmitters and small cell size permit reuse of allocated frequencies; cells separated by a sufficient distance can use the same frequencies. Moreover, small cell sizes are compatible with the limited range and unfavorable propagation characteristics of the frequencies now being made available near 900 MHz for mobile communications. The base station within each cell is connected to the wireline telephone network through a switching system, thus providing a direct link with any fixed telephone.

In addition to the basic components of MCTS, a cellular system requires facilities for determining in which cell a mobile unit is located. It also must be able to switch receiver/transmitter frequencies and transmitter stations automatically as the mobile unit moves from one cell to another. The major advantage of the cellular system is its enormous potential capacity, measured in hundreds of thousands of users per system and thus millions nationwide might provide the means for deployment on the scale of the fixed telephone. The capacity is proportional to the number of channels assigned, but it is also inversely proportional to cell area; as cell size is reduced, the number of cells in a given service area is increased, thereby increasing frequency reuse. Additional advantages are privacy and virtually unlimited effective range through interconnection with the standard telephone network.

As a result of recent regulatory decision, the Federal Communications Commission favors using FM cellular technology to accommodate widespread demand for more mobile services, but the use of other techniques has been suggested that also may increase subscriber capacities. For example, narrowband, digitized voice systems constitute an alternative approach. Such systems are technologically feasible because of recent advances in signal processing and solid-state technology. Technology also is being developed that may utilize computation to reduce present spectrum requirements for conveying both analog and digital information. Linear predictive coding techniques make intelligible communication of speech possible over digital systems, with lower bandwidth requirements. Yet some authorities argue that synthesizing speech systems would create voice transmissions inferior to those utilizing telephone channels.

Digital voice equipment is more expensive than FM equipment, but it is likely that the cost differential will decrease. Because new digital techniques are expected to reduce bandwidth requirements even further than presently possible, it has been argued that installation of digital transmission systems would allow new advances to be accommodated more easily. Such technological alternatives will have to be considered as land mobile systems evolve.

Thus, a number of methods being proposed would increase the capacity of land mobile systems significantly, including some more speculative methods based on satellite transmissions that we have not mentioned.[3] Some argue that the cellular method may not be needed because the alternative methods may be more economical in the long run, resulting in a controversy over whether we even should deploy this advanced technology and, if so, in what form. However, there is no controversy surrounding its ultimate capacity to accommodate a very large number of users.

The Services

For clarification, the following distinctions are made among the four kinds of mobile communication services that can be offered, because each has different characteristics and requires specialized technology. These include:

- Paging service: One-way communication service that permits an extremely brief message to be transmitted to a

recipient. Frequently this message is no more than a signal to the recipient to call a prearranged number, though simple voice messages are possible. Some paging systems are connected with the wireline telephone system, permitting direct access to pagers from a telephone.[4]
- Dispatch service: Service requiring a simple two-way system and frequently used by taxis and ambulances, as well as various units of commercial organizations. The messages are brief and often consist of instructions concerning deployment. A few channels can accommodate a very large number of users.
- Mobile telephone service: A means of communication with characteristics comparable to those of the fixed telephone. Interconnect with the fixed telephone system is essential. The conversation can be extended, and users have a higher degree of privacy than with other two-way services.
- CB service: Service based on a broadcast system of limited range. Use of the service is not restricted to a particular class of users. Essentially, it provides a large number of broadcasters with a "party line."

The amount of spectrum required per user varies greatly among the services. If the service is characterized by short message lengths and long wait times for access to a channel, the spectrum needs will be less. Consequently, of those systems that limit access to the service (i.e., all but CB), paging service requires the least spectrum per user and mobile telephone service the most.

The differences in spectrum demand are substantial. For example, in noncellular systems, a block of 1 MHz of spectrum could accommodate about 200,000 five-digit paying units, nearly 7,000 conventional dispatch users, but only 650 mobile telephone users with trunked systems and present-quality parameters inferior to those of the standard telephone. If one wished to provide mobile telephone service with characteristics comparable to that of the fixed telephone, the number of possible users per MHz of spectrum would be even less. CB radio accommodates a very high number of users per unit spectrum, but the bands often are highly congested.

The amount of spectrum allocated to a given service not only determines whether the service can be provided, but also affects the cost of the units. More spectrum can be exploited to serve users with simpler technology, and this can lead to lower direct costs. The trade-off between spectrum conversation and system cost should be considered in the assignment of spectrum.

The Uses

The services are used for many purposes and it is not possible to define a set of functions that are mutually exclusive. Nevertheless, it is useful to make some partial distinctions. The following utilization categories are self-explanatory: paging, vehicle monitoring, emergency beaconing, data transmission, and general conversation. The use of mobile communication devices to redirect personnel and vehicles after they have left their base of operation, thus reducing response time of field units, is particularly important and is known as dynamic routing.

The Service and Manufacturing Industry

About 700 independent radio common carriers supply mobile telephone service through interconnect with the wired system. Many of these companies are small and have typically provided service for only 90 subscribers per system. Wireline common carriers such as AT&T and GT&E also operate 1,300 mobile telephone systems, and these systems are also small; about half of them are operated by AT&T companies. Paging service also is provided by wireline companies and the radio common carriers.

If it were not for regulatory constraints, the mobile communications service industry could be competitive. However, firms providing two-way service have had insufficient spectrum available to be able to exploit potential economies of scale. The growth of common carrier dispatch service has been limited; most companies desiring this service have had to operate their own systems. Privately owned paging systems are also very common.

In the equipment manufacturing sector (excluding CB equipment), four firms (Motorola, GE, RCA, and E.F. Johnson) accounted for 90 percent of the output in 1973. Motorola dominates the domestic industry; it produced nearly two-thirds of the equipment in 1973.[5] One cannot conclude, however, that there are economies of scale inherent in the industry that account for this high degree of concentration—several small firms with market shares of less than 3 percent have been able to survive. Much of the CB equipment is manufactured by foreign companies.

The industry (particularly the service sector) has faced a number of difficulties in responding to increased demand.

Limitation of spectrum has been a serious constraint, and the FCC's new allocation of spectrum only recently gained final judicial sanction. Technical problems with the new high-capacity systems remain to be solved. Also, uncertainty about the effects of recent regulatory decisions probably has made the industry and its investors cautious about committing major resources.

Demand and Growth: The Social Pull

Demand is growing for the various services. In some cases this demand is being met by increasing the number of systems, but in others the level of unsatisfied demand is rising.

The convenience and inexpensiveness of paying systems is leading to an increase in their utilization. The number of units in operation increased from 240,000 in 1972 to more than 512,000 in 1975. Of the latter, 11 percent were used in services provided by a wireline carrier and 62 percent by a radio common carrier, while 27 percent were associated with private systems. Projections suggest that 1.5 million units will be in use by 1980 and nearly 3 million by 1985.[6]

Use of dispatch systems for police, fire, public utility, business, and local government communications also has grown rapidly since the basic frequency allocation was established in the 1940s. The total frequency allocation was approximately 40 MHz, distributed among the number of bands, until 1972 when it was increased fourfold by the FCC. A combination of social and economic factors has contributed to an extremely large and even explosive demand for land mobile services. Excluding CB, the number of transmitters in operation grew from 150,000 in 1949 to nearly 3,000,000 in 1970 and 6,000,000 in 1974. As mentioned earlier, use of the frequencies has intensified as a result of more sophisticated technology, but the growth in demand (projected by the FCC to reach more than 12 million units by 1980) continued to increase congestion within the allotted 40 MHz.

Demand for mobile telephone service has also been strong, but in many respects less effort has been devoted to this demand than to other systems. The spectrum allocated to this service has been modest (a total of 3 MHz); such limitations and the consequent restrictions on the number who could be served have made it difficult to respond in some areas other than by assigning priorities. For these reasons, mobile telephone service has not been promoted by the vigorous salesmanship characteristic of other services, and in many areas there is a long waiting list for service. Because of the high loading of channels, the quality of the service provided has been inferior to what is technologically possible if more spectrum were used. Widespread mobile telephone service of a quality comparable to that of the fixed telephone is now possible with the additional spectrum allocated in recent regulatory decisions. Estimates of the number of mobile telephones in use range from 100,000 to 130,000, totals that have not changed appreciably during the past few years; estimates of recent growth rates vary from 1 to 6 percent per year. The market is shared about equally by the wireline companies and the radio common carriers.

The most explosive increase in demand has occurred within the CB service. Its extraordinary growth has attracted national attention; journalists, businessmen, and policymakers eagerly are attempting to understand the phenomenon. They are asking why 980,253 persons applied to the FCC for CB licenses in January 1977, and how we arrived at an estimated 17 million sets with perhaps 35 to 40 million users by late 1976. Is CB a fad, or does its tremendous growth reflect a more fundamental need or desire for electronic communication while on the move?

The growing use of CB has implications extending beyond the service itself. One must ask whether it is not, in fact, a precursor of the public's demand for more sophisticated systems. Although much of the broadcasting on the CB can be characterized as entertainment and hobby use for which the more advanced technologies are not suited, certain patterns of use suggest that those employing CBs eventually may want more sophisticated systems. For example, in addition to recreational uses, CBs also are being utilized for more professional and instrumental purposes, including emergency use. Because the numbers involved in the CB phenomenon are so great, if only a small fraction of these users decide they wish to "upgrade" their equipment, a significant increase in the demand for the more complex two-way systems could result, even including mobile telephone service.

It also should be noted that, despite tremendous growth, CB has reached only about 20 percent of the population. Potential for further expansion is still very great. And because devices normally are not used by trained personnel, the CB service presents special problems of regulation and control.

Technological Advances: The Industrial Push

The appetite of the public for new communication systems is being matched by the ingenuity of the technologists. The revolution that began with the first solid-state devices is continuing, and these devices are becoming smaller, less expensive, and more versatile. The evolution of large-scale integrated circuits—permitting the most extraordinary miniaturization of entire segments of the receiving and transmitting units—has been a major stimulus in this field. Although the principles involved in channel sharing, trunking, digital transmission, and frequency reuse are not especially new, the feasibility and the economics of application have been transformed by developments in semiconductor technology. Improvement in the performance of high-frequency transistors and the evolution of microprocessors will ensure exploitation of the higher frequencies. Also, power requirements have been reduced steadily, which is especially important for the development of portable units. As a result, the fully portable telephone is now feasible.

The Cost of Systems

As a result of regulatory decisions to be described later in this chapter, most of the growth in land mobile communications systems (excluding CB and paging) will occur within the newly allocated spectrum near 900 MHz. In this frequency range, as at lower frequencies, costs will be determined by a number of factors, including service parameters such as message length, wait time, and signal-to-noise ratio, as well as range and capability for random access. The cost to the user depends greatly upon the number of users that can be accommodated, and this number is affected by the type of service provided. For example, because requirements for providing dispatch service are less stringent than those for telephone service, costs for the former are lower.

For technical reasons, one would expect some 900-MHz systems would be more costly (perhaps by 10 to 50 percent) than those operating at lower frequencies if they are produced at the same scale with the same manufacturing techniques. However, some future 900-MHz systems may provide some services at less cost than present lower-frequency systems because the equipment is likely to be produced in larger quantities using more advanced techniques. Were there sufficient spectrum available at lower frequencies, this cost advantage would not occur.

Because we are dealing with systems that have yet to be deployed, estimates of costs vary considerably, but *Table 8-1* exhibits some characteristic costs. In the past, telephone companies' annual charges for service have been equal to approximately one-third of their total investment per user. This would mean monthly charges of $67 for cellular mobile telephone service (MTS) and $57 for cellular dispatch. Increased competi-

Communications for a Mobile Society

Table 8-1
Costs of 900-MHz Systems
(Approximate estimates in 1976 dollars)

System and Service	Number of Users	Total Base Station Investment	Base Station Investment Per User	Cost of Mobile Unit	Total Investment Per User
Cellular MTS	25,000	22×10^6	$ 900	$1,500	$2,400
Trunked MTS (20 channel)	500	900×10^3	1,800	1,500	3,300[a]
Cellular Dispatch	25,000	16×10^6	640	1,400	2,040
Trunked Dispatch (20 channel)	2,000	900×10^3	450	1,150	1,600
Conventional Dispatch	50	$5,000	100	1,150[b]	1,250

Notes: (a) R. Lane ("Spectrual and Economic Efficiencies of Land Mobile Radio Systems," *IEEE Transactions*, Vol. VT-22, November 1973, pp. 93-103). Estimates a value that is less than that of cellular MTS. (b) Conventional units are simpler than those needed for the more advanced systems, but they are likely to be purchased in smaller quantities.

Entrees are adapted from an analysis by H. Ware (Cornell University) and E. A. Blackstone (Temple University) that was based on many sources. Many of the estimates cited above have been rounded to within 10 percent, and do not reflect the significant deviations that exist in current projections of cost.

---

tion in providing this service may result in monthly charges that are a smaller fraction of the total investment being adopted. Also as the market expands, the cost per user can be expected to decline.

For purposes of comparison, we give the following figures for a lower-frequency paging system: the base station equipment for a system accommodating 10,000 users will cost roughly $230,000, or $23 per user; and the portable unit, about $250. Significant reductions are expected in the cost of these units if the market continues to expand at its recent rate.

The Benefits

Substantial social and private benefits are likely to result from increased deployment and use of mobile communications. Extrapolation from our present experience makes it clear that the devices can help to increase productivity in commercial activities, improve public transportation through better control of vehicles and development of passenger-responsive systems, and improve public safety services by reducing response times and increasing coordination. The growth of CB demonstrates a significant desire for systems that permit more private and recreational uses. Although such uses are less amenable to quantitative analysis, they should be included in the list of public benefits.

Increased Productivity

Mobile communications can substitute for labor and capital, enabling a production unit to have the same output with fewer resources or a greater output with equal resources. For example, a study in the late 1960s found that without two-way mobile communications, the police would have to double the number of patrol cars to provide the same level of service. Although this is not typical of the opportunity for increased efficiency, it does suggest that deployment of mobile communications can have major economic implications. Moreover, the benefits can be both private (accrued by the possessor or purchaser of the device), or social (enjoyed by a broader public through such things as more prompt delivery of services and shorter delays in response to emergencies). Mobile communications should have special significance within the large and expanding service sector of the economy, where productivity gains have lagged behind those in the manufacturing sector. Even small productivity increases in this area could result in large net economic gains.

Quantifying potential benefits is a highly speculative endeavor because the magnitude of increases in productivity that might result from improved mobile communications is quite uncertain. Nevertheless, some rough calculations are made in the subsequent discussion to indicate the scale of the possible benefits.

Mobile communications systems are particularly useful for increasing the efficiency of the delivery of services and goods by vehicles. For example, using mobile communications to determine optimum routes and schedules for a fleet of trucks can increase their productivity by a substantial amount; estimates suggest that this increase could be as large as 20 percent in many circumstances.[7] Because such savings have been achieved in specific cases, one might expect that dispatch service already is used widely in trucking. In fact, its use is significant but limited because the costs of private systems are heavily dependent on fleet size, and only the large operations find it economical to use such devices. Approximately half of the firms with gross revenues exceeding $500,000 per year use mobile radio, while only 5 to 8 percent of the smaller firms use them.

New technological developments and recent increases in the allocation of spectrum will make adoption of mobile communications economically feasible for smaller operators, especially if the service can be provided through a common carrier. Some statistics indicate the large savings that could result: In 1972, approximately 20 million trucks were in use; 84 percent of these, or almost 17 million, were in fleets of 5 vehicles or less. It is likely that almost half of these trucks, or about 8.5 million, would be able to benefit from low-cost, common-carrier dispatch service which could spread fixed costs among many subscribers. If we accept $20,000 as the annual cost of operating a truck and assume an average increase in productivity of 10 percent (half of the figure cited above), the annual savings could amount to $400 million for each percentage point of the total number of trucks that utilize the technology. If in the long run use by trucks becomes common, the benefits are measured in the tens of billions of dollars. It might take decades to reach the latter level of use and realize the potential benefits. Although these figures are speculative, they are not unrealistic. The cost of providing the communication technology is likely to be significantly less than $1,000 per year for each truck. Thus, the net benefits will amount to at least half the sums we have calculated.

An example also can be presented from the service sector. About 13 million service workers are potential users of mobile communications, including public administration officials, foremen, medical personnel, electricians, etc. If the use of mobile communications could accrue savings of merely 2 hours per week for each worker at an opportunity cost of $15 an hour, annual gross savings would be roughly $200 million for each percentage point of the total number of service workers using the technology. With a 25 percent penetration of this market (or 3.3 million users), annual gross savings could reach more than $5 billion. Of course, the benefits will be reduced by the cost of the communication device, but even if the most costly service, the mobile telephone were used, the net benefits still would be about $2 billion. Or, to look at it another way, it is not unreasonable to assume that improved communication could increase the productivity of many service workers by 5 percent. Given that the service sector now accounts for more than $600 billion of the gross national product, each percentage-point increase in productivity could save $6 billion.

The magnitude of the benefits, even on this rough scale, indicates that improving mobile communications would be economically worthwhile because of the impact on very large sectors of the national economy. One would have to assume implausibly low figures for penetration and productivity gains and unreasonably high equipment costs to expect overall economic loss. Large savings appear to be attainable in one or two decades.

Vehicle Monitoring

Automatic vehicle monitoring, whether associated with or independent of the new mobile communication systems, appears to have significant social benefits. Some studies have suggested that applying this technology to urban bus service could lead to more regular service and a reduction of approximately 5 percent[8] in the number of buses required for a given level of service. An examination of the technical details indicates that the cellular system may be able to provide the required display and coordination equipment, as well as the more usual communication service, for costs below the figures quoted for a fully independent system. The new technologies, including computer-assisted dispatch systems, are important for the further development of passenger-responsive public transport, such as dial-a-bus systems.

Another benefit of the locational ability of cellular systems may be a reduction in truck hijackings, which is a serious problem with substantial economic consequences. The value of goods lost in this manner was estimated at about $1 billion in 1972. Moreover, associated costs such as claim processing are said to exceed the value of lost merchandise. Thus, even using 1972 figures, automatic vehicle monitoring or automatic beaconing could save $20 million for each percentage-point reduction in losses. If these technologies could decrease the losses by one-third, savings to society could exceed half a billion dollars, when the costs of claims processing and insurance premiums are taken into account. Such results are not implausible: A much larger reduction has been reported using a combination of methods based on monitoring and beaconing.

It should be noted that a serious obstacle to realizing this kind of benefit—especially with current mobile technology—is the "public goods" aspect of the advantages we are considering. That is, the benefits to one firm do not justify the cost of establishing a network for monitoring and beaconing on its own, or even in collaboration with a few other firms. But it may benefit society to have such a network. Certain fixed costs could be shared by the association of specialized networks with a widely deployed cellular system, thereby reducing the barriers to implementation. Moreover, if the social advantages were sufficiently large, government subsidy might be desirable.

Public Safety

Response to vehicular accidents can be improved. Each year there are roughly 50,000 vehicle fatalities, in addition to roughly

1.2 million people who suffer at least moderate injury. Although the extent to which mobile communications can reduce deaths and injury by shortening response time is subject to dispute, one study suggests that both fatalities and serious disabilities could be reduced by 18 percent.[9] Even if this estimate is too high, providing faster medical assistance could have considerable social and economic benefits.

One cannot express the value of saving a human life in purely economic terms; the most fundamental consequences of lives lost by accident are not quantifiable on any scale, least of all financial. However, for our purposes it is instructive to compute some economic consequences based on the present discounted value of the income lost due to the premature curtailment of an individual's earning capacity—$200,000 being a reasonable figure.[10] If the previously cited report is correct, the economic savings implied by avoiding fatalities amount to more than $2 billion; including serious disabilities at an appropriate level of income loss would produce a figure many times higher. These dollar figures provide a measure of lost production and are related to sums paid by insurance or public funds for compensation.

Similar benefits can be obtained by shortening the response time in other medical emergencies such as heart attacks, the leading cause of premature death. There are approximately 1 million heart attack victims in the U.S. annually.[11] More than half of the 650,000 who do not survive these attacks die before reaching a hospital.

Improved mobile communications also should deter crimes, such as robbery, vandalism, arson, drunken driving, homicides and assaults. The cost of these crimes in 1974 has been estimated at nearly $14 billion.

We emphasize again that the social and economic benefits which may result from improved deployment of mobile communications are uncertain and even controversial. Our understanding of the economic effect of communications is still too primitive to permit precise estimates of the impact that can be expected. Yet it is certain that these communication technologies will influence a range of large-scale commercial and public activities. The magnitude of the "leverage coefficient" must be a matter for speculation however.

Nor should we overlook the possibility of extensive problems from utilizing sophisticated mobile communications, especially in the public-service sector. For example, reducing response time may save lives in medical emergencies, but it also may increase the number of those who are legally alive, though incapable of life in any practical sense. Also, public safety agencies easily could resort to advanced technology in circumstances where more manpower and increased human interaction are called for. The more sophisticated and complex dispatch systems, especially those requiring advanced computer systems, could become the political pawns of the public safety bureaucracy without really benefiting potential clients. Finally, if these new technologies are expensive and beyond the financial resources of small communities, resorting to them almost certainly will increase dependence on the federal government.

Private Uses

The fixed telephone has a wide variety of private uses. Maintaining contact with family, seeking information, making appointments, and calling for assistance are just a few examples. Until recently, few people could transact this kind of private business from a vehicle, but in many situations a mobile communication device could help citizens conduct their personal affairs.

The explosive growth in the demand for and use of CB radio speaks for itself: The public wants two-way mobile communications. Although scholars can argue about why people want it—and even whether they should want it—the public has not waited for answers to these questions. The CB is being used for specific instrumental and professional purposes. Users are seeking emergency help, asking for information, making appointments, and conducting business by means of this service—uses similar to those that characterize the more business-oriented and sophisticated mobile communication technologies discussed earlier. However, many of the uses of CB radio are quite different from those of other communication technologies. CB provides an immense party line, and "electronic tavern" where people can find company, even anonymously.

Attempts have been made to understand this latter phenomenon by asserting its connections with the increased complexity, transience and anonymity in American life, because CB radio appears to facilitate peoples' desires to extend their social bonds and form new kinds of communities within our highly mobile culture. In terms of public policy, the critical issue is how to respond to this need. Although these recreational uses are not related directly to the economic issues of productivity emphasized earlier, they are related to other needs, such as the allocation of resources for public parks and other leisure-time facilities. Indeed, the lack of connection with productivity in the narrow sense may mean that these private needs do not receive adequate attention and resources from the political sector. It is extremely easy, for instance, to rationalize the lowest priority for private and recreational uses when allocating spectrum. We must guard against such a tendency.

Direct and Indirect Costs

In extending mobile communication services, society will incur a number of costs. The most easily quantifiable is the direct economic cost borne by those who adopt the technology. This cost will vary depending on the type of system, and will change as new technology is incorporated and new manufacturing processes are adopted. The basis for these costs has been described briefly earlier. We can summarize costs to the consumer. CB radio requires purchasing a unit that costs $75 or more; there is no monthly charge for use of the service. One-way paging involves a relatively inexpensive technology with monthly charges of about $15-$25. The cost of conventional dispatch service is two or three times higher. Mobile telephone service is much more expensive, in the range of $90 to $120 per month. This price is high because the number of users is limited by spectrum constraints, thus preventing economies of scale; the character of regulation of MTS service is probably also a factor in some areas. It is thought, however, that mobile telephone service provided through a cellular system will be considerably less expensive, perhaps half the cost of current systems. Although such costs are high when considered as consumer items, they are well within the reach of commercial and public service organizations.

Many believe that the high cost of the more sophisticated sytems precludes their adoption by consumers. But it is difficult to find reliable indicators for assessing the potential level of diffusion of the more advanced mobile communication technologies. Although cost is obviously important—and even may be the most important factor—we must be wary of making that the sole determinant; the commercial history of color television, high fidelity equipment, and the "fully equipped" automobile suggest the American consumer is motivated by more than a desire for the lowest-cost device that will perform a given function. Furthermore, it should be kept in mind that rates for telephone service in the 1880s were also high. A call from a pay telephone cost 15 cents, and charges for subscriber services were approximately $150 for 1,000 messages. Translated to contemporary values, this is equivalent to a charge of 50 cents for a pay telephone call and $500 for the subscriber service—costs that did not prevent the rapid growth of telephone service.

As already mentioned, introducing the cellular system should reduce mobile telephone service costs from their current level by about a factor of two. Because the technology of mobile telephone service is less mature than that of the fixed telephone and is deployed on a much smaller scale, one can expect the disparity in costs between the two services to keep decreasing. This surely will contribute to increased adoption. Moreover, the technology we are analyzing combines two technologies that have had a profound influence on our society—the automobile and electronic communications; one can almost say that the American is in love with both the automobile and the telephone. It is the subjective judgment of this author that mobile telephones are likely to proliferate over the next several decades.

Industries involved in the deployment of new systems will have to raise substantial capital, the amount of which will depend on the deployment level. To roughly indicate the scale 10 cellular markets, each serving 250,000 dispatch users and 100,000 mobile telephone users would involve a total capital investment of about $7.5 billion (in 1976 dollars). This example assumes a medium level of deployment—3,500,000 units. Such capital requirements would be spread over several years and, while substantial, they are within the range of new capital investments commonly made in the communications industry.

In principle, the direct costs we have been considering can be quantified, although the precision and reliability of the resulting data may be questionable. But there is a broader range of potential costs or "disbenefits" that are much harder to analyze and virtually impossible to quantify. The extension of communications devices to transportation vehicles and even to our pockets surely will affect the social environment. Privacy and solitude will decrease, and we will have taken a further step in the evolution of the communicating and chattering society. The devices are potential nuisances in public places. The concept of the wired man, a favorite topic of science fiction, becomes a little less fantastic; only the purist will notice that the means will involve wireless devices.

Although many citizens need more information and a higher level of communication, some feel unable to cope with the levels to which they are already subject. In the case of a busy executive, for example, a pocket or vehicular telephone might improve decisionmaking within an organization, but possessing the device may subject the executive to more stress. The phenomenon of "information overload" has been addressed by others, but its effects are not irrelevant to the private citizen.

Some of the new technologies—particularly those involving vehicle monitoring—potentially can provide the state with new means for surveillance of its citizens. For instance, the cellular system will require a centralized computer that knows in which cell a mobile unit is located. To be sure, even the most advanced systems currently contemplated will be able to come only within one or two miles of the exact location, and so it might be easy to conclude that this has little bearing on surveillance. Nevertheless, past experience should make us extremely cautious in this area; the potential for abuse and misuse clearly exists.

There is also much concern about the increasing size of organizations in our society, and the new technologies are likely to be agents of organizational change in those industries and public agencies where they are adopted. In many respects, these devices can provide the worker who uses them with a greater level of freedom, but they also provide the means for greater centralization within the organization and consequently of the decisionmaking process as well. As a result, some of the gains in efficiency and coordination might be offset by the reduced autonomy of the field worker. Surely the level of job satisfaction experienced by these people will be affected, not always for the better.

Increased deployment of mobile communication devices will raise the ambient level of electromagnetic radio. What is a safe level of nonionizing electromagnetic radio is a very controversial question. In the past, U.S. authorities have based their safety standard on the contention that the only important physiological consequences resulted from heating tissues, and this led them to establish a standard (of ten milliwatts per cubic centimeter) for extended exposure. On the other hand, one can find in the literature data suggesting that more subtle physiological effects occurring at levels too low to produce extensive heating may be significant. Much of the early work on these "athermal" effects was performed in eastern Europe, leading the Soviet Union to adopt much more stringent standards than those of the United States—limiting long-term civilian exposure to a standard, a thousand times lower than the U.S. (ten microwatts per cubic centimeter).

We cannot review here the controversy associated with the safety issue, but the following is clear. Unlike the situation with ionizing radiation emitted by the nuclear power industry, we already have increased the level of nonionizing radio above that of the natural environment by many factors of ten. Further deployment of personal communication devices will increase it further. An analysis of the radiation emissions expected from mobile units themselves suggests that such emissions will be inconsequential if judged by the American safety standard and small in terms of the Soviet standard. The one exception is hazard to the user of a portable device held close to the head during operation; this requires further examination. Other safety and environmental concerns, such as using these devices while driving, also will require careful attention.

Allocation of Costs and Benefits: Values and Priorities

Regulatory decisions will play a major role in allocating costs and benefits. When decisions such as allocating spectrum are being made, the choices reflect, either explicitly or implicitly, our values and preferences. Although the alternatives rarely will be clear-cut, we can give priority to communications for emergency and public services, or to uses that will increase the efficiency of our systems of private production, or we can emphasize recreational and private uses. In oversimplified terms, the choice will not be one of "guns or butter," but rather "safety, work, or recreation."

If we emphasize the development of mobile telephone systems, we will be giving priority, at least in the initial stages, to the managers of our production and service organizations and to the affluent citizen. Favoring the development of two-way dispatch systems would reflect the value we place on productivity. In the past, such priorities have been prominent in the area of mobile communications. But the growth of CB, the "peoples'" mobile communication device, presents us with a different kind of choice.

The use of the airwaves for private and recreational purposes is not new, of course. Nevertheless, although some similarity to the standard broadcasting situation exists, important differences are also obvious. For instance, in standard broadcasting, the citizen plays a passive role and the medium is often under the control of large commercial enterprises. The dominant characteristic of CB, however, is "participation," and its rapid growth clearly reflects an unfilled need in our society.

In appraising developments in this area and attempting to assess their social impact, one is tempted to ask whether these advances are being stimulated by a perceived and important social need or by what has become known as the "technological imperative." Certainly, the technologies that we are examining can serve many useful purposes. As is true of most new technologies, they were not developed as a direct response to some of our most basic social problems such as unemployment, the high cost of medical care, or the state of the physical environment. Yet perhaps by indirect means they will be able to help to ameliorate these problems. Instead, the technologies we are analyzing are products of a highly industrialized society that emphasizes productivity and the saving of time and, with the exception of CB devices, they largely are oriented to serving business and governmental ends, rather than those of a cultural, private, or social character. It would be erroneous to suggest though that these technologies are being thrust upon an unwilling public and that they reflect an artificial stimulation of wants. To be sure, many of these devices are manifestations of our highest technological skills, but the public has demonstrated many times its large appetite for them. The quality of service and dispatch service, has been far from satisfactory, essentially because of spectrum constraints. Still, waiting lists grow, and the public has made it plain that it prefers poor service to none at all.

The "push" of the technology and the industry is unmistakably present, but there is also substantial public "pull." The new technology, however, also will exact social and economic costs; foresight in deploying and controlling the new systems is necessary. Nevertheless, it is difficult to sustain the argument that a national policy should be developed to discourage the continued proliferation of these devices.

The Regulatory Response

Spectrum Allocation

Since the 1940s, growing demand for mobile radio service has been met by increased allocations of the spectrum to that use. As the technology developed, operation at higher frequencies became economically feasible and allocation of higher bands was necessary; increased utilization of lower bands for a variety of purposes prevented expansion of mobile services at those frequencies. Consequently, the FCC set aside frequencies near 150 MHz and 450 MHz to complement the original allocations below 50 MHz. At these frequencies, the total spectrum allocated for public safety, industrial, and transportation use is now 39 MHz. More than 3 additional MHz have been allocated for domestic public radio (radio telephone) service.

But during the 1960s it became plain once again that the spectrum that was provided for conventional systems could not meet the demand for such services with an acceptable level of service quality. Because there was little sharing of channels among different organizations, spectrum congestion had become severe in certain geographic regions, particularly urban areas.

Pressures on the FCC to improve the situation grew from users, industry, and Congress. These led to the proceedings associated with Dockets 18261 and 18262, initiated in 1968, proposing transfer of some UHF spectrum to land mobile services. The proposed reallocation was contested vigorously by major broadcasting organizations, as well as smaller independent stations. They argued that the existing land mobile spectrum was utilized poorly, and that such a reallocation was prejudicial to the established commitment to localized television for both commercial and educational purposes. But ultimately the land mobile interests prevailed and, in 1974, the FCC transferred 115 MHz to land mobile use. The order ensuing from Docket 18262 also promulgated a number of rules that will have considerable influence on the deployment of the new technologies. These will be discussed later.

Citizens Band (CB)

The extremely large increase in demand for CB service has presented the FCC with many problems that could not be handled solely by increased spectrum allocation, and that required modifying rules governing the use of the service and new licensing procedures.

In 1959, which marked the end of the first twelve years of CB, only 49,000 licenses were in effect. This was one year after the FCC reassigned a portion of the spectrum from amateur radio operators and inaugurated a new class of CB service (Class D) in an effort to promote private citizens' personal use of radio. Reallocating the spectrum made it possible for manufacturers to build good-quality CB sets at a price that the average consumer could afford, in the range of $150 to $250. In 1962, four years after Class D service was introduced, the number of licensees had increased to 300,000—a sixfold increase. By 1965, the number of licensees had increased another two-and-one-half times to 745,000, and the FCC was beginning to express considerable alarm. In its *Annual Report,* the FCC stated that "widespread rule violations threaten the continued usefulness of the service" and that "illegal practices are seriously impairing the legitimate use of the service." For some reason there was no significant growth in the number of CB licensees during the 1967-1973 period, with the number continuing to hover at about 800,000; there is still no satisfactory explanation for this stabilization.

The rules in force from 1958 through late 1976 permitted any citizen 18 years or older, or any business owned and operated by citizens, to obtain a Class D license. Unlike the amateur service, no tests were required. Rules were adopted limiting use to the transmission of substantive and purposeful business or personal messages; hobby use, which is frequent within the amateur radio band, was prohibited. The FCC intended CB to be a readily accessible, two-way personal communication technology for the average American.

From 1971 through late 1973, the number of applications for CB licenses approximated 20,000 a month. After the truckers' strike in late 1973, however, the number of applications increased to about 45,000 per month in 1974, and then climbed dramatically throughout 1975 to reach a level of 443,000 in December. During 1976, the number of applications being received showed signs of stabilizing and even exhibited a decline in mid-year, no doubt partly in response to the confusion over new types of CB units and the FCC's regulatory response to these options. But growth resumed in 1977; more than 2 million applications were received in the first three months of the year.

Not surprisingly, keen dissatisfaction, among both CB users and the FCC, accompanied this extraordinary growth. Experienced CB users and those who employed CB in their businesses were especially outraged when their channels became overloaded and were misused. The FCC found itself under strong political pressure to do something about the "CB mess." Deteriorating service was attributed not only to the explosive growth but also to poor enforcement by the FCC.

The FCC responded to the complaints and political pressure by partially deregulating the CB service. In late 1975 the Commission promulgated a series of more permissive rules that, among other things, removed the restrictions on hobby use and "chit-chat" conversations, while still limiting duration of transmissions. The FCC is relying on liberalization and deregulation to improve the service. The unresolved question is whether the situation will get better or worse as a result of the FCC's new approach. In fact, the FCC may have deregulated the service because there was no effective means of, and perhaps no compelling reason for, regulating it.

Docket 18262

Because of their special significance, the FCC decisions associated with Docket 18262 require more extensive discussion. These decisions, rendered in 1974 and 1975, and subsequently upheld by the Court of Appeals, resulted in the transfer of 115 MHz of the 900-MHz band (most of which had been allocated to television) and land mobile use. Of the 115 MHz, 40 MHz were allocated to high-capacity, common-user systems utilizing cellular technology, 30 MHz to dispatch use employing either conventional or trunked systems, and 45 MHz were held in reserve. There were also some other important and precedent-setting aspects of the decision. These include:

- Allocation of spectrum by technological system rather than by type of use—a departure from previous practice.
- Concession to cellular operators of an effective monopoly over mobile telephone service.
- Creation of a special class of common-user systems in the 30 MHz allocated to dispatch and reliance on competition rather than regulation to protect consumers within this class.
- Encouragement of competition by preventing vertical integration by cellular operators, requiring complete separation of the cellular subsidiary from its parent if the latter is a wireline telephone company, granting cellular operators the right to offer dispatch service as well as mobile telephone service, but prohibiting cellular firms from providing fleet calls by which all vehicles are called at once, and prescribing an open-entry policy for the dispatch market.

These decisions have major implications for the structure of the mobile communications service industry. Although they have been controversial, they are consistent with trends in regulatory policy during the past two decades. The FCC gradually has reduced the scope of the monopolies of the wired telephone companies by its decisions on interconnection, private microwave systems, and specialized microwave common carriers.

The transfer of spectrum from television to land mobile use was particularly controversial, and parts of the television industry regard it as an ominous precedent. In evaluating the merits of this transfer, it should be kept in mind that as late as 1973 only 42 MHz were allocated for all land mobile uses. At the same time, 492 MHz were allocated for television channels, including VHF (lower frequency) and UHF (higher frequency) spectrum. Each UHF television channel requires 6 MHz, which is twice the total allocation for all public mobile telephone service prior to Docket 18262; one television channel occupies the same spectrum space as roughly 4,000 mobile telephone users, or 40,000 conventional dispatch users. This large allocation to television resulted from policies adopted by the FCC in the early days to encourage diversity and local programming; in particular, the FCC decided that a large number of UHF stations was necessary to achieve these objectives. The latter policy was not successful; UHF frequencies often went unused and the stations frequently were unprofitable. Even today there is substantial excess capacity in the UHF allocation.

It is difficult to make a quantitative analysis of the costs and benefits of this spectrum transfer because one is forced to compare uses that are very different. Only rough estimates can be made of the potential economic benefits of land mobile radio, and revenues from entertainment and trade stimulated by commercials are even more difficult to assess. Furthermore, allocating spectrum to UHF television in order to expand programs and increase diversity involves a number of issues that are not primarily economic. Nevertheless, estimates made earlier suggest that large savings can result from using mobile communications, substantially greater than the costs of the equipment. Additionally, the low utilization of UHF channels suggests that benefits from the large UHF television allocation are quite small now. Moreover, cable is a good substitute for UHF television; it is already economically viable, and improvements in technology will make it more attractive. It seems probable that the cost of transferring to cable in areas where there is a shortage of spectrum would be significantly smaller than the expected long-term benefits from more widespread deployment of mobile communications.

We must emphasize again that the issues involved in reallocation are not solely economic. Cultural, regional, and consumer issues also are involved, and there is no formula that can take into account all these aspects. Ultimately, the decisions must be political ones, reflecting our values and priorities. A significant question that can be raised is whether present decisionmaking processes are adequate to reflect these values and priorities, and whether the interests of affected parties receive sufficient consideration. Because of its resources, industry is assured representation, even if the process is cumbersome, costly, and time consuming. Representing broader public views is more problematic, however. It should be noted that in the seven years of proceedings associated with Docket 18262, there was virtually no substantial public participation and no input from public interest groups.

The decisions in Docket 18262 contain several specifications, including restrictions on equipment manufacturing and operation of services, that are intended to ensure competition in the industry. Whether they will be effective is still subject to dispute. The characteristics of mobile communication systems are sufficiently different from those of the fixed telephone network to render inapplicable many arguments for a "natural monopoly."

Perhaps the best case for a fully competitive environment is the dispatch market, where the service parameters are such that small-scale systems will not be at a severe disadvantage with respect to cost and spectrum efficiency. For systems interconnected with the wireline network, a balance must be struck between the efficient and diverse services that competition might foster on a regional level, and the desirability of providing a nationwide mobile telephone system capable of serving many users. The issues are complex. If wireline companies and equipment manufacturers are permitted to operate mobile services, the potential for cross-subsidization exists and should be monitored carefully.

As mentioned earlier, four firms account for 90 percent of the output in the mobile equipment manufacturing industry (excluding CB equipment), and one firm dominates the industry. It is not known if the provisions in Docket 18262 will ensure that the mobile communications equipment industry will be fully competitive. However, expanding demand for units will favor competition, and firms manufacturing other forms of communication equipment may enter the market.

Apart from regulatory constraints, the structure of the mobile communications service industry is conducive to competition, though in the past the firms had insufficient spectrum to be able to exploit potential economies of scale. Decisions by the FCC in Docket 18262 may increase concentration in the mobile communications service industry. For example, cellular companies probably can service more than one-third of the total market (dependent, of course, on demand). They have been allocated approximately the same amount of spectrum as the entire mobile communications industry had prior to Dockets 18261 and 18262. Trunked companies probably will control a substantial share of the dispatch market in many areas. Because the FCC has chosen to encourage more efficient use of spectrum by promoting complex systems with high fixed costs, this policy also will tend to concentrate output in fewer firms.

On the other hand, in some markets the vast increase in spectrum made available for commercial use can offset the impact of this FCC policy. Excessive concentration is not likely to be a problem in large markets with sufficient demand to support several trunked dispatch systems, many conventional systems, and a cellular system. In smaller markets, however, concentration may be high in the dispatch field. Finally, mobile telephone service provided by cellular systems will be highly concentrated in all markets, with the only direct competition coming from radio common carriers at lower frequencies. Consequently, nonbusiness users will face little choice when purchasing mobile telephone service.

The structure of the mobile communications industry will be quite similar to the long-distance segment of the fixed telephone system used by businesses. These consumers can establish a private microwave system, or use either a specialized common carrier or the telephone company. Small firms that previously did not use mobile communications because they had too few mobile units to make a system profitable now will be able to employ common user or common carrier systems; this should enhance the ability of small firms to compete with their larger rivals. Moreover, the rates for dispatch (except for cellular) will be unregulated by the FCC, which argued that easy entry and a competitive structure will assure reasonable prices. The FCC also believes, though, that the absence of similar constraints on cellular firms make it mandatory that traditional economic regulation be applied to those companies.

In conclusion, there are two general observations to be made on regulatory practices. First, the history of telecommunications provides many examples that should make us cautious about expanding "over-the-air" transmissions if the communications can be accomplished using wired systems. One of the merits of the cellular system—as well as others involving interconnection—is that a substantial part of the transmission is carried by the wireline system. In a sense, this situation is analogous to one we face with television broadcasting: Many wish that earlier allocations of the spectrum for this purpose had been more frugal because the same ends now often can be achieved by using cable television. Thus, as we look to the future of mobile communications, we not only should emphasize efficiency in the use of the broadcast spectrum, but also favor systems that carry part of the message traffic over wires or cables. Principles of conversation and option preservation justify such a policy; we must leave room on the airwaves for future technologies.

Second, it should be noted that allocating the spectrum by technological system rather than by category of use represents an important departure from previous practice. It is hoped that this method will avoid the wasteful situations that exist when, within the same geographical region, one band is highly congested, and a neighboring one is used very little.

The Future: An Agenda of Issues

As we attempt to foster the evolution of land mobile communications in a manner that will be most beneficial to our society, we will confront numerous problems, many of which are implicit in the preceding discussion. Unfortunately, it is not possible to translate these problems into a set of questions that, if answered, will solve the problems for the indefinite future. Rather, there are issues that will require continuing attention and reappraisal. I shall conclude this chapter by specifying the more significant ones.

Coordination Within the Government

Rapid growth of land mobile use is likely to continue and even accelerate, permeating the commercial, public, and private sectors. Although the FCC will continue to be the focal point of regulation, the characteristics of land mobile systems are such that a high level of coordination with other regulatory agencies and executive departments will be of crucial importance.

For example, if Interstate Commerce Commission rules that trucking routes and service areas not be modified, the benefits of more extensive use of mobile communications in the trucking industry are unlikely to be realized. Nor are the possibilities in the public safety area likely to be realized without some coherence in the decisions and activities of the FCC and Departments of Health, Education and Welfare, and of Justice. The Department of Transportation is vitally involved in renovating urban public transportation; new mobile communication technology has a significant role to play in this process as well.

In no area are the consequences of poor coordination likely to be more serious than in the energy field. We already have drawn attention to the fact that new communication technologies may save energy in some situations and increase energy use in others. Giving further advantage to the automobile as a mode of transportation is a matter of considerable concern. Therefore, it seems essential that electronic communications, especially of a mobile character, be considered in the evolution of an energy plan.

Federal Purchasing Power

The federal government alone operates more than half a million vehicles. This constitutes a massive market, and government procurement decisions can influence development of the technology, the level of deployment, and safety standards. If, for instance, the General Services Administration decided to equip its automobiles with CB radios for safety reasons, there could be a major impact on further adoption of such devices.

The Military and Space Programs

Military and space programs are continually extending the limits of mobile communication technology. They are the source of many advanced techniques relevant to the civil sector. But does the private sector derive the maximum possible benefits from military and space research in these areas? This is the old problem of "spin-off."

In analyzing this problem, it is important to recognize that many of the firms performing communications research and development for the military and space effort transfer information from one sector to another, and considerable "leakage" of technological methods often occurs. This author does not have sufficient information to judge the need for further explicit efforts to increase the interchange. Extensive direct transfer of technology between the sectors should not be expected, because the kind of economic and performance specifications involved in choosing military and space technologies differs markedly from those in the civilian sector.

FCC Decisions

The FCC continues to make decisions that will determine the diversity of services being provided and the degree of competition among them. This is done by allocating the spectrum, promulgating rules governing use of the services, and regulating entry into the service market. Although many recent decisions, particularly those associated with Docket 18262, are intended to maintain diverse services and to stimulate competition, it is still unclear whether they will be effective. The industry requires a stable regulatory environment in which to operate, but earlier decisions must be reevaluated as conditions change.

If land mobile communications continue to expand at the present rate, or even more rapidly, new strains will be placed on the FCC. By the end of this decade, we could have 12 million users of the more sophisticated systems and many times that number of CB users. Despite the present clamor for a reduction in government regulation and spending—not all of which is informed or devoid of self-interest—it is difficult to see how the FCC will be able to deal with this area effectively without increased resources.

Because the policy issues implicit in this brief discussion require continuing attention, the central question remaining is whether our capability for policy analysis is commensurate with the need. Clearly, some capability does exist within the Legislative Branch, the regulatory agencies, the Executive Branch, the universities, and private-sector research organizations. But it is hard to persuade oneself that these are adequate given the large resources that industry can bring to bear in support of its interests.

Of course, such analysis is only a means to an end: to better inform the actions and decisions of government. Whether our governmental agencies can cope with the new demands that rapid growth of land mobile communication will place upon them is an even larger question. Improved analysis is a necessary, but not a sufficient, condition for improving governmental actions. In this broader context, land mobile communication deserves no unique consideration; it takes its place alongside many other segments of the communications field discussed elsewhere in this volume.

Notes

1. The author wishes to thank colleagues at Cornell University who are members of a project assessing the future of land mobile communications. The results of this project will be published as a book in early 1978 by Sage Publications, Inc. The contributions of Alfred M. Lee and Harold Ware of Cornell University were especially helpful. Research for this article was supported by funds from the National Science Foundation (Grant ERP 74-20555); the Cogar Foundation; and Cornell University, derived from foundations and industry. The opinions expressed here are those of the author and do not reflect any of these organizations.

2. Frequency reuse is employed by all systems when the same set of frequencies is allocated for land mobile purposes in widely separated service areas. But in the subsequent discussion, we shall consider a much higher level of frequency reuse, involving repeated use within the same service area.

3. See Ivan Bekey and Harris Mayer, "1980-2000; Raising Our Sights for Advanced Space Systems," *Astronautics and Aeronautics*, July/August 1976, pp. 34-63.

4. The reader should note that the nomenclature used in this field is confusing. Words such as "paging" or "dispatch" can refer to a service, a technical system, or a use. The context in which the word is used is often the only means of clarifying which of these meanings is intended.

5. Paul F. Kagan, *Barrons*, July 10, 1974, p. 72.

6. Arthur D. Little, Inc., "A Study of the U.S. Radio Paging Market," prepared for the Office of Telecommunications, Contract No. OT-117, Washington, D.C., April 30, 1975, pp. 4-5.

7. Harry A. Plotkin, "Mobile Radio—A Pickup and Delivery Tool," *The Logistics and Transportation Review*, Vol. 9, no. 4, 1973, pp. 357-366. Also, M. J. Fogarty, "Closing the Mobility Gap," *Electronic Weekly*, October 9, 1974, pp. 21-22.

8. Systems Applications Inc., *Land Mobile Communication and Public Policy*, Vol. I, National Technical Information Service, U.S. Department of Commerce, Springfield, Virginia, 1972, p. 163.

9. *Ibid.*, p. 221.

10. Brian C. Conley, "The Value of Human Life in the Demand for Safety," *American Economic Review*, March 1976, p. 54.

11. William F. Renner, "Emergency Medical Service," *Journal of the American Medical Association*, October 14, 1974, p. 253.

Selected Readings

Daniel E. Noble, "The History of Land Mobile Radio Communications," *Proceedings of the IRE,* Vol. 50, May 1962, p. 1405.

U.S. Advisory Committee for the Land Mobile Radio Services, *Report,* U.S. Government Printing Office, Washington, D.C., 1967.

Richard Lane, "Spectral and Economic Efficiencies of Land Mobile Radio Systems," *IEEE Transactions on Vehicular Technology,* Vol. VT-22, November 1973, pp. 93-103.

William C. Jakes, Jr. (ed.), *Microwave Mobile Communications,* John Wiley and Sons, New York, New York, 1974.

"Second Report and Order," Docket 18262, Land Mobile Radio Service, *Federal Communications Commission Reports,* 2nd series, Vol. 46, 1974, pp. 752-803.

"Memorandum, Opinion and Order," Docket 18262, Land Mobile Radio Service, *Federal Communications Commission Reports,* 2nd series, Vol. 51, 1975, pp. 945-1010.

Carlos Roberts, "Two-way Radio Communications Systems for Use by the General Public," Master's Thesis, University of Colorado, 1975.

Raymond Bowers et. al., *Communications for a Mobile Society,* Sage Publications, Inc., Beverly Hills, California, 1978.

9.
Electronic Alternatives to Postal Service

Henry Geller and Stuart Brotman

The chairman of a recent study commission on the U.S. Postal Service said succinctly, "Our long-range outlook for the Postal Service is dismal."[1] From July 1, 1972 to September 30, 1976, Postal Service costs exceeded total income by $2.78 billion. The Service estimates a loss of $991 million for FY 1977, and anticipates deficits totalling $2.7 billion in FY 1978 and FY 1979.[2] If postal rates remain at present levels, the cost to the Treasury and to the taxpayers will increase to an annual cost estimated at $12 billion by 1985.[3] The problem within the Postal Service is basically a consequence of a structure of uneconomic services coupled with extraordinary labor costs, which constitute 86 percent of all costs. Since 1971, total expenses have risen from about $9 billion to roughly $14 billion, with labor costs accounting for about $4.7 billion of this increase. In short, quite apart from external factors, the Postal Service is facing monumental financial problems.

At the same time, the Postal Service faces internal problems. Competition, long absent because of the Private Express Statutes, is emerging in the form of electronic alternatives, which are not subject to the statutes. The concept of electronic alternatives to physical delivery of mail is not new. The telephone has long been such an alternative, and its use has cut significantly into the use of mails. This is reflected in the closing gap between the cost of sending a letter and of making a three-minute telephone call from New York City to Los Angeles—a ratio of 83:1 in 1950, down to about 6.5:1 today. Recently, there has been an explosion of new technologies and services in the telecommunications area, spurred by such developments as the computer and its digital links, satellites, cable television and fiber optics. Such technological innovations are bound to siphon major revenues from the conventional mail activities of the Postal Service at a time when that organization already faces serious financial problems.

Approximately 80 percent of today's first class mail is business mail; 70 percent of this is payment transactions (the mailing of bills or checks, and so on). The Postal Service will lose a large part of this desirable business over the next two decades.[4]

There are thus two converging trends—the internal crisis of the Postal Service and the external competitive environment. Public attention has been largely focused on the internal problems—deteriorating service, increased postal rates, labor problems. But there has been relatively little public awareness that the internal problems are exacerbated by the external trends.

This chapter therefore explores present problems of the USPS, the emerging new electronic technologies or services, their probable impact on the USPS, and the possible responses of the USPS to these developments, and the regulatory and jurisdictional implications of such responses. The discussion also briefly touches upon social values that may be affected by the interaction of advanced technology with mail delivery.

Present Problems of the USPS: An Overview

In 1971, the U.S. Post Office Department stopped delivering Aunt Minnie's mail in Podunk. Replacing it was the newly formed U.S. Postal Service (USPS), a corporation chartered by Congress and wholly owned by the federal government.

The formation of USPS was the end result of a process started by the President's Commission on Postal Organization, which argued that the then-existing Post Office Department, headed by a cabinet member, was nearing rigor mortis:

Electronic Alternatives to Postal Service

The United States Post Office faces a crisis. Each year it slips further behind the rest of the economy in service, in efficiency and in meeting its responsibilities as an employer. Each year it operates at a huge financial loss. No one realizes the magnitude of this crisis more than the postal managers and employees who daily bear the staggering burden of moving the nation's mail. The remedy lies beyond their control.[5]

The department's failure, in part, was thought to be a result of its strong political orientation. Since the New Deal, it frequently had been used for patronage appointments and was the source of lucrative construction contracts that Congressmen used as political plums in their local districts.

By 1970, Congress was faced with a fundamental choice: retain the department or redesign it along the lines of private enterprise so that it could operate on a self-supporting basis. Despite the burgeoning problems that had been heavily documented, Congress was still reluctant to give up a department that enhanced the political power of its members. It required gathering broad, bipartisan support to secure adoption of the Postal Reorganization Act of 1970. Unfortunately, as a result of the necessary compromising entailed in passage of this legislation, the Act emerged as an all-purpose magic potion. It promised lower, fairer rates for mailers, better working conditions and wages for workers, and continued service to rural areas and small post offices, as well as managerial efficiency. It proposed to attain these objectives and to put the mails on a self-sustaining basis within ten years. The deep-rooted conflict between operation as a public service and operation as a business was neither fully acknowledged nor satisfactorily treated.

Instead of a growing legacy, the six-year-old Postal Service has established a record that is more likely to appear in an obituary. Although reorganization was intended to reduce postal costs by at least 20 percent through utilization of practices modeled on those of private industry, the USPS has incurred mounting operating deficits which, by 1979, are estimated to total more than $2.7 billion. Since 1973, current liabilities have consistently exceeded current assets. It has avoided insolvency only by borrowing heavily from the U.S. Treasury, by entering the private financial market, and by receiving a Congressional subsidy of $1.5 billion a year. Although these numbers may seem abstract to Aunt Minnie in Podunk, the dramatic rise in rates for first-class mail (FCM) affects every American citizen. In 1971, the FCM rate was 6 cents; today it is 15 cents; and by 1985, if the Postal Service continues to operate as it has, a 28-cent stamp and correspondingly higher rates in all other classes of mail can be expected.[6]

The financial problems of the USPS will not disappear. The overall cost of providing traditional postal services (i.e., door-to-door delivery six days a week) will continue to climb dramatically, and more revenues will have to be raised through additional congressional appropriations, additional borrowing and, of course, higher postage rates.

The USPS problem has been aggravated by several external forces. Since World War II, the role of the mails as a primary medium of personal communication, news, political commentary and commerce has declined in favor of increased dependence on the telephone, radio and television, and automobiles. The national recession of the early 1970s caused a substantial loss of revenues.[7] And the growth of the public sector labor movement has generated higher-priced contract settlements, partially as a response to spiraling inflation. In turn, wages and benefits for postal employees have risen dramatically, and these costs have been passed on to Aunt Minnie in the form of higher prices for postal services.

The USPS has attempted to maximize efficiency by streamlining operations. Approximately 80 percent of its delivery routes are now mechanized (motorized)—approaching the practical limit of the process. The Postal Service has deployed electronic letter-sorting machines and solid-state cancelers, and has increased the percentage of letter mail handled by machine from 25 percent in FY 1971 to 63 percent in FY 1976. It has reduced personnel by over 50,000 in the past two years without materially affecting service. All these efforts have had a short-term impact by improving productivity (13.1 percent higher in 1976 than in 1971). However, such productivity has had no effect on a number of postal services, including door-to-door residential delivery, rural free delivery, and the maintenance of small post offices. In effect, the race for productivity is now being run on a treadmill, as ever-rising costs quickly cancel out any efficiencies achieved.

The USPS has expressed the resulting frustration:

> The Postal Service today is expected to behave like a business by bringing costs and revenues into balance and satisfying the needs of its paying customers. It is also expected to behave like a public service by continuing to provide all of the traditional services and by honoring all of the constituencies that evolved through two

centuries of political rule. As a long-run proposition, it is impossible to reconcile these expectations.[8]

Scores of postal critics have emerged to barrage the USPS with a variety of reasons for the current situation: misguided capital investments, management indecision, excessive labor settlements, and general inefficiency. While some or even all of these may be contributing factors, it appears that the core problem remains a simple failure to operate within the basic economic restraints of supply and demand.

Although one-third of total postal costs (about $4 billion) are used for delivery services, this allocation does not reflect the demand of the mailer for such services. Extensive public-opinion surveys have shown that individual mailers are largely indifferent as to whether mail is delivered at curbside or at the door. Businesses have indicated that less frequent residential delivery would be acceptable. Additional research has shown that alternate-day residential delivery would meet the demand for over 90 percent of the present market using the mail service.

Mail volume itself is stagnating, and FCM volume is entering a projected decline caused by price increases and the growth of technological alternatives. As a result, the lack of FCM competition may be another factor that may have contributed to the financial plight of the USPS. For over a century, the U.S. Post Office and its successor, the USPS, have operated a monopoly service under the Private Express Statutes. These statutes confer upon the Postal Service an exclusive right to carry "letters" for others over postal routes, principally to secure revenues for providing needed postal services to all areas of the United States, as mandated by Congress. Today, it may be argued that the statutes, representing the public service philosophy of the Post Office Department are harmful to a Postal Service designed to achieve economic self-sufficiency. In achieving the goal of postal delivery, the costs have been passed on to the postal rate payer, taxpayer, or both. There has been no competitive incentive to resist economic inefficiencies.

As a result, the Private Express Statutes have maintained the USPS as a labor-intensive public monopoly. Without the stimulus of competition from private enterprise, the USPS has no incentive to adjust its rates and services to reflect market conditions. The historical record is clear: Since 1850, the Postal Service has operated in the black for only seven years. But the competition which the Postal Service itself might welcome would not be "under all of its present political regulatory constraints. Competition in that case would be less competition than cannibalism."⁹

Postal services continue to be tailored to the residential correspondence market even though this represents less than one-fifth of total mail volume. Postal costs and postal revenues have become grossly overbalanced on the deficit side, so that costs continue to rise, while revenues decline. In the next decade, unless prevailing policies change, the USPS can be expected to provide increased delivery services to more points at greater cost, while receiving relatively less revenue. Aunt Minnie may continue to get her mail six days a week, although if asked to purchase this service in the marketplace, she might clutch her purse while ordering Fido to bite the postman.

In 1974, the USPS tried to close several hundred post offices—out of a total of 40,000—and found that it had initiated a Congressional crisis. Constituents simply want the local post office to remain open, even if it cannot balance costs and revenues.

Although it is easy to criticize inefficient or irrational decision-making in this area, there may be countervailing social values at stake. The rural post office is often a symbolic meeting place in small villages and townships; it also may be the sole presence of the federal government in particular communities. Social values may also support door-to-door mail delivery. For shut-ins and for people living alone, the postal carrier may be the only daily visitor, a guarantee for regular face-to-face conversation. As all other home delivery services have faded, the postal carrier remains the last hold-out resisting the social costs of personal isolation, albeit an economically inefficient one.

Yet the fact remains that Congress will have to consider how much it must charge the nation as a whole to meet such social needs, especially as the population increasingly moves toward major urban centers. Perhaps the quality of rural life might be better improved by expenditures in other fields (health, education, and so on), or in low-cost loans for broadband telecommunications service. For rural areas to be compensated for the loss of full local service, it might be equitable to redirect a portion of the saved postal subsidy toward such rural development projects.

First-class mail volume contributes the majority of the USPS' total operating revenues, and eight out of every ten pieces of FCM are business mail (including invoices, bills, payments, statements of accounts, purchase orders, financial papers, and business letters). Business mailers now are forced to pay for postal services whose costs exceed those they utilize. In view of rapidly developing technology which will be able to meet business demands with greater efficiency and economy, it is a virtual certainty that a mass exodus from the mailstream will result. Former Postmaster General Benjamin F. Bailar described the potential situation:[10]

> If the public elects to continue the postal system in its present form, it will have to pay a steep price. It may find the first class stamp becoming a luxury item in the next decade and the Postal Service a ponderous and costly left-over from simpler, more affluent times.

Aunt Minnie herself was a product of such times, and it seems inevitable that she will have to adjust to the Postal Service's reduced services. Even if technological development remains at a standstill, the USPS will have to make aggressive changes in its operations if it ever hopes to conform with the legislative intent of the Postal Reorganization Act. Otherwise, honesty will compel the public and its representatives to acknowledge regress to the former model of the Post Office Department, relying on subsidies to bolster mounting deficits in order to maintain a universal mail service.

At the turn of the century, in a nation largely of small towns and farms, the post represented the basic means of personal business and news communication, and was also an important source of entertainment and education. Today, with the advent of the telephone, broadcasting, and the auto, and an increasingly larger number of urbanized areas, the structure of the USPS, which has remained unchanged, is challenged. The Congress must ask what kinds of postal services make sense today. In the words of an astute former postal officer, John McLaughlin:[11]

> Does the nation really need to pay $12,000 or $14,000 a year to have someone standing around in a Post Office at every forgotten crossroad on the chance that someone might want to buy a single 13 cent stamp? Is it really a public service to spend hundreds of millions annually to bring mail to your front door instead of to a curbside box? Or to spend a billion or more delivering mail daily to households when the people who receive the mail and the people who send it tell us most of it is not urgent?

In particular, Congress must ask what kind of postal service is needed in the face of the telecommunications revolution. We turn now to that revolution.

Telecommunications Technological Development: 2001 Arrives in 1985[12]

The House Subcommittee on Postal Facilities, Mail and Labor Management, accurately summarized the challenge that the USPS is facing:

> In the next few years postal managers will be called upon to make vitally important decisions which will determine the direction which the USPS will take when technology in the form of electronic transfer and other advances change the role of the organization and the very concept of postal service as we know it today.[13]

The development of communications technology today is moving by giant steps, as private enterprise has become fiercely competitive in the production of communications "hardware" that can be integrated into business operations. The technology discussed here does not refer to the long-established electronic alternatives to the mail—the telephone and well-known TWX and Telex provided by Western Union. TWX/Telex teleprinters compose, correct, transmit and receive text that originates and is sent to some distant point; 94 percent of the nation has telephone service, while there are still only 100,000 subscribers to Western Union TWX/Telex. The recent telecommunications revolution that is the subject of this section, however, began with the introduction of the digital computer, accompanied by enormous strides in transmission.

The Computer—Interactive Word Processing

The post-war development of the computer is an appropriate starting point for any discussion of present and future developments in communications technology. The computer created a major revolution in office management as an increasing number of routine functions were switched from office personnel to computers. The computer became the technological rallying point that stimulated further research and development utilizing computer technology.

The microprocessor, the so-called computer-on-a-chip, has been a vital cog in the development of such technology. The development of low-cost, integrated circuits has produced a fingernail-size square of silicon that can duplicate the data storage and retrieval functions of the gigantic computers that are now being phased out. Since the invention of these microproces-

Electronic Alternatives to Postal Service

sors in 1971, their price has declined sharply, and they are now widely used in minicomputers and in office machines that can be programmed with intelligence functions, such as the "smart" typewriter in word processing systems. This is a form of electronic typewriter in which an electrical or magnetic record is created at the same time that an operator keyboards the text. These typewriters are used in the offices in which the text is originated or read—an important advantage—and they can be readily connected by high-speed telephone lines. Although the majority of today's 200,000 word processors are not used for communications (only about 8 percent are currently connectable to networks), all new production comes with optional communications features, which can facilitate writer-to-reader communication from one office to another. Word processing, as a form of electronic mail, appears to have marked cost and convenience advantages over rivals like TWX/Telex or facsimiles. It has been estimated that about 800,000 word processing machines will be used by businesses in the early 1980s, although many of these machines will not be interconnected.

There are other important forms of electronic text transmission that are facilitating the telecommunications revolution—in particular, elaborate business communications networks established through computer terminals, which produce computer mail. (Mail in this sense does not necessarily mean "hard copy"; the message may be displayed on a cathode ray tube for visual inspection only, and then placed in the memory unit.) For example, thousands of businesses have established private networks to connect outlying activities with computer-controlled operations. While used for many purposes, these systems are also utilized for administrative "traffic" that might otherwise be sent by mail. And new carriers or companies have come into existence offering similar services.[14]

Communications Brokers and Value-Added Networks

The Federal Communications Commission has embarked upon a policy of permitting the entry of specialized common carriers into telecommunications markets that previously were solely the province of the established communications common carriers. The convergence of telecommunications and computers has, in part, led to the creation of "value-added" services where, for example, the telecommunications service is enhanced by applying computerized switching instructions to the distribution of electronic information (i.e., packet switching).[15] This convergence has also encouraged "brokerage"—acquisition of communications services or facilities from established or specialized common carriers and offering their use to third parties—and "shared use"—a form of brokerage in which the broker also satisfies a communications need of its own.[16] It is these services that have created in the telecommunications industry the real potential for economical large-scale replacement of USPS physical delivery of messages by electronic message distribution systems. Established carriers have also penetrated this market with responsive offerings, including AT&T's digital "data under voice," pilot "transaction network" project transmission services, and the "DataSpeed 40/4" and "Transaction" telephone.

Narrowband Transmission: ViewData—Combination of Computer, Television and the Telephone Line

There is a tendency to ignore the capability of the ordinary telephone line in light of the dream of a broadband "pipeline" (discussed in this section). But the twisted-pair, narrowband telephone is used today in 94 percent of American homes, and can effectively deliver non-video information. In a development called ViewData, the British Post Office (which runs both the telephone and postal services) proposes to marry the telephone, the computer, and the television screen. This is an advanced and versatile system whereby data from the computer is ordered by the telephone subscriber over the regular telephone line and exhibited on the television screen by means of an attachment. The British Post Office provides the means by which a growing number of "information providers" can make available a wide range of information for users to retrieve according to their needs. The British system will thus provide the physical resources necessary to store and access this information, but it will be the task of the information providers to collect and edit their own contributions before input to the data base. The system is designed to appeal to both the residential and business communities, and offers a subscriber access to virtually unlimited numbers of information pages.

One of the most significant features of ViewData for telecommunications is its coded, individually addressable "mail" service. Either by selecting pre-formulated messages or, if the user is provided with a fully alpha-numeric keyboard, by generating complete messages, ViewData will accept messages from one user for collection by another. In short, it provides electronic mail using two elements found in virtually all U.S. homes (the telephone, with its 94 percent penetration, and the television set, in 98 percent of these homes). Thus, this system could make message transmission relatively inexpensive as a simple extension of telephone service.

ViewData is currently being pilot tested. At the end of the trial period (in March 1978), the service, including an extensive data base, is to be ready for test marketing (i.e., a "public trial" involving 500 or more selected users to access information at three or more sites). If the results are satisfactory, full commercial service could commence in 1979. Similar experimental systems are being inaugurated in West Germany and Japan, with several other countries (such as Canada) conducting feasibility studies.

Broadband Transmission

Within the next ten years, there will be a number of broadband transmission systems able to accommodate more input and output of messages than the narrowband telephone network at low cost. These include satellite, coaxial cable, wave-guides, and optical fibers, which are described in the following paragraphs.

Satellites

Broadband connections between businesses are possible via satellites, which enable the interchange of all types of data, including of course the mail. To give one example, the Federal Communications Commission recently authorized the Satellite Business System (SBS).[17] This system, to be built by a Comsat-IBM-Aetna consortium, promises a switched broadband digital service providing two-way voice, data and image service through small (16-23 feet in diameter) and relatively inexpensive transceiver antennas located directly on the user's premises. Significantly, the system would also entirely bypass the telephone landline switched network.

Costs for satellite communications systems are decreasing as new series of satellites are introduced. At present, satellites, when used for point-to-point circuits, are economically competitive with other communications systems only when the ground stations are located more than about 500 miles apart. By the late 1980s, however, satellite circuits may be cost-effective, compared to terrestrial systems, at distances of about 50 miles. It is estimated that a full-scale electronic mail system (EMS) utilizing satellites would require at least 100 ground stations which, for the newer Intelsat (international satellite) systems, range from $3 million to $6 million per installation. The cost projections for earth stations with small antennas (15 feet in diameter) range from $60,000 to $150,000. Nevertheless, the advantages of satellites in an electronic mail system are numerous. In comparison with microwave relay systems, satellites could accommodate the larger volume of messages that any large-scale electronic mail service would produce. Costs per page of a letter produced digitally (such as by computer, Telex, or "smart" typewriter) and transmitted by satellite could be extremely attractive to large business mailers who initiate the bulk of present and projected first class mail.

"Time-sharing" arrangements for satellite transmission of home electronic mail could be arranged. Under time-sharing, electronic messages could be transmitted at off-peak hours, after business users have left their offices. The messages could then be stored in home delivery terminals and be delivered electronically the next morning.

Satellites will also serve to create the "switched-data network in the sky"—as any earth station could connect with any other earth station. This could provide greater coverage for EMS, and the greater access capability could also help reduce transmission rates.

Coaxial Cable

The use of coaxial cable as a distribution medium for television programming is now over 20 years old. Its principle use has been in the development of cable television systems. Originally, such systems were developed as community antennas to improve television reception in rural areas. Today, cable television includes importation of distant signals, multi-channel capacity, and specialized channels for public access, first-run movies, and news service reports. Twelve million persons nationwide have coaxial cable connected in their homes by 3,500 cable systems. Besides providing entertainment, cable television has enormous, though unrealized, potential as a home information service.

The big question mark about cable is its ability to operate in the core area of urban markets, where the majority of the U.S. population lives. Warner Cable Corporation's recent announcement of a massive experimental system in Columbus, Ohio, if successful, may result in a dramatic breakthrough of "blue sky" cable development in urban areas. Warner describes the experiment as a computerized, interactive cable system. It will offer 30-channel programming, pay-by-program selectivity, a hookup that can serve as both a fire and burglar alarm, and two-way capability between a viewer (using a remote control device) and cablecaster (hooked up to a computer in its studio). Viewers will be able to take tests, play games, and select pay or broadcast programs through the two-way device. Because of cable's multi-channel capacity, specialized services, such as video shopping, could be added for a relatively low cost. Such a system may eventually be tied to a keyboard/hard copy printer that could transmit mail electronically at a cost substantially lower than projected first class mail rates. Keyboard technology is developing rapidly, and portable lightweight models that can be coupled with a telephone are already being marketed.

The TICCIT system, developed several years ago by the Mitre Corporation, a research and development firm, suggests how another type of electronic mail system (via cable television) might work. The TICCIT system devotes one channel of a cable system to the transmission of still images in a half frame (a unit lasting about 1/60 of a second), with a digital code addressing the frame to its intended recipient. When the transmission and reception code match, the receiver then "holds" the frame; stores the image on magnetic tape, disk, drum or electronic storage tube; and flashes the message on the television screen at a rate of 133 words every 30 seconds. Such a system presently would cost a cable subscriber $33 per month, and the picture quality is fair to poor. However, it is undoubtedly a matter of time before engineering improvements refine production and costs are lowered. Such home terminals will be feasible only when prices are sufficiently low for consumer acceptance. Given the multibillion dollar market potential of the home terminal, there is substantial incentive for further research, innovation, and competition in the production of home terminal hardware.[18]

Optical Fiber

The coaxial cable and existing telephone wire could be supplemented, or even supplanted, by transmission through optical fibers. These fibers, about 2/100th of an inch in diameter, use modulated light beams generated by a laser source to carry digital data at transmission rates of millions or billions of bits per second; at the same time, the fiber eliminates frequency assignment and interference problems. The fibers are now being tested by AT&T and others in trunking operations, and in time could become the broadband "highway" into the home, greatly lowering the cost of data transmissions (including electronic mail). Indeed, because of its great capacity, there could eventually be virtually instantaneous low-cost transmission of the kind of information currently found in magazines, books, and lengthy documents.

Interim Systems—Facsimile; Interconnection

Before advanced telecommunications systems are fully developed, interim systems may be introduced for transitional purposes. Facsimile systems may be the bridge to future all-electronic systems using digital (i.e., keyboard) input and output. The USPS has expressed interest in utilizing facsimile to offer more efficiency and economy by reducing costs of long-distance manual transmission of letters. In the early 1970s, it established an experimental facsimile service between post offices in Washington, D.C. and New York City, and a six-city experimental facsimile system is under consideration.

The facsimile system can copy a complete page, including figures and pictures, and transmit it to a terminal that reconverts the original signal back to hard copy. Telephone lines are currently used to connect facsimile terminals, which now rent for as little as $30 a month, and transmission time per page is 4 to 6 minutes over dial-up lines. These terminals may not be cost effective for home use; however, costs will drop when broadband transmission is utilized. For example, with a terrestrial microwave relay a facsimile page can now be transmitted for 3.3 cents, whereas satellite transmission could reduce that to 0.3 cent per page if a more expensive terminal were utilized. This facsimile system may appear attractive now, but once all-digital systems are in use, facsimile transmission will not be able to maintain itself as viable competition. It takes at least 250 times the number of electronic signals required for a one-page digitally coded message to carry a similar facsimile message. Because transmission rates are based upon numbers of signals delivered, facsimile seems destined to be the bridesmaid.

Facsimile, despite the high cost of terminals for rent or purchase, may be more promising in the home because it provides hard copy and does not require the hardware or skill necessary to operate word processing keyboards. This, too, may be a short-lived advantage that will no longer exist once the home terminal market is reached. Resistance to technological adaptation should be minimal as new communications technologies are introduced to consumers accustomed to using pocket calculators, computer terminals, the telephone, and various learning machines. Massive education at the secondary school level, similar to present driving lessons and typing instruction, will create a new generation of consumers that takes for granted processing of electronic messages. Facsimile could remain as an alternative to home users who refuse to work with or are incapable of using word processing equipment.

Electronic Alternatives to Postal Service

The market for facsimile in business is small, inasmuch as most business messages can be digitally coded by computer; this leaves only correspondence for facsimile transmission. Hard copy messages may decrease as more business communications become electronically projected by cathode rays to electronic screens. This will be especially economical in the home where the television set can be a message center, and an "order button" can retrieve messages from computers for repeated display.

In the meantime, there are a variety of hardware systems besides facsimile that produce hard copy. The most familiar system is the teletype, which is relatively inexpensive. Unfortunately, it only prints a line every few seconds, and it would not be compatible with sophisticated computer systems. Digital line printers, presently costing $10,000 to $20,000, may be better suited to developing communications technology. These printers can handle 1,000 lines per minute at a cost of 0.5 cent per page of prepared letter copy. The main disadvantage is print quality— digital line print lacks the sharpness and clarity of the office typewriter. Moreover, because these printers have an impact printout system, the resulting noise would prevent use in a home terminal system.

Perhaps the best hard copy system for home terminal use will be the thermal printer. Present costs for a printing unit ($2,000) and per page (one cent) are too high for the consumer market but if prices drop, such printers would be adaptable for home use. The system is silent because it applies heat to specially treated paper that changes color to record the printed character.

It has been suggested that facsimile might be the easiest way for the USPS to introduce an electronic mail system—namely, a facsimile service between post offices that could be initiated with a relatively small capital investment with the possibility of favorable economies of scale. Such a system would have to be designed to serve business-to-business mail traffic in geographically concentrated areas, where a postal carrier could deliver the final hard copy. The benefit would be a reduction in transportation costs and delivery time, not a reduction in letter-carrying personnel. (See the discussion in this chapter under the heading "USPS Entry into EMS: Policy and Pragmatism.")

Initially, this system would probably be part of a partnership arrangement that enabled use of electronic common carrier facilities during off-peak periods. Such a productive partnership of communications technology and hard copy postal delivery is already in existence. The Western Union/USPS Mailgram (which can be delivered to a post office, telephoned to a telegraph office, or telexed directly to a receiving terminal) has become a popular, though limited, mechanism for transmitting postal messages more efficiently. The Mailgram's digital input, transmitted either over a data link or on magnetic tape, is printed out on terminals at selected post offices.

The devices used to originate and reproduce electronic mail—word processing machines, computer input and output machines, facsimile, TWX/Telex—are not alike. A Western Union teleprinter, for example, is not as easily connectable to a computer as is an IBM word processor. It might be argued, therefore, that to become operable, electronic mail must be limited to a single, designated carrier, which in turn will create rigid specifications for terminals and their use.

An alternative exists, however, which allows the innovative forces of the market to prevail and yet ensures universal compatibility with all terminal devices. The alternative rests upon the word processing sector and the use of "computer on a chip." The inexpensive chip connected to a terminal can perform the necessary interconnection; this chip can be attached to all word processing devices located in the offices of senders and receivers. Thus, available technology can convert existing devices already in place as well as machines of the future into a functioning system of electronic mail.

Toward the Cashless Society

The effects of technological development on mail service can be seen most clearly when both software (services) and hardware (machines and transmission systems) are examined.

An important element in the software revolution is the Electronic Funds Transfer System (EFTS), the innovation that could lead to a "cashless" society. EFTS is designed to replace conventional payment systems, such as cash and checks, with electronic systems, rather than paper instruments, to record financial transactions.[19]

EFTS is an outgrowth of the public's increasing acceptance of checks and credit cards as methods for financial transactions. Today the typical American is involved in over 1,000 cash transactions yearly, and a large majority have checking accounts. Total check output is staggering: Individuals write or receive about 19 billion checks a year, and business and government produces over 8 billion checks a year that do not even involve individuals.

The growth of major national inter-bank cards, such as Master Charge and VISA, also serve to stimulate the development of EFTS. These cards have found consumer acceptance because of

Electronic Alternatives to Postal Service

their convenience, production of accurate purchase records, and reduction of the threat of cash robbery.

EFTS would make these credit systems more profitable because of significant cost savings relative to the current system of handling checks. The present profit structure of inter-bank cards arises from interest on unpaid consumer account balances and from income acquired through merchant discounts. A major portion of expenses involves steadily rising clerical and postage costs associated with billing and rebilling. EFTS could eliminate this intermediary step, as money would automatically be transferred at a specified date from an output terminal at the consumer's bank to an input terminal at the collection site via a computerized clearinghouse. In turn, the clearinghouse would distribute funds to credited accounts.

A fully established electronic funds transfer system would also perform a variety of other tasks. An EFTS card could be used for automated banking transactions, such as withdrawals and deposits which can be handled by computer. Such systems are already in use in several major banks, including Union First in Washington, D.C., and Citibank in New York City. These systems provide convenience and flexibility—many terminals are open 24 hours a day. It is estimated that banking institutions have already made a $1 billion investment in automated banking equipment. Because all the technology necessary to develop a full range of electronic payment services is now available, it will be relatively easy to interconnect the present machines to form larger EFTS networks.

Retail establishments would be another important part of any nationwide EFTS network. Many stores already have electronic check verification systems, and these could be upgraded to provide electronic transfers at the time the sale is transacted (point-of-sale system). Such a system would replace or supplement the present methods for retail payment: cash, checks, and credit cards. The electronic point-of-sale system could capture data relating to the transaction (e.g., retailer, customer, price, date, merchandise), and then make an immediate transfer from the purchaser's bank account to the business account.

The promoters of EFTS predict it will have significant social benefits, such as a reduction of cash-related crimes (e.g., robbery, assault and battery); increased ability to attribute liability for products and services (leading to greater consumer protection); and changes in the use and value of credit, making it more widely available to lower socioeconomic groups, and more convenient and probably cheaper to all groups.

Such optimism should be tempered by several considerations. Although street crime may be reduced, it is possible that white-collar crime will increase as criminals devise new schemes to "tap" the computers handling the financial transactions. Most computer experts agree that currently it is easy to commit massive financial embezzlement through fraudulent electronic transfers.

Tampering with large data bases that would be the core of EFTS also raises serious questions concerning privacy for both individuals and businesses. Appropriate technological safeguards and enforceable privacy statutes would be necessary components of any nationwide electronic funds transfer system.

To make EFTS cost effective, it must be operated on a large scale. Its success may depend on how well it is marketed to American consumers and businesses. EFTS is a system created as a result of technological development rather than from any dissatisfaction by consumers with the current methods of payment. Consequently, it may be difficult to "sell" the concept to the poor and uneducated, who would be forced to deal with unfamiliar procedures and with abstractions that they might find difficult. The introduction of a full-scale point-of-sale system may also cause problems for retail buyers, who may resent the elimination of long-accustomed choices for payment. For example, with an EFTS retail system, Aunt Minnie could no longer "float" funds in her checking account, since the omnipresent computer would made an immediate transfer.

Probably the most acceptable electronic funds transfer system will involve pre-authorized deposits and payments conducted through automated clearinghouses. Presently, there are over 2.5 billion deposits made per year, including payroll, Social Security, pension, dividend, and welfare payments. In addition, there are over 4 billion payments made per year, including rent, mortgage, utilities, insurance, credit cards, and installment loans. Since most of these payments pass through the mail, the costs of conducting transactions are reflected in additional clerical, accounting, and banking services.

Under a "paper" transaction system, when a paycheck is received or a utility bill payment made, the check itself must be moved physically, if only to the bank on which it was drawn and then returned to its originator.

By contrast, EFTS would deposit a paycheck or a utility bill payment electronically at an agreed-upon date. Even though paper records would still be provided to the individual as a receipt, the volume of paper messages would be reduced and the sequential return process, including the sending of follow-up

Electronic Alternatives to Postal Service

notices, would be abbreviated. Pre-authorized payments are projected to increase in many areas of the commercial sector, for example, among landlords, mortgage holders, and large-volume billers (e.g., utilities, insurance companies), who may offer incentives to consumers for agreeing to pre-authorization (e.g., a one-percent discount). Unlike point-of-sale systems, there should be less resistance to accepting paychecks by pre-authorization because the primary benefit of physically receiving a payroll check is the psychic satisfaction of depositing it in the bank.

The federal government is and will be a major force in the switch to pre-authorization. It is now a major distributor and recipient of income to and from individual employees, taxpayers, and welfare beneficiaries. Today approximately six million monthly payments, such as Social Security checks, are electronically transferred to recipients' banks via automated clearinghouses, and the figures are expected to rise to 18 million by 1980. Within five years, over one-third of these checks will be converted to EFTS, which will result in diversion from the first-class mail. Moreover, by 1980 the federal government will have converted nearly half of its 62 million monthly payroll and pension mailings to EFTS, which itself will result in a potential annual revenue loss to USPS of $46 million per year (based on the current rate for first-class mail, 15 cents per check). The reason is obvious: it now costs the Treasury 19 cents per check (15 cents for postage, 4 cents for processing), while it can transfer the funds directly into the bank accounts of recipients at a cost of 2 cents a transaction.

This may just be the tip of the iceberg. Similar massive business mail diversion through EFTS is also expected within five years.[20] Arthur D. Little, Incorporated, a prominent consulting firm, has estimated that as many as three-quarters of all checks written might eventually be replaced by EFTS. A nationwide automated clearinghouse system could be established in the 1980s.

By that decade, the annual volume for first-class mail is projected to be 55 billion. Six out of every ten pieces of first-class mail will involve business or governmental financial transactions that could be diverted to pre-authorized electronic transfer. In economic terms, nearly half of today's total postal revenues generated by those financial transactions could potentially be lost to EFTS, primarily through pre-authorization. Point-of-sale systems will have little impact on mail revenues because many retail checks are not mailed.

EFTS is likely to remain a competitive, private enterprise. It is possible that the USPS may try to integrate EFTS technology into its own operations (e.g., electronic money orders). On a larger scale, the USPS could upgrade its current electronic equipment to provide automated clearinghouse services for pre-payment transfers. An arrangement whereby the federal government provided a special subsidy to the USPS to handle all pre-authorization would only be possible if the USPS could offer competitive EFTS services. The questions involving the USPS role in EFTS are analogous to those that must be faced for electronic mail as a whole: Can and should the USPS enter the competitive electronic marketplace?

Impact on USPS

In light of the previous, little more is needed to establish the seriousness of the problem confronting the USPS. First-class mail constitutes the most profitable type of mail because of the way in which USPS charges for it. The electronic transfer of funds will decrease the transaction segment of the first-class mail, which currently comprises 35 percent of *all* mail, particularly transactions among businesses. These transactions constitute the majority of first-class mail and are the most profitable part of the mail stream. The USPS cannot predict how long it will be before the EFTS erosion will occur and therefore estimates losses ranging from 5 percent to 25 percent of first-class mail within ten years.[21] Other estimates are from 9 to 16 percent of current first-class mail by 1980, and 25 to 40 percent by 1985.[22]

It is even more difficult to estimate the impact of the other EMS devices. But as noted, 80 percent of first-class mail is clearly identifiable as business mail (44 percent business-to-business), and probably about 30 percent of all government and business transactions are amenable to transmission as electronic messages. It has been estimated that within the total "message market" the USPS share will decline from 22.7 percent in 1970 to about 13 percent in 1980.

Even if these estimates appear to be high, it is clear that there will be substantial erosion of the USPS by competitors in the message market, and the implications for the Postal Service are evident. If the USPS experiences substantial reductions in mail volume, especially in the first-class mail and is still required to maintain traditional services, the cost for the remaining volume of mail will increase because, as previously discussed, a large part of its costs are relatively fixed. Under the present rules governing the USPS, it cannot reduce its costs in proportion to volume losses.

The Role of the USPS in the Electronic Communications Revolution: Probable Scenarios

Historical Background

The history of the USPS may well contain some answers to the USPS problems concerning electronic alternatives to conventional mail. That history indicates a clear and consistent dichotomy: Postal Service control over the physical delivery of material (letters) from sender to recipient but the reservation to the private enterprise sector of the transmission of information or messages by electronic means.

As Charles Jackson has pointed out,[23] the Postal Service has considered the problem of electronic mail for over 130 years, ever since the invention of the telegraph. The 1845 telegraph line between Washington and Baltimore was operated by the Post Office Department, which urged at that time that the government run the telegraph system. While the Congress did not then respond, a provision in the telegraph legislation of 1866 (Post Roads Act) authorized the government to purchase existing telegraph plants after 1871. Postmaster General John Wanamaker in 1892 proposed, in effect, a Mailgram service: postal telegraph messages would be collected and delivered by the Postal Service, with the long-distance telegraph service provided by private companies under contract to the Postal Service. (Wanamaker even suggested that the postal telegraph system might someday offer a facsimile service.) None of these proposals made headway, however, because of Western Union's then-powerful private monopoly (with revenues in 1890 of $20 million, one-third of the postal system's revenues). And in 1947 Congress repealed the Post Roads Act.

The Post Office Department did have one brief period of electronic operation in 1918 when the government, as a wartime measure, took control of the telephone and telegraph systems. But this episode ended with the war, and the next experiment did not occur until 1959, when the Post Office tested "Speed Mail." This was a facsimile experiment between Washington and Chicago in which government agency mail was transmitted using facilities supplied by private carriers. Western Union strongly protested against this intrusion of government into the private electronic sphere, and in early 1961 the new Kennedy Administration terminated the experiment because, "...the very limited research and development funds should be applied on other high priority projects."

The telephone and telegraph eventually established their own scheme of regulation. In the 1910 Regulatory Enactment, Congress gave the Interstate Commerce Commission jurisdiction over these interstate communications, although the Commission largely failed to exercise this control. And in 1934, Congress enacted a comprehensive statutory scheme for the regulation of interstate and foreign communications services by wire, cable, and radio and of all persons engaged therein. The Communications Act of 1934 sets forth the broad policies and standards governing the regulation of these services and entities and specifies the processes to be followed in implementing its authority. The Act reserves to the states the regulation of intrastate electronic communications matters. The Act buries the last vestige of postal control over telegraph companies, with its provision in Section 601(b):

> All duties, powers and functions of the Postmaster General with respect to telegraph companies and telegraph lines under any existing provision of law are hereby imposed upon and vested in the Federal Communications Commission.

The importance of this history is obvious. It appears that the Congress, while making the traditional mail services the sole responsibility of the Post Office Department, has historically relied on private enterprise to serve the needs of the public for the transmission of intelligence by electrical means; this policy was codified in the 1910 Regulatory Enactment and the Communications Act of 1934. In the following we address the question of whether the 1970 Postal Reform Act changed that legal position.

The USPS Authority to Engage in Electronic Mail Services Under the 1970 Postal Reform Act

The argument for USPS authority to engage in EMS stems largely from the rather broad provisions of the Postal Reform Act (PRA) of 1970. This Act contains provisions supporting the application of new technologies to increase the benefits that the nation derives from the operation of the Postal Service:[24]

> (1)...provide prompt, reliable, and efficient services to patrons....

Electronic Alternatives to Postal Service

(2) ...give the highest consideration to the requirement for the most expeditious collection, transportation, and delivery of important letter mail.

(3) ...receive, transmit, and deliver...written and printed matter...and provide such other services incidental thereto as it finds appropriate to its functions and in the public interest.

(4) ...provide for the collection, handling, transportation, delivery, forwarding, returning, and holding of mail....

(5) ...provide, establish, change, or abolish special nonpostal or similar services...

(6) ...promote modern and efficient operations and should refrain from...engaging in any practice...which restricts the user of new equipment or devices which may reduce the cost or improve the quality of postal services....

(7) ...have all other powers incidental, necessary, or appropriate to carrying on its functions or the exercise of its specific powers.

These provisions support a dual approach to USPS authority in electronic mail: (1) it is a more modern and efficient way to accomplish the Postal Service's existing mission; or (2) it is a new "special nonpostal or similar service" to postal service.

There is no question concerning the authority of USPS (the first approach listed above) to substitute electronic transmission of information for a segment of the mail stream (a function analogous to the World War II "V-Mail" service). Under this scheme, letters would be converted to electronic signals at the post office nearest the sender for transmission to the recipient's nearest post office, where it would be converted to hard copy for delivery (called Generation I EMS in the recent National Research Council [NRC] Report).[25] There would be savings in using technology to bring about this more efficient transmission but the savings do not come in areas of the greatest cost to the USPS—the physical processing and delivery of the hard copy output.

There would also appear to be little question about cooperative ventures between the USPS and the electronic carriers or large mailers, whereby the electronically transmitted information would enter the mail stream at a post office close to the recipient (called Generation II in the NRC Report). Mailgram is of course an obvious example of such a cooperative venture—a most successful one (as shown by its growth from 370,000 messages in

1971 to about 24 million in 1975). Such joint efforts could be extended to other electronic carriers.

Further, the NRC Report estimates that about one-third of all letter mail would be capable of input in electronic form; and that the mail volume from large corporations is so concentrated that it would make electronic inputs practical and economic.[26] (There are also, of course, disadvantages—e.g., the inability to insert advertising or other material, but the cost savings could be substantial.) The most likely candidates for such electronic inputs are utilities, banks, department stores, gasoline, credit card and insurance companies. The companies themselves could supply the electronic input from their computers, through carrier lines to the postal installation, or could employ value-added carriers to do the job.

Thus far, there has been no serious legal issue because the USPS is not engaged in electronic message transfer itself but rather would be entering into joint enterprises with others who supply the electronic input and make use of the Postal Service's existing strength—universal physical delivery to homes and businesses.[27] But what if the USPS itself enters the electronic message-transfer business?

First, the 1961 "Speed Mail" experiment was an attempt to enter the all-electronic era (although delivery by the postman was still part of the service). The legality of this endeavor was vigorously challenged by Western Union. A decade later the USPS 1971 experimental offering facsimile mail service between selected post offices in New York and Washington, D.C., aroused no electronic carrier opposition. The different responses may be a result of a practical consideration: Electronic carriers do not consider such facsimile activity to be a real threat. (See discussion in this chapter under the heading: "USPS Entry into EMS: Policy and Pragmatism: Generation I".)

A very different issue would arise if the USPS sought to emulate the electronic carriers, including the value-added networks (e.g., Telenet or Tymshare) and provide duplicating electronic message services directly from the sender to the recipient (the so-called Generation III mentioned in the NRC Report). Here, a serious legal issue would be present: While the quoted sections of the Postal Reform Act of 1970 are broad in scope, surely if the long-settled, historical dichotomy between physical and electronic message delivery were to be set aside, there would have been some mention of this in the Congressional floor debate or in the Congressional hearings. The Commerce Committees would be intensely interested, since the centralized concept of the Communications Act of 1934—an FCC

Electronic Alternatives to Postal Service

with full authority over all interstate electronic carriage—would be undermined. But the legislative history contains no indication of such a drastic turn—evidence that the USPS was not given authority to challenge the value-added networks or AT&T and its competitors.

In many ways the situation is similar to that in Canada where a high-level Canadian postal official, after examining portions of the Canadian Postal Act containing very broad provisions, stated that his "first impression was to congratulate our legislators on their amazing foresight." However, his legal advisor explained that electronic mail "would represent such a quantum leap of progress that it could not possibly have been envisaged by the legislators," and that therefore the Postal Act would have to be changed to allow electronic services.[28]

The most important question is what policy should Congress adopt in its pending reconsideration of the role of the Postal Service? We discuss that question in terms of both policy and pragmatism in the following.

USPS Entry into EMS:
Policy and Pragmatism—Generation I

The USPS should be encouraged to use electronic means for inter-city transfer of mail, in the interests of policy and pragmatism. However, transportation between post offices represents only a small percentage of the USPS costs (about 6 percent). The USPS would still have to pay for the delivery portion (over one-third of total costs) and for the actual mail processing—the movement through the Postal Services facilities after it is received and before it is delivered—a little less than one-third of total costs). And it would face the increasing loss from telecommunications competition.

It seems appropriate at this point to discuss a USPS facsimile service. One strength of the Postal Service is the thousands of post offices, many with convenient parking spaces in suburban areas. The obvious EMS development utilizing these locations is a facsimile system. As noted, there has been an experiment to provide a facsimile service between Washington, D.C. and New York City, and a larger facsimile system experiment is under consideration.

While competitive with firms like Graphnet, this facsimile service is, on analysis, a Generation I system. The sender can deposit a letter at the nearest local post office and send it to the recipient's post office for physical delivery or the sender can use the facsimile method to move the material between the post offices. Thus, the fact that the electronic carriers are now offering facsimile services should not, *per se*, rule out this type of USPS operation.

As a practical matter, however, the USPS would face a difficult obstacle because of growing competition. The electronic common carriers would not only capture the high density markets but, more importantly, they would be able to provide electronic output on the user's premises, thus eliminating the cost and delay of physical delivery. The USPS can only succeed if it fills some need for low-volume users that is not currently met. The Postal Service has been reluctant to compete with private firms for this business. Thus, it recently rejected an unsolicited proposal from Xerox, that entailed a postal investment of about $30 million for field testing a facsimile system in postal installations. The USPS decided that this was too risky, based on unsuccessful market tests of similar concepts by postal authorities in Sweden and Great Britain.

Generation II

The cooperative ventures between USPS and electronic carriers such as Western Union (Mailgram) have been promising. The delivery system, as noted, represents the largest cost factor. But since the letter carrier delivering the mail cannot be eliminated, the factor should be viewed as a strength, and exploited to the largest degree possible in cooperation with industry. Hence, services like Mailgram should be expanded. But again, success of such ventures would not significantly reduce the cost burdens, nor would it stem the loss of business to EFTS and the other electronic competition involving business-to-business communications.

Thus, although Generation II ventures are helpful, they cannot solve the problems of the USPS. Indeed, if the USPS constructed a larger-scale mail system based on electronic input/physical delivery, it would represent a considerable risk in terms of both technology and market acceptance. It appears that such a system would face the eventual development by the electronic carriers of comprehensive electronic input/electronic output schemes, eliminating the cost, delay, and uncertainties involved in physical delivery. Clearly, the key is the ability of the USPS to engage in Generation III endeavors (electronic input/electronic output).

Generation III—Business-to-Business

At this stage the USPS would compete against electronic carriers and service firms (e.g., AT&T; Western Union; satellite and specialized common carriers; value-added networks) in providing electronic message services. As noted, these firms, using primarily computer and data transmission systems, now offer business-to-business messages of all types, including EFTS. Both policy and practical considerations are strongly against USPS entry here.

First, the USPS comes very late to this telecommunications scene. The computer-communications explosion has occurred, and private enterprise—including giants such as AT&T, IBM, and Comsat—is pushing the limits of the expanding technology steadily outward. As J. T. Ellington Jr., Senior Associate Postmaster General, acknowledged,[29] there is a substantial question of the USPS..."capability, from both a technological and managerial standpoint, to launch (an EMS) effort associated so much with advanced technology"; to put the pragmatic case bluntly, the electronic business-to-business message services are probably already beyond the reach of the USPS.

It might be argued, however, that this is a challenge for the USPS to enter the competition and let the marketplace determine the outcome. To enter this market, however, the USPS would need, in practical if not explicit political effect, congressional authorization to increase substantially the level of research and development and to channel such efforts in the wholly new direction of outright competition with the electronic carriers or service firms.[30] The USPS requires each year a very substantial congressional subsidy and, whether or not earmarked for electronic message delivery, the subsidy facilitates the overall postal operation. Thus, if USPS should seek congressional authorization to engage in electronic business-to-business message operations, it might expect to meet several objections as follows.

First, what purpose is being served by such USPS operations? This is not an area where the nation lacks full or innovative services. On the contrary, with the pro-competition policies being pursued by the FCC, the field is filled with new entrants pushing the emerging technology and services as rapidly as possible. Why then should the federal government subsidize the governmental entry of USPS? It is no answer to say that USPS might use the revenues to subsidize the traditional mail operation. If Congress decides to continue that operation for the national good, the monies can and should come from the general Treasury. (In any event, it is more unlikely that the USPS could reap substantial profits by competing with AT&T, IBM, ITT, and other growing service firms.) In the words of former Postmaster General Benjamin Franklin Bailar:[31]

We don't have a clear, legislative mandate to get into electronic funds transfer, and we don't have the capital resources or the know-how to get involved in that. Apart from all of that, I question whether we should be getting involved in activities so different from traditional mail service. If private industry is willing and able to provide a service, why in heaven's name should the Government get involved?

Second, because there is the issue of whether the USPS could be regulated by the FCC, USPS entry into Generation III enterprises might well dilute the centralized control over telecommunications envisaged by the Communications Act.[32] In any event, such entry would raise difficult problems. If the USPS entered as a carrier, it would become the second largest carrier in the country, altering the entire character of competition in electronic carriage. Such entry by USPS, a partly subsidized, quasi-public entity, would result in extraordinary problems of cost allocation, e.g., cross-subsidization.[33] If the USPS entered as a value-added network, it would be a *broker* and would not be in the electronic transmission business. But its massive subsidized presence might drown competitors like Graphnet and hence force regulation of a service that should be largely unregulated. Could the FCC deregulate the brokerage markets because it believes the competitive market forces will protect the public interest, *if* the USPS with its great resources, including Congressional subsidy, enters this market?

In short, the nation has a domestic telecommunications system that has served the people well and compares advantageously with government-owned foreign systems. It is most unlikely that the Congress would opt now for government entry into a well-established and well-performing private enterprise system, particularly in view of the regulatory and economic disarray that would be produced.

The Private Express Statutes (PES)

This section presents a brief discussion of the Private Express Statutes (PES) because of the fear that has been expressed that these statutes, together with 1974 regulations promulgated by the USPS, might be used to prevent electronic message developments by carriers other than the USPS. As noted, the statutes confer upon the Postal Service an exclusive right to carry "letters" for others over post routes, principally to secure revenues for providing needed postal services to all areas of the United States, as mandated by Congress. In 1974, the USPS adopted new PES regulations defining "letters" with certain exceptions to include every tangible or "hard" mailable object except parcels and unaddressed circulars specifically including data-processing materials like recording disks and magnetic tapes.[34] This has aroused fears concerning extension of the postal monopoly to include the transmission of virtually all commercial information.[35]

As a practical matter, however, the PES regulations could not be used to curtail the provision of business-to-business telecommunications services. The Postal Service denied such intention in the 1973 Report of the Board of Governors on Restrictions on the Private Carriage of Mail (p. 10). And former Postmaster General Bailar reiterated this pledge in September 1976, when he stated (in a speech to the National Postal Forum):

> Let me stress that we have no interest or desire to intrude in any way into existing telecommunications as they are conducted by American business. Not only would that be a practical and financial impossibility but it would also run directly counter to the notions of free enterprise that have made the United States the world leader in communications.

In any event, the PES regulations have no applicability to the heart of these telecommunications services. These services do not include delivery of "hard" objects over mail routes; indeed, the time and cost savings from the electronic process are severely diminished if there is a need for physical delivery. Under these services, the message is transmitted in seconds and read on a television screen (cathode ray tube) or comparable device, or printed in the office or place of the final destination.

Finally, the applicability of the PES regulations has not been determined for the Graphnet-type of physical delivery operation. Graphnet, one of the new value-added service firms, provides a facsimile service that accepts input in a variety of forms, uses electronic storage and transmission, and provides output in either electronic or hard copy form; the service offers the option for ultimate physical delivery of a piece of paper bearing a message and a written address, which can be delivered via the usual postal routes. USPS claims that it has a monopoly of this final physical delivery, and states that Graphnet must either use its services or affix the proper postage. Graphnet uses private contractors for the delivery, and leaves it to them to settle this issue with the USPS.

The USPS claims the legal right to adopt its 1974 regulations under the broad rulemaking power given to it in Section 401(2) of the Postal Reform Act. But opponents argue that this rule-making power does not extend to criminal statutes (the PES is contained in the Federal Criminal Code Title 18) and cannot be used to redefine the parameters of the postal monopoly. These adversaries rely on the long-established exception for telegrams, showing that it would be arbitrary to extend the monopoly to other electronically originated messages and, indeed urge, on the basis of old cases, that the law be restricted to current and personal correspondence.[36] If not reconsidered by the Congress, the matter will eventually be settled by the courts; reportedly, the Department of Justice, which must prosecute for violations of the PES, is not enthusiastic about instituting cases, so the court test may be somewhat delayed.

Generation III—Electronic Mail to the Home

If the foregoing analysis is correct, the business-to-business electronic message services are beyond the reach of the USPS. That leaves for consideration the more difficult issue of the electronic message service to the home (either business-to-home or home-to-home). The USPS does not encounter private enterprise pre-empting this area at a rapid rate. On the contrary, while there are developments in this area, the process appears slow simply because the home represents a low-volume user of written messages. Because of this consideration, there is the opportunity for USPS services. We discuss below a possible scenario and the countering arguments.

Electronic message service by the USPS to the home would be based upon two related factors: (1) there is a gap in the present telecommunications drive by the electronic entities (a lack of interest or plans to provide extensive household coverage), and (2) the USPS has a broad charter to provide nationwide household delivery service. The USPS, with its mandate, would thus seek to fill that gap by cooperating with existing carriers and electronic equipment manufacturers to become the manager of a household electronic message system. It would gradually phase out its high-cost physical delivery services, while a nationwide electronic message system would be constructed along lines such as the following:

First, the USPS would work with equipment manufacturers to obtain a terminal, in the beginning perhaps for reception only, which would sell in the neighborhood of a few hundred dollars, or rent for less than $10 a month. This terminal would be coupled with the telephone line, and would have a visual display and hard-copy option for an additional small rental fee. Such ter-

Electronic Alternatives to Postal Service

minals might be later integrated into television sets or home video entertainment units. The consumer would inspect mail on the terminal screen, and if there were no hard-copy attachment, could request the USPS to deliver hard copy (for an additional charge). The system would also indicate to the consumer when more mail was waiting if the memory unit was full.

Second, the USPS would accumulate messages, sort and route them for transmission over common carriers, sort again, and then deliver them for storage and electronic distribution during off-peak operating hours (e.g., midnight to 6:00 a.m.) for telephone company lines (or cable television lines or radiowaves).

The USPS could argue that a household electronic message delivery system requires management with a broad charter, since there is a need for universal coverage and a central focus to bring about nationwide distribution. By acting as manager, the USPS would be drastically shortening the otherwise long and difficult process of standardizing initially multiple, competitive, and often incompatible systems. It could not be substantially competing with the electronic carriers, which have shown little interest in the household market because of the high costs. On the contrary, the USPS would be assisting those carriers using their facilities in a way that generates revenue in off-peak hours. It might well need the help of the value-added networks for this large undertaking. Moreover, the USPS would assist the terminal equipment manufacturers by using the large-scale flow of mail as a basis for universal, high-quality home terminal systems, which would in turn encourage development of other private services. Indeed, if successful, the USPS would probably initiate the development of local broadband systems for immediate, rather than overnight, delivery of messages. And this might lead to the development of subsequent systems through which the home resident could send as well as receive messages.

In sum, it is argued that this approach would allow the USPS not only to escape its labor-intensive problems but to do so in such a manner that both carries out its charter and fulfills an unmet telecommunications need.

The countering considerations may be briefly stated. The recent NRC study noted that the average household receives only about ten items of mail per week, of which about six are first-class. The NRC panel concluded that this relatively small amount makes it "quite improbable that the average household will install a terminal solely for the receipt of electronic messages.[37] The report asserts that if a household terminal becomes available for other applications, the marginal cost of adding an electronic message service might be economically acceptable. In short, contrary to the above scenario, this view, shared by postal authorities in Canada and Great Britain, is that electronic mail will follow, not lead, the household terminal process.

A second point is that the USPS entry into the telecommunications area might well be opposed by the established carriers, who wish to render the service themselves. They would argue that without government entry, out of the competitive clash of private enterprise will come innovation and standards that reflect the best possible marketplace solution. In any event, as noted previously, computer technology already available resolves the standardization problem by allowing full interaction among technically dissimilar terminals.

A third consideration is that the home electronic message delivery market is admittedly the hardest one to penetrate, and thus calls for the greatest technical competence and managerial skills. The USPS has not demonstrated such competence or skills. Indeed, in light of this deficiency, it may already be too late for the USPS to enter this field; to be ready for service by the mid-1980s, it must begin to plan now. Yet the USPS has no present concrete plans, and it seems quite likely that one or more telecommunications entities might well be ready to pre-empt the home electronic message field through marketing strategies that come to fruition by 1985. Finally, as a practical matter, is it feasible to undertake an electronic message service, without the critically important business-to-business facet (now largely foreclosed to the USPS)? Could AT&T have succeeded if it had proceeded in this fashion?

Evaluation

The scenario described above, projecting a large USPS role in household electronic message deliveries, is not without substance. There is a present gap in the newer electronic *household* service; the USPS has a broad charter for universal service; and it would made sense for it to assume a leadership and managerial role in effecting a nationwide system, particularly in light of its present labor-intensive problems.

But pragmatically it is not likely to occur. The USPS has not developed the skills or initiative to capitalize on whatever its charter may allow in the telecommunications area. Further, a move in this area would represent an enormous switch for the Postal Service—and for the country. The United States has relied on the private sector to promote the national interest in telecommunications. It seems most likely, therefore, that we shall continue to let private enterprise lead the way to electronic mail

Electronic Alternatives to Postal Service

service to the home. Such private competition has the promise of the most important benefit—the fullest possible play for innovative technology and services.

The above judgment is necessarily speculative, because it is difficult to foresee the home telecommunications environment of the next two decades. It is easy to speculate about the technology because the technology is known; the economic and the marketing factors remain difficult to assess.

Experience abroad may be helpful here. Thus, the Japanese experiments such as Tama New Town and the British effort with ViewData will provide further marketing insight, upon which surer judgments may be possible. As noted, ViewData has a coded "mail" feature, individually addressable, that could operate with television screens and telephone lines already in 94 percent of all U.S. homes. But the mail feature of ViewData is one part of a complex system, being marketed by the British Post Office which controls both the telephone and mail system. Assuming that the package is commercially successful, what does that portend for this country, where the telephone service and related services are committed to private enterprise?

The above has been a discussion of a large USPS role in household electronic message delivery. There is also the issue of a smaller, perhaps transitional, role for the USPS. At some point, there will be significant numbers of households with electronic message capacity, but substantial numbers without such capacity. Clearly there must be some interim arrangement whereby the mail can be directed either electronically or by physical delivery. Because of the latter aspect, the USPS must have an important role in this transitional system, perhaps the manager, or, in effect, a "value-added" communications service operating as a resale carrier. Thus, the USPS could receive message traffic, route what it could to the electronic firms, and deliver the remainder by its own letter carrier system. The USPS would be permitted, and indeed required, to accept messages for transfer to the electronic carriers.

This scenario raises the issue of interconnection of electronic telecommunications service and the Postal Service. Today Mailgram offers an efficient form of interconnection between Western Union's electronic network and the USPS physical distribution network. Tomorrow, the interconnection issue will be multiplied greatly as the household electronic delivery capability grows. Thus, two basic questions are presented: (1) whether the USPS should be required to offer interconnection to all communications carriers on a standard basis; and (2) whether the Private Express Statutes should be revised to integrate the roles of all information carriers, and specifically, if the USPS were allowed to offer "value-added" communications services, would private value-added networks be allowed to provide physical distribution services?[38]

The general conclusion reached here is that the USPS will not have a dominant or large role in the electronic message household delivery services; rather, it will play a significant part in the transitional scheme. It this conclusion is correct, the USPS must plan for a declining revenue base because of growing message losses to the electronic carriers. This does not mean, of course, that the USPS should not be forward-looking and efficient as to its continuing, already assigned mission. It should continue its new technology program, and where economically sound, move ahead with cooperative and other ventures (e.g., Mailgram; other Generation II services) which make use of its strengths; its delivery system and postal offices. However, the increasing loss of volume from the large business mailers, who generate much of today's postal revenue, raises a question: Will Congress continue subsidizing the traditional postal services, with their ever-increasing deficits, or will it cut back services (e.g., to delivery three days a week; closing of marginal postal installations)?

Whether the USPS operates as a public service or as a business, Congress must face the problem of revenue loss to the operators of new telecommunications technologies, and must plan for a scaled-down system. The planning should begin now, particularly in light of the fact that more than half of the present full-time workforce of the Postal Service is expected to retire within the next decade. This means that no major dislocation of present employees would be required, if the Congress should conclude that major changes in the Postal Service should be initiated within the next few years.

Congress should also consider another issue that is basic but beyond the scope of this chapter, namely: Should the USPS attempt to develop its labor force asset? There is a wide array of services that are still labor-intensive. Physical delivery of most forms of mail may no longer be such a service in the future. However, physical delivery of a wide variety of items is still a valuable service. Thus, rather than enter a service in which its labor force would be a liability, the USPS might attempt to compensate for its losses in mail by subsidiary services that are labor intensive. Some examples of these services are local delivery for merchants or libraries; conducting surveys; reading utility meters or home security devices; or checking on the status of elderly and handicapped persons. These suggestions, if implemented, could eliminate home delivery by other businesses, supplement home services of government entities, or reduce the number of trips that home dwellers must make.[39]

We recognize that there has been opposition to the Postal Service providing services other than classic mail services. But we urge that this issue of more effective utilization of the USPS work force, including ways of retraining that force to undertake assignments such as suggested above, is basic to the future of the USPS, and should at least be given serious consideration and accepted or rejected on the merits.

The Congress must also address the issue of competition to such scaled-down postal organization. Wholly apart from EMS considerations, the Antitrust Division of the Department of Justice and the Council on Wage and Price Stability has urged the repeal of the Private Express Statutes.[40] The argument is based on two points: There are no economies of scale in postal services, and the present monopoly does not serve the public interest. The statutes are, it is asserted, a device to gain revenue for the service, and such revenues can be obtained by direct subsidy (general taxation) without sacrificing the benefits of competition. The issue must be faced and decided upon in the context of the emerging postal situation in the next decade.

Impact on Society

First, to the extent that electronic message service is just one part of the computer-satellite, new technology communications revolution, there is little need to single out its impact on society. It is just one element in the information revolution that is affecting America and other advanced societies.

Electronic message service, an important element of increased business efficiency, can assist in coping with the energy crisis facing the nation in the last decades of this century. The service is basically energy conservative: communications and computer technologies are not heavy users of energy and do not pollute. Thus, EMS contrasts sharply with USPS mechanization, which means increased use of motorized transportation. Electronic transactions and message transfer can also contribute to a reduction of energy-consuming transportation.

There is also the issue of the impact of EMS on letter writing. The physical delivery of mail will not only continue for decades but will probably be available for a lengthy period, as long as the writer is willing to pay the additional charge. But the majority of these writers will learn word processing techniques in the same fashion that they learn handwriting. And there is no reason why letter writing in this manner cannot remain an art, to the same extent it is today.

There are also issues of privacy and security. The EMS (and EFTS) creates the potential for increased abuse of privacy, with special safeguards needed on a national level to prevent such abuses. Federal law must declare unlawful unauthorized use of electronic information, and these laws must be vigorously enforced. This important issue has been treated extensively in recent documents issued by governmental study groups.[41]

Persons living in rural areas generally have telephone service and could obtain EMS via telephone (with satellite transmission for the long-distance link). Their participation in the broadband (EMS) services depends on a Rural Electrification Act-type program of low-interest loans to facilitate the construction of these rural "pipelines." The poor also have telephone service in light of the 94 percent figure; to the extent EMS becomes a prevailing means of communication, a subsidy (either specifically as part of welfare or as a guaranteed minimum income) might well be needed. This problem, in sum, is the same as the provision of telephone service for the poor.

Institutional Considerations

The problem of the USPS, vis-a-vis the new telecommunications services, are acute and will obviously call for careful and informed examination by the government entities involved. Unfortunately, the governmental process is fragmented and presently appears inadequate.

Legislative jurisdiction in the Congress is divided between the Postal Subcommittees (which oversee the USPS) and the Commerce Subcommittees (which oversee the FCC), with no joint hearings or even staff exchanges contemplated. Thus, neither committee is examining the entire issue. The House Subcommittee on Communications has embarked on a review of the Communications Act but none of the groups established to assist in this review will examine a comprehensive postal plan or effects—an understandable omission in light of the subcommittee's jurisdiction. Congress created ad hoc study committees on the future of the USPS and EFT; however, they had short deadlines for issuing their reports. It is difficult to perceive what Congress expected could be accomplished with such short-term approaches. The Privacy Protection Study Commission had a longer time span but it, too, was an ad hoc commission.

The USPS, of course, has been pressing for examination of the problem by the Executive Branch, but its top management is engaged, to a large extent, in responding to the current financial crisis. The agency primarily charged with long-range telecommunications planning for the President, the National Telecommunications and Information Administration (NTIA) has neither the budget nor personnel to properly study the subject and make recommendations. Further, there appears to be a lack of any formal bridging mechanisms between "electronic" and "mail" jurisdictions in the Executive Branch (NTIA and USPS).

Finally, the FCC, which should also be engaged in long-range policy planning, has all it can do to keep slightly ahead of its own regulatory business, without taking on the perils and promise of future mail services. The other regulatory entity, the Postal Rate Commission, appears much too restricted to be effective in this overall planning area and is confined largely to passing on rate requests of the USPS.

Possible solutions to the USPS problems include close consultation (joint or sequential hearings) between the interested congressional committees, with technical assistance from the Congressional Office of Technology Assessment. Also needed is strengthened and coordinated long-range policy planning, both in the Executive Branch (led by an organization like NTIA) and at the FCC.

John M. McLaughlin recently observed:[42]

> The Post Office is the archetype of public service institutions in a society that is troubled by the performance of public service institutions. Frankly, I despair of our chances of resolving macro problems like health, education and welfare if we are unable or unwilling to deal with the institutional and political problems of something so prosaic as delivering the mail. This is not to say that delivering the mail is unimportant; it simply does not have the emotional content or the life and death nature of many other social institutions.

It might be more accurate to say that the emotional impact of declining Postal Service has not been fully realized by the public because of an institutional process that has not fully considered the principle policy options and implications. Americans complain about rising postage stamp prices and delayed services but they have not been faced with the possible future developments such as the three-day week delivery, the 34-cent stamp, the $15 billion taxpayer subsidy, or the drastic reduction of delivery service. It is the imperative of institutional reform to develop measured policy alternatives to such possibilities for the future.

The recent report of the Commission on Postal Service, under Chairman Gaylord Freeman, has made a marked contribution to public awareness of these possibilities. It has underscored the problem, making clear that "there is no free lunch." And it has called for resolution within the next two years of the issues posed by electronic message service for the USPS. This, then, is the challenge and the time frame confronting the institutional process.

Notes

1. Statement of Gaylord Freeman in Hearings before the Subcommittee on Postal Operations and Services of the House Committee on Post Office and Civil Service, 95th Congress, 1st Session, 1977, pp. 2-7.
2. Report of the Commission on Postal Service, U.S. Government, Washington, D.C., 1976, pp. 10 and 12.
3. See note 1 *supra*, p. 5.
4. See note 1 *supra*, p. 4.
5. Report of the President's Commission on Postal Organization, "Towards Postal Excellence," U.S. Government, Washington, D.C., 1968, p. 1.
6. For detailed discussion of the figures set out in this section, see Report of the Commission on Postal Service, note 2 *supra*; Donald E. Ewing and Roger K. Salaman, "The Postal Crisis: The Postal Function as a Communications Service," Office of Telecommunications 77-13, January 1977; Postal Service Staff Study, "The Necessity for Change," House Committee Print 94-26, 94th Congress, 2d Session 1976; J.T. Ellington, Jr., Senior Assistant Postmaster General, Address to the Postal Customers Council, New York, February 16, 1977, pp. 2-7; John F. McLaughlin, Remarks before the National Association of Manufacturers, November 15, 1976, pp. 3 and 5 (unpublished).
7. The rising cost of fuel, resulting from an energy crisis, has also been passed on to the USPS. Transportation costs increased $271 million between 1971 and 1976.
8. Postal Service Staff Study, "The Necessity for Change," House Committee Print 94-26, 94th Congress, 2d Session, 1976, p. 12.
9. Postal Service Staff Study, *Ibid.*
10. Benjamin F. Bailar, Remarks before the Economic Club of Detroit, Detroit, Michigan, March 8, 1976, (unpublished).
11. John F. McLaughlin, Remarks before the National Association of Manufacturers, November 15, 1976, (unpublished).
12. For a more complete discussion of the figures and concepts in this section, see the discussion by Walter Baer, Chapter 2;

Remarks of M. Rogoff on "Electronic Mail," Communications Law Workshop, Federal Bar Association, Washington, D.C., January 31, 1977, (unpublished); R.J. Potter, "Electronic Mail," *Science*, Vol. 195, pp. 1160-65; Roy M. Salzman, "The Computer Terminal Industry: A Forecast," *Datamation*, November, 1975, pp. 46-50; Harry R. Karp and Gerald Lapidus, "What the Future Has in Store for Data Communications," *Data Communications*, May/June 1975, pp. 25-40; *Facsimile and Electronic Mail*, International Resource Development, Inc., New Canaan, Connecticut, 1976; *Communicating Word Processors*, International Resource Development, Inc., New Canaan, Connecticut, 1976.

13. Staff Report on National Bulk Mail System, House Subcommittee on Post Office and Civil Service, House Report No. 94-12, 94th Congress, 2d Session, 1976, p. 12.

14. Portable computer terminals are now advertised that can be used, via an acoustic coupler, with any telephone. Indeed, because of their inability to use the ordinary telephone, the deaf are using these and inexpensive home terminals, together with acoustic couplers, to receive printed messages through a central computer in cities like Washington, D.C.

15. This is not to be confused with circuit-switching, in which a connection requires a number of distant nodes to piece together a continuous path from end to end; and, for the life of the connection, its constituent circuits are dedicated to carrying a conversation. Very short conversations are not economical because they require a large number of bits to be transmitted. For a conversation in which there are substantial periods of inactivity, the numbers of useful bits transmitted is small in contrast to the number that might have been transmitted if the constituent circuits were fully utilized. Circuit-switching makes poor use of communications facilities when the conversations being carried are either short or interspersed with periods in which no message is sent.

Packet-switching, on the other hand does not dedicate circuits to set up connections; rather, the messages that form a conversation are injected individually at the exact moment of input. Each packet is transmitted with a complete specification of the communication: desired destination, source, size, sequence number, priority. Because there is no connection set-up to amortize over a conversation, short conversations are not seriously disadvantaged relative to long ones; because a packet-switching system allocates its resources to messages rather than to conversations, the inactive periods in one conversation can be used to support other conversations. Packet-switching thus fits well with computer "conversations," which are shorter and intermittent.

16. *Regulatory Policies Concerning Resale and Shared-Use of Common Carrier Services and Facilities,* 60 F.C.C. 2d 261, 1976; 62 F.C.C. 2d 588, 1977.

17. *Satellite Business Systems,* 63 F.C.C. 2d 997, 1977, appeal pending *sub. nom., United States v. FCC,* 77-1249 (Dist. of Columbia Cir.).

18. Another video teletext system has been developed for over-the-air broadcasting (called "Ceefax" or "Oracle") and is presently used in Great Britain by the BBC and ITV. Teletext utilizes unused lines in the television picture to transmit coded messages, which can then be decoded by special equipment attached to or built into the television receiver. Thus, the BBC's Ceefax service offers news headlines, shopping guides, stock market data, weather maps, sports, and so forth on four spare lines of television signal, and can accommodate up to 400 pages on each channel, each page displaying 24 lines of characters and simple graphics.

The Japanese are also conducting experiments in Tama New Town and Higashi-Ikoma with the new communications technology for home or business delivery of information (e.g., news and stock information; facsimile newspaper; individual memoranda; automatic reading of household utility meters; safety and fire alarm messages).

19. For a more thorough discussion of the EFTS, see the Report of the National Commission on Electronic Funds Transfer, "EFT and the Public Interest," 1977; Arthur D. Little, Inc., *The Consequences of Electronic Funds Transfer,* Contract NSF-C884, National Science Foundation, Washington, D.C., June 1975; Robert E. Knight, "The Changing Payments Mechanism: Electronic Funds Transfer Arrangements," *Monthly Review,* publication of the Federal Resource Bank of Kansas City, July/August 1974; James B. Rule, *Value Choices in Electronic Funds Transfer Policy,* Office of Telecommunications Policy, Washington, D.C., October 22, 1975; Joann S. Lublin, "'Checkless' Banking is Available, but Public Sees Few Advantages," *The Wall Street Journal,* November 18, 1975, p. 1; Ernest Holsendorf, "Postal Service and Electronic Transfer," *The New York Times,* March 10, 1977, pp. 49 and 57; R. K. Jurgen, "Electronic Funds Transfer: Too Much, Too Soon," *IEEE Spectrum,* May 1977, pp. 51-57.

20. EFTS is not the sole source of diversion from transaction mail. Utilities are experimenting with a system whereby meter readers deliver monthly bills which would be outside the jurisdiction or the Private Express Statutes (although, under the law, those "private postmen" cannot place the bills in the mailbox

itself). Other firms are conducting similar experiments, billing less frequently, or consolidating their first-class mail.

21. Actual mail volume in 1976 was 89.8 billion, 20.2 billion fewer pieces than had been projected in 1968 by the Kappel Commission on Postal Organization (see note 5 *supra*). The economic impact of this decrease was described in the 1977 Report of the Commission on Postal Service (p. 15):

...A decline in mail volume immediately affects postal revenue, resulting in losses because postal costs do not change in proportion to volume changes. A one dollar change in revenue changes costs by about 50 cents, so a volume loss of 100 million pieces of first class mail, while costing the Postal Service more than $14 million in revenues, will only reduce costs by about $8.4 million...

22. These estimates are described in fuller detail in the Report of the Commission on Postal Service, Commission on Postal Service, U.S. Government, Washington, D.C., April 1977; "Electronic Message Systems for the U.S. Postal Service," National Research Council (hereinafter "NRC Report"), Washington, D.C., 1976; Ewing and Salaman, "The Postal Crisis: The Postal Function as a Communications Service," Office of Telecommunications, publication 77-13, January 1977; Hearings before the House Subcommittee on Postal Operations and Services, April 6, 1977; Arthur D. Little, Inc., *Electronic Mail Systems*, August 1976.

23. For a discussion of the history of postal service vis-a-vis electronic communications, see Charles L. Jackson, "Electronic Mail," No. 2197, prepared for the Office of Applications, National Aeronautics and Space Administration, Washington, D.C., April 1973, (unpublished paper).

24. The specific sections of the Postal Reform Act of 1970 from which these benefits derive are as follows: No. 1, 39 U.S.C. 101(a); No. 2, 39 U.S.C. 101(e); No. 3, 39 U.S.C. 403(a); No. 4, 39 U.S.C. 404 (1); No. 5, 39 U.S.C. 404(b); No. 6, 39 U.S.C. 2010; see also 39 U.S.C. 101(g); 401(5), 403(b), 404(3); No. 7, 39 U.S.C. 401(10). Provision No. 5 is further clarified in H.R. Rep. No. 91-1104, 91st Congress, 2d Session, 1970.

25. "Electronic Message Systems for the USPS," NRC Report, 1976.

26. *Ibid.*, p. 32.

27. See *United Telegraph Workers, AFL-CIO v. FCC*, 436 F. 2d 920 (District of Columbia Cir.), 1970.

28. Statement of C.F.C. Vandergeest, Director of Systems Research and Development of the Canadian Post Office, Report to the 1975 Airlie House Telecommunications Policy Research

Conference, Airlie, Virginia, April 1975, pp. 15-16.

29. Speech delivered by J. T. Ellington, Jr., Senior Assistant Postmaster General, to the Postal Customers Council, New York, February 16, 1977, p. 12.

30. The USPS currently devotes about .025 percent of its annual revenues to research and development (R&D). The office equipment and computer industry, R&D represents an average of 5.4 percent of sales; 1.9 percent of sales for the communications industry; and 0.5 percent for service industries. See *Business Week*, June 27, 1977, pp. 62-84.

31. *The New York Times*, March 10, 1977, p. 49.

32. The USPS argues that it is not a "person" within the definition of that term in the Communications Act, and is subject to an entirely different regulatory scheme in the 1970 Postal Reform Act (e.g., by the Postal Rate Commission). There are counterarguments, but the issue is beyond the scope of this chapter.

33. When a carrier serves both monopolistic and competitive markets, the FCC is confronted in its rate-making and other activities with the problem of preventing revenues from monopoly services from subsidizing its competitive services, which would distort or inhibit the free play of market forces to establish reasonable prices. Thus, a major FCC concern has been to create and maintain an environment in which all participants may compete "fully and fairly" in the specialized communications market when a monopoly such as one with the size and pervasiveness of the Bell System is a participant. For almost a full decade, the FCC has been wrestling with the problem of establishing principles for pricing and costing the competitive operations of AT&T, which is predominatly a monopoly enterprise. Thus far, the problem is unresolved.

34. 39 Fed Reg. 33211-13; 36114, 1974; 40 Fed Reg. 23295, 1975.

35. See, for example, L. L. Hillblom, et. al., "The Recent Expansion of the Postal Monopoly to Include the Transmission of Commercial Information: Can it be Justified?" *University of San Francisco Law Review*, Vol. 11, p. 243, 1977.

36. *Ibid.*

37. NRC Report, *op. cit.*, p. 35.

38. See, for example, Statement of Chairman Lionel Van Deerlin, before the House Post Office and Civil Service Committee, April 6, 1977, pp. 5-6.

39. See H. M. Watson, Jr., *The Post Office, Business Opportunities and Technology for the 1980's*, National Research Council, Washington, D.C., PSPS-31, September 1975.

40. See "Changing the Private Express Laws: Competitive Alternatives and the U.S. Postal Service," U.S. Antitrust Division, January, 1977; Comments of the Council on Wage Price Stability Concerning the Private Express Statutes, in Docket No. R 76-4, Postal Rate Commission, January 16, 1976.

41. Report of the Commission on Electronic Funds Transfer, "EFT and the Public Interest," U.S. Government, Washington, D.C., 1977, pp. 24-30.

42. John F. McLaughlin, Remarks before the National Association of Manufacturers, November 15, 1976, pp. 8-9 (unpublished).

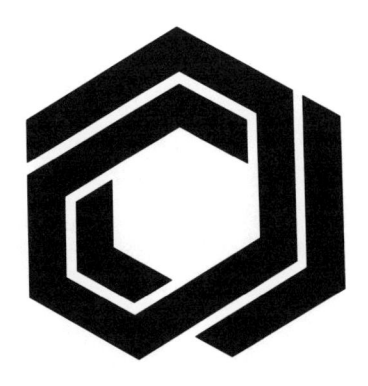

Part 4.
Government Institutions and Policymaking Processes in Communications

ABC

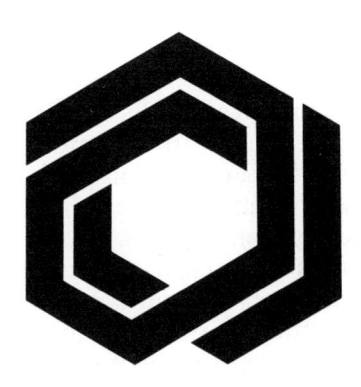

10.
The Federal Communications Commission

Glen O. Robinson

The increasing importance of electronic communications in our society, together with accelerating advances in communication technologies and services, have raised new issues of public policy; these new issues have, in turn, raised questions about the adequacy of public institutions that regulate the system. Given its central role both in the formulation and implementation of communications policy, the Federal Communications Commission is rightly the first focal point of critical attention.

To some degree, the following critique draws on personal experience as an FCC Commissioner (which is to say that rein has been given to subjective opinions that cannot be objectively verified). What follows is not, however, an attempt to portray the "inner life" of the Commission, or of its members. In truth, there is less to tell of the inside workings of the Commission than is commonly supposed and what there is can be found in the trade press reports, from which little seems to be withheld. This essay focuses on the institutional capacities and incapacities of the agency to cope with its present and anticipated future responsibilities, and is both a personal evaluation of the agency and its processes and a survey of critiques and proposed reforms that have been made by others.[1]

Even independent of the increasingly important role of communications and apart from the rising interest in communications policy issues, the FCC is an object of interest as one of the several federal regulatory agencies whose role and functions have come under fire from different public and private critics. On the strength of news accounts of attacks on the regulatory agencies and their activities in the past five or six years, it would appear we have entered a new era of government reformation; more precisely, what appears is a counter reformation against the reform tradition of the 1930s. To be sure, the current critical mood is not unprecedented. Criticism of the regulatory agencies and proposals for their reform have been a well-practiced occupation of investigative commissions, legislative committees and individuals and institutions since their creation. Among its "sister agencies" (for some reason the feminine form is customary even though most of the agencies are overwhelmingly staffed with males), the FCC has not been preeminent as a target of criticism or reform, but it has been a prominent one. To date, not much has come of such institutional critiques or the proposed reforms which have accompanied them. Despite persistent criticism of the regulatory agencies in general, they have grown in number, in size, and in power. So too the FCC: it has endured over a score years of more or less continuous and often trenchant criticism (beginning at least as early as the mid-1950s), yet— unreformed in any significant manner—it has expanded its range of influence substantially.

Not sharing the common faith in regulation as an instrument of social and economic control, I approach the problem of reform with skepticism about many of the aims that are promoted by regulation. Though it might be unrealistic to hypothesize an unregulated marketplace for electronic communication, there is nevertheless a large area of communications where government regulation is neither required nor beneficial. In such cases, reforms in administrative mechanisms are insufficient and may be counterproductive (on the sensible notion that we should not encourage efficiency in pursuit of mischief).

For those matters which remain appropriate subjects for administrative control, the useful procedural and institutional reforms are simple in concept and limited in number. Several of my predecessors at the FCC have advanced ingenious and ambitious proposals for reorganizing the FCC (not all of which are special to this agency), such as by restructuring various component offices of functions of the agency into new and separate agencies. Some of the main ideas are reviewed later, but such reorganization proposals are unlikely to yield significant improvements in administrative performance, information, and public participation. The old verities about the need for good people, simplified procedures, and public participation still hold. This is not an exciting conclusion, and certainly it demonstrates no creative imagination. But inventiveness has its limits. In the field of administrative reform, what is most needed is inventive ways of bringing about uninventive reforms.

The discussion that follows is organized broadly into six parts. First, it briefly discusses the legislative mandate. Second, some major problems and reforms in FCC processes are addressed. Third, internal organization is investigated, with a principal focus on staffing. Fourth, the commissioners themselves are observed as well as the endlessly controversial problem of agency appointments. Fifth, several basic organizational reforms are critically examined. Finally, some brief comments on administrative regulation conclude the critique.

Legislative Charter

The role of Congress in making communications policy, and the relationship of Congress and the FCC, are explored by Ernest Gellhorn in Chapter 13 and do not require examination here. However, to understand the character of the FCC it is necessary to review the character of its mandate from Congress. It is not necessary to trace in detail the substance of the Commission's statutory mandate. The two most striking things about the Communications Act of 1934 are that it is generous in its grant of powers to the agency and sparing in its limitations thereon and that it is out of date.

Delegation of Powers

At least as it has been interpreted by the courts since Justice Frankfurter's celebrated dictum in the *NBC* case, the Communications Act could be described with only modest exaggeration as a grant of power to regulate electronic communications as the FCC deems necessary.[2] Indeed, even that description does not do full justice to the scope of the Commission's substantive discretion; the Supreme Court has several times affirmed that the powers are not only broad but flexible—capable at least of being *expanded* to meet needs as they are perceived by the agency—independent of any needs imagined by the legislative draftsmen.

Broad, ill-defined and self-enlarging powers are an attribute the FCC shares with many other federal agencies. This delegation of legislative powers to agencies, without limiting standards of guidelines as to their exercise, has become an accepted norm of modern administrative government. It is accepted, that is, by Congress, the courts, and many scholars; but a few eccentrics (myself included) still question both the legitimacy and practical wisdom of wholesale delegation. This is not the place to pursue this issue at length. However, it is worthwhile to mention briefly some of the effects and implications of delegating broad, discretionary legislative powers.

The wholesale transfer of legislative power to administrative agencies has several obvious political effects. For one thing it places a strain on traditional democratic principles insofar as they presuppose that a politically accountable legislature is the primary maker of laws. Of course, no one sensibly supposes that the legislature can or should make all the law, and no one expects it to draft such statute in words so plain and detailed as to leave no room for administrative (or judicial) invention. But a legislature that does no more than enact platitudes of public interest purposes and direct administrative agencies to carry them out, hardly fulfills its democratic purpose as a maker of laws. The transfer of legislative power to agencies does not necessarily mean that lawmaking is removed from political influence; political influences are rather redirected from legislature to the agency. Certainly this is the case with the FCC, which has supplanted Congress as the primary target of lobbying efforts. Indeed, the chief purpose for lobbying in Congress today is not so much to obtain legislation but to gain legislative leverage over the agency to force the adoption of some particular policy.

Because administrators (and their staffs) are not directly accountable to the electorate, they could be considered less susceptible to general popular concerns than would congressmen. However, the behavior of the FCC over the years belies any influence that the agency is indifferent to popular clamor of the kind that usually moves a legislature. Indeed, an inspection of the kinds of activities to which Commissioners individually and the Commission institutionally devote their time shows a fair correspondence between agency interests and popular concerns. At the same time, it would be ingenuous to say that the Commission is closely and directly responsive to the general public, as opposed to particular, active segments thereof. In the FCC, as in Congress, results depend on organized, sustained and concentrated efforts by interested persons. Not surprisingly, this gives private industry groups a decided advantage vis-a-vis less

The Federal Communications Commission

organized groups purporting to represent the interests of the general public.

While the shift in the focus of political influence may not have greatly changed the character of political influence, it has altered somewhat the ground rules for its exercise. Except for inconsequential registration requirements and poorly enforced restrictions on favors, gifts, etc., there are virtually no legal restraints on the type or manner of political influence in the legislature. In agencies such as the FCC where policy is often made in adjudicatory proceedings, such influences are essentially forbidden by a requirement that all communications in adjudicatory cases be made on the record, with notice to all parties to the proceeding. Even allowing for the well-known occasional violations of these restrictions, they have, by and large, served to constrain the exercise of political influence in ways that it is not constrained in the legislative arena. A recent court of appeals ruling—if ultimately sustained—promises to broaden this constrain by restricting *ex parte* communications in informal rulemaking as well as in adjudicatory proceedings. More will be said on this later; here it is necessary to point out that whether or not this new restriction is sustained, the decision reflects a trend toward introducing judicial-type formality in agency rulemaking, the effect of which is to limit the scope and manner of lobbying by private groups before agencies.

As this *ex parte* ruling illustrates, the delegation of legislative power ultimately transfers power not only to the agencies but to the courts that supervise the exercise of agency power. Theoretically this may seem contradictory because the broader the terms of agency power and discretion, the more limited the occasion for judicial interference will be. However, when the statutory terms are (or, what is the same thing, construed by the courts to be) so broad as to lose all practical significance, there is ample room for both agency and court to share the power that the legislature has relinquished. Students of judicial realism might discern in this a special self-interest motive for the judicial penchant for construing legislative mandates broadly. Whether or not such a conclusion is warranted, it is reasonably clear that the courts have shared in the lawmaking power and they have been liberal in construing legislative intent to transfer.

In addition to political effects, there is room for endless debate on the social and economic consequences of broad delegation. On the one hand, it provides the opportunity to create and shape policies without the very cumbersome machinery of the legislative process. In particular it provides a way of avoiding political stymie, which stems from the diffusion of legislative power and fragmentation of interests. On the other hand, the combination of power and discretion has the same liability imputed by Macaulay to the American Constitution: "It is all sail and no anchor."

For those liberal free spirits who imagine that the ship of state ought always to be in full motion, the lack of an anchor is not a problem. For them, the incapacity of Congress for swift and resolute action is merely a frustrating impediment to social progress. For landbound conservatives on the other hand, Congress' incapacities are more a virtue than a vice; they discourage facile legislative solutions to social and economic problems—solutions that often prove short-sighted and ultimately mischievous. If the rallying cry of the former is "do good," the slogan of the latter is "do no harm."

The difference between these polar outlooks is real and substantial, but it only partly reflects a conflict in fundamental political theory. To a significant degree, these outlooks reflect different perceptions of the underlying substantive problems and differing assessments of how probable it is that government will respond sensibly. We can test this by asking whether those who decry the absence of specific legislative standards for, say, the regulation of competition in the telephone industry would prefer (a) congressional enactment of legislation to reverse the FCC's policies allowing competition in the supply of telephone terminal equipment or (b) the status quo. Any number of such choices can be proposed to test the apparently reliable proposition that where you stand depends on where you sit. As a personal predilection, I would probably choose option (b) in this case, but a democratic conscience counsels that in the long run it is healthier to abide the costs of legislative error (or inaction) than administrative abandon.

Modernizing the Act

Most of the text of the Communications Act dealing with broadcasting is 50 years old—derived from the Radio Act of 1927 and incorporated into the 1934 Act without change. Many of the common carrier provisions are even older—largely derived from the Mann-Elkins Act of 1910. In a field that has evolved as rapidly as electronic communications, it is a natural presumption that any regulatory scheme constructed so long ago must be somewhat out of touch with the times. The general presumption is easily supported by specific cases. The commonest example is that, except for one small provision added in recent years, cable television is not mentioned in the Act; yet it has been at the very

The Federal Communications Commission

center of regulatory controversy in the field of radio and television for over a decade. That a particular technology or service is not mentioned in the statute would not render the statute out of date, but when we realize that the nature of cable television presents regulatory issues quite outside the basic legislative objectives underlying the statute, one must wonder about the pertinence of the statute of modern regulation. The Commission, with acquiescence of Congress and the Supreme Court, was able to patch together snippets of statutory language and prior judicial dicta to stretch its statutory jurisdiction and permit the creation of an entire scheme of regulation untouched by legislative hands; the achievement of this task does not prove the statute's modernity—merely its irrelevance.[3]

The case of cable is illustrative of a more fundamental anachronism that pervades the statute. One of the most vexing and important problems in the field of common carrier regulation is defining the regulatory boundaries between communications and general computer services in cases where both can be provided by the same facility.[4] When the "relevant" provisions of the statute were written, electronic computers did not exist even in the imagination of science fiction. The question must then be raised: What guidance can be found in the statute to resolve the perplexity of regulatory policymakers as to how to treat computer-communications?

Although it might be obvious that the statute is out of date, it is not obvious what should be done about it. As this is written, there is considerable effort by the House Communications Subcommittee to attempt a revision of all or at least major parts of the 1934 Act. On the premise that the Congress should not surrender its prerogative entirely to agencies such as the FCC, that effort seems admirable (not to mention ambitious), providing that it seeks to impose real standards, directions, and limitations on agency discretion. However, if the result is simply to blow more wind into the FCC's sails, it will serve little real purpose. A modern irrelevant statute is not necessarily preferable to an outdated irrelevant one.

Commission Processes

Against this background on the FCC's broad legislative powers, we can assess the processes by which they are implemented. The Commission's processes (as well as its policies) have been the object of severe criticism over the years from numerous sectors, both public and private. The FCC is not unique in this respect, although it has been one of the most consistently attacked of its sister regulatory agencies. The honor is not entirely deserved. Indeed, in some ways the Commission deserves special commendation for procedural innovation. For example, it has been a leader in the use of informal rulemaking processes—so admired by most modern administrative law scholars—for the promulgation of general policies. And in its use of formal procedures, the Commission has been receptive to the use of flexible hearing procedures, delegation of decisional authority, and other procedures that are commonly regarded as innovative. Nevertheless, the Commission has not been entirely guiltless of the charges made against its processes as even a casual review of some prominent problem areas will reveal. It is impossible to explore here all of the different processes by which the Commission renders hundreds of formal decisions and rules and takes an uncounted number of informal actions each year. The discussion is limited, therefore, to a few of the salient, and recurrently controversial, elements of Commission processes.

Procedural Inefficiency

Of all the procedural inefficiencies, there is one that has attracted such attention that it has become almost a symbol of administrative inefficiency. This is the problem of delay. The problem is not, of course, special to the FCC;[5] nor is it characteristic of administrative proceedings, it should be noted in passing that the delays are often no greater than those encountered in analogous judicial and legislative arenas. This observation does not require further elaboration. It is pertinent as a matter of critical perspective, but it is not intended to be either an explanation or a justification for administrative delay.

In broadcast and cable proceedings, the most common and most protracted delays have occurred in broadcast license hearings, particularly those involving comparative hearings, where five to ten years are often spent in dubious battle for the prize of a broadcast license. These more notorious cases are not fair representatives of general Commission activity, but they illustrate how bad the situation can be. And in all hearing cases, a serious delay problem has existed. It is difficult to identify any single element of procedure that is significantly responsible for this or any simple procedural reform that would correct it. The Commission recently made a number of modifications in its procedures aimed at streamlining the process of adjudication. But the changes are not especially important in themselves; they make only marginal changes in the adjudicatory process, none of which seems likely to yield savings in time that are substantial

The Federal Communications Commission

in relation to the five- to ten-year schedule of the cases that cause concern. Whatever significance the changes have lies more in the attitude they reflect toward the problem than any specific reform made.

Regulatory attitude seems plainly the heart of the matter. None of the formal processes that underlie Commission decisions in the area of broadcast and cable are intrinsically so time consuming as to account for the delays that are frequently encountered, and modifications in the procedures themselves are unlikely to correct the problem. On the other hand, a simple, strong commitment by the Commission to expedite agency action at all stages can effectively eliminate the problem. (A commitment by the chairman is particularly important: as chief manager of the agency, the chairman is the only person in a position to shape the agency's calendar.) Such a commitment in cases that do not involve hearings has made remarkable strides in the past few years. The surge of challenges to broadcast license renewal filed by public interest groups in the past five or six years has created such a backlog of undecided renewal cases that, whatever other adverse effects the backlog might have had for licensees and challenges alike, it was creating a serious administrative problem for the Commission. By reassigning some personnel, prescribing deadlines and scheduling special meetings to review cases, the backlog has been virtually eliminated.[6]

The FCC has also been frequently dilatory both in the disposition of requests for initiation of rulemaking and, more seriously, in the disposition of rulemaking proceedings after they are initiated. In some cases, rulemaking proceedings have been so protracted that the policy issues have become mooted by events. The still pending proceeding on a table of allocations for noncommercial FM stations is a case in point, as are the erstwhile proceedings on clear channel allocations. While such instances are atypical, many cases can be found where the proceedings took longer than could be justified by the nature of the issues, even giving due regard to political considerations (e.g., the desire to find a politically expedient time for decision).

Again, there is no simple procedural "fix" beyond a commitment to move forward with all deliberate speed. However, with respect to the initiation of rulemaking, it might be useful to promulgate a general rule prescribing a time period for acting on requests for rulemaking. Except in unusual cases, six months is sufficient time to dispose of such requests. It may not be practicable to establish a similar *general* time schedule for conducting the proceeding itself, but the Commission could for most cases set (and publish—say, in the Notice of Proposed

Rulemaking) an estimated deadline for final disposition on a case-by-case basis. Such a deadline would not be unalterable, of course; reasonable flexibility must be allowed to accommodate unforeseen complexities in the proceeding—or changes in staff resources. Nevertheless, even an adjustable and flexible deadline could provide a benchmark for internal and public scrutiny of the agency's progress. It might also provide at least a partial check on the disturbing tendency of the Commission to institute rulemaking proceedings to eliminate a problem or to delay acting on a controversial issue until the heat of controversy has dissipated. It is unrealistic to expect an agency never to employ such tactics; indeed, in some instances deferral or "benign neglect" is not merely political expediency but sound social policy as well. However, there are proper limits. Decisions on controversial issues cannot be deferred indefinitely without damaging the public interest at stake and the credibility of the agency as an instrument of public policy.

If a simple commitment to celerity is the key to the problem of delay in most broadcast and cable matters, the problem in the field of common carrier regulation is somewhat less amenable to such a solution because of the nature of the issues involved. Delays in some common carrier cases have been inexcusable and could have been avoided by a more forceful dedication to the task. The ten-year history of the Private Line Rate proceeding is one case in point; the ten-year history of the Comsat Rate case is another. However, the complexity of the issues, which has greatly increased as a consequence of new technology, new services and new competition, together with the limited resources available to the Commission (which have not grown in proportion to the increased complexity and burden) are reason to be somewhat more tolerant of lengthy cases here than in the case of the usually simpler broadcast and cable matters. As in broadcast and cable cases, relatively little savings in time are likely to be accomplished by marginal procedural adjustments. In fact, some timesaving measures have in the past proved counter-productive. For example, efforts to expedite rate hearings by eliminating all oral testimony and cross-examination have sometimes been frustrated when the abbreviated proceedings yielded a record insufficient (and insufficiently scrutinized) for a decision—necessitating a supplemental hearing or a complete new proceeding.

Some of the delay problem has proven to be the difficulty in obtaining adequate information in utilizable form from the parties involved (most notably from AT&T). Insofar as the difficulty stems from willful or neglected failure to comply swiftly and responsively to Commission demands, the problem would seem to admit only one solution: Those issues that turn on withheld or missing information should be decided against the carrier. However, few cases warrant such action. In most cases, it would be difficult to support a finding of willful, or even neglectful, failure by the carrier. In some cases, the information simply does not exist—or is not readily available—in the form demanded by the Commission. In others, the Commission is not clear or precise about what it wants or why it wants it; and ambiguity is always an invitation to delay, particularly where, as is often the case here, the carrier has an interest in prolonging the investigation. More will be said later of the information problem which transcends the immediate issue and in fact pervades all aspects of the communications regulation.

Licensing: Standards and Purpose

It is often impossible to separate procedural reform from substantive policy issues. A notable illustration is the comparative hearing process. Probably the most maligned of all agency processes, this process has become for critics a modern symbol of administrative inefficiency and ultimate futility.[7]

According to many critics, the central problem of the comparative hearing process has been the FCC's inability to develop clear and meaningful (in terms of licensee performance or public interest concerns) criteria for choice among otherwise qualified competing applicants. The result has been confusion and inconsistency in results as well as inefficiency in procedure. It has also arguably led to improprieties—the absence of clear, rational standards for influencing decisions leading to other forms of influence. The problem has been alleviated somewhat since the 1950s and early 1960s, when broadcast licensing decisions were the core of public scandal for the FCC. The alleviation is perhaps attributable in part to some clarification of the choice criteria but in greater part it is simply a consequence of the reduction in the number of broadcast comparative hearings in the past decade.

Although comparative hearing cases still account for a significant portion of broadcasting cases docketed for hearing, in recent years the focal point has shifted from licensing of new applicants to the renewal process. Here the comparative hearing process has played a rather small role for the simple reason that, with one exception that is now chiefly of historical interest, the Commission has always favored incumbent licensees seeking renewal over challenges unless the incumbent has been disqualified by serious wrongdoing. Nevertheless, if the comparative hearing process has declined in importance, it continues to have relevance. And the relevance is not confined to broadcast licensing; the comparative choice problem arises in any context where, for economic or technical reasons, there are mutually inconsistent applications for authorization. Thus, comparative hearings have been designated for competing applications for common carrier certificates for multipoint distribution systems, and it is possible that the occasion for comparative choice will arise in a number of other areas in the future—for example, in satellite authorization, private-line telephone services, mobile radio services (such as paging), etc. Thus, the necessity for comparative choice is likely to continue in one context or another.

There is almost unanimous agreement both within and outside the agency that the comparative choice process should be improved, but little agreement on what action should be taken. The Commission itself made some modest efforts in 1965 to clarify its standards on comparative broadcast licensing but this accomplished rather little. Modifications in the hearing process and in the standards have since been proposed but not adopted. While some changes could perhaps be made that would simplify and clarify the standards for choice, it is doubtful that enough could be done in this direction to justify the effort. Basically what is wanted is not merely clarity but relevance. It would be relatively simple to devise criteria by which one applicant could be separated from another. What is difficult is to match such criteria with some demonstrable public purpose that will be furthered by the choice indicated.

For the most part, the quality of the proposals the Commission examines is meaningful only in terms of thresholds. An applicant's engineering proposal will be either sound or unsound. The technical and business staff an applicant hires to manage a station will be either acceptable or unacceptable. An applicant's ascertainment of the community will be either adequate or inadequate. Programming proposals probably should not be examined at all, for First Amendment related reasons; but, in any case, if the proposals meet some minimum level of acceptability, nothing further should be required. To go beyond an inspection of basic qualifications is productive of nothing but senseless waste of applicant and agency resources. Yet confining the agency to evaluating threshold qualifications almost invariably proves insufficient for making the choice between competing applicants. Particularly where only paper records are involved, it is unlikely that many who survive the initial staff scrutiny will be ruled out on the strength of a basic deficiency. And it is even more unlikely that all applicants but one will be eliminated by this process.

If there are not meaningful distinctions between applicants, then the choice between them will perforce be arbitrary. Arbitrariness *per se* is not necessarily a bad thing. It is, in any event, often unavoidable. But if a government agency is required to make an essentially arbitrary choice, it is important that the arbitrariness equates to randomness rather than personal whim. The wheel of fortune—a lottery—is much to be preferred to that different class of arbitrary criteria, the capricious preferences of bureaucrats.

A simple lottery, in fact, might be a sensible method of choosing among qualified license applicants (those meeting minimal, threshold standards), but an even better mechanism would be to "auction" licenses. An auction combines the simplicity of the lottery with two additional virtues. First, it would allow the public to recoup the economic value of the benefits conferred upon private licensees. Second, unlike a lottery, an auction measures the intensity of individual preferences, in accordance with the prevalent standard for allocating resources in our economic system.[8]

Proposals to auction licenses to the highest bidder have been criticized on grounds of social policy. It is said, for example, that an auction (1) would be an abandonment of "public interest considerations," (2) could lead to the view that the winning bidder has a property right in the license, and (3) would favor the wealthy. None of these objections is convincing.

The first objection is a little vague but it seems to harbor both a misunderstanding of how the auction system would work as well as a misunderstanding of how the present system does work. An auction system need not, and would not, eliminate all inquiry into licensee qualifications. Minimum standards of fitness or eligibility need not be waived for anyone who has sufficient cash. On the contrary, only those applicants who meet the basic legal and character qualifications for licensees would be eligible to bid. Thus, under any circumstances, successful bidders at auction would be at least as well qualified as licensees whose applications are uncontested. In the final analysis, that is all we can realistically expect from the comparative hearing process also. The idea that comparative hearings really select the "best" candidate is naive. Seldom is there any single "best" candidate; more often the candidates are, for all practical purposes, fungible. In any case, history shows that the comparative process, despite adjustments made in it over the years, is not designed to discover "the best," and even where the best might be selected, there is no mechanism (other than the general rule against "trafficking" in licenses) that prevents the winner from ultimately selling out to a less qualified candidate.

It is sufficient to respond to the second objection—that an auction would support the view that licenses are a property right—by observing that such a property right does in fact exist as a consequence of the historic practice of routinely renewing licenses. The question is then raised: What follows once we recognize this reality? What follows, I think, is a reconsideration of the basis on which such property rights are acquired.

The third objection—that an auction would favor the wealthy—hardly seems a decisive objective because the same can be said of any pricing system. Why is it, for example, any more pertinent to broadcast licenses than land or mineral leases or the use of dialysis machines? In any case, particularly in the context of a comparative hearing, it is unlikely that an open-pricing system would lead to greater control of licenses by the rich than does the present system, given the current expense of obtaining a license.

Comparative licensing illustrates an important point with regard to licensing and regulatory functions generally. Samuel Johnson once said of a dog walking on its hind legs: "It is not done well; but you are surprised to find it done at all." So it is with many Commission functions. Without attempting to make a catalog of them, it may be worthwhile to offer two examples to reinforce the point that only satisfactory reform of some activities has about the same overall effect of not performing them at all. The illustrations are limited to relatively minor activities simply to avoid an extended discussion of substantive regulatory policy; it should not be implied that the only cases for regulatory reform—and regulatory retreat—are minor ones.

The first illustration is Citizens Band (CB) licensing. The explosive growth of CB radio has brought with it a corresponding increase in license applications—currently over 5 million are received a year, a two-fold increase since 1975—that swamped a Commission unprepared to handle such a workload. Perhaps needless to say, the FCC's inability to handle the enormous new workload expeditiously spawned criticism from CB users and Congress. And this, in turn, has stimulated considerable effort to find ways of streamlining the applications review process. To its credit, the Commission appears to have now more or less caught up and the problem of long delays has been solved both by management efficiencies in processing and by granting interim licensing authority from the time the equipment is bought by the user. However, the transitory procedural problem focuses attention on a more basic question, as yet unanswered: Why are CB users licensed at all?

The only procedural inefficiency of concern in the procedures is that they are essentially purposeless. There are virtually no public interest criteria administered or enforced through licensing. Unlike other radio licensees, CB users receive no protection against interference. The only basis for licensing is to assist in protecting non-CB radio users from CB interference, but licensing provides little assistance in this regard. Enforcement of restrictions on CB operations relies basically on field monitoring or responses to complaints. (The commonest problem in these operations is the illegal use of linear amplifiers to boost power output beyond the prescribed four watts.) Licensing is of marginal assistance in this task. Nor is licensing justified in terms of obtaining information about CB use or CB users, since millions of CB users never obtain a license. It is argued by some Commission staff members that the license is an important element of acculturation of CB users to comply with legal restrictions: licensing makes CB users more responsible. But there is no solid evidence behind this sociological speculation and it seems a very weak base to support the use of scarce government resources that are needed elsewhere. In the final analysis, the only believable reason for CB licensing is that it enhances the size and importance of the Safety and Special Services Bureau staff to whom the function is given.

A second illustration is the certification of cable systems. Technically the FCC does not "license" cable systems as it does broadcast stations; this function is left to local, community, or state agencies. But the Commission has since 1972 required cable systems to obtain from the FCC a "certificate of compliance" as a prerequisite to carrying any broadcast signals in addition to whatever instrument of authorization (typically a city franchise) is required by local authority. In contrast to broadcast licenses, the certificate of compliance is not subject to periodic renewal, though a new certificate must be filed if any change is made in the broadcast signals carried by the system. Thus, the certificate is not intended as a mechanism for continuous oversight; rather, it was conceived basically as a means to ensure initial compliance with Commission rules and regulations.

In recent years, the Commission has become burdened with a growing backlog of certificate applications. In contrast to CB license applications, the problem arises not simply from the growth in applications but from the increased complexity of the FCC rules compliance with which is the purpose of reviewing applications, and from the opposition to applications (from local broadcasters claiming economic injury). There do not appear to be any managerial efficiencies that would substantially ease the problem, except to increase the size of the staff. The Commission has done so up to a point but present staffing constraints limit the effectiveness of this remedy. The one "remedy" that has not been adopted (though it has been considered) is to eliminate the process itself. It seems doubtful that the process of prior clearance is any more necessary or useful here than it is in the case of CB radio.

It may be unfair to cite these particular examples. As mentioned earlier, the activities themselves are rather minor examples of regulatory excess. In particular, the cable certification process is only part of a larger scheme of cable regulation, the purpose and character of which are extremely dubious.[9] As for the propensities of staff proponents of these particular functions, it should be noted that they are not much different in character than those of bureaucrats elsewhere—everywhere. However, important as it surely is to keep a critical eye on the "larger picture," it is useful periodically to notice the smaller parts that comprise it. There is no better place to begin than with a critical look at the Commission's broadcast, cable, and special services licensing functions.

The Information Problem

Just as, according to Napoleon, an army marches on its stomach, so administrative agencies march on the information they digest. In contrast to the problem that inspired Napoleon's dictum, however, the administrator's problem is seldom one of quantity. If anything, quite the reverse is more likely true; to use Alvin Toffler's phrase, the problem is one of "information over—load." No one supposes, of course, that individual agency members themselves digest all or even a large proportion of the material submitted to them. That, after all, is what a staff does: it reviews, digests, summarizes and makes recommendations from this huge volume of information. This response is not fully responsive to the problem, however. So far as the Commissioner's personal staff is concerned, two professionals (the number allotted to each commissioner except for the chairman who has three) cannot be expected to digest more than a small part of the information received. In fact, most commissioners' staffs are occupied simply in assimilating material for their commissioner and advising him on the week's agenda of agency staff reports and recommendations. (Typically this agenda is received less than a week in advance of the Commission meetings for which the matters are scheduled to be heard.) Thus, with rare exception, the commissioners and their staffs are greatly dependent on bureau and office staff for information and guidance. Under the circumstances, one can appreciate the truth of political scientist James Q. Wilson's remark that if agencies are captured by anyone, it is by their staffs.[10]

Nevertheless, the "information problem" is troublesome not because it causes agency members to be dependent on bureau staff. Assuming that the information is relevant to the functions to be performed, it is presumably not the overwhelming information that is ultimately responsible for this dependency but the functions themselves. Certainly no one would seriously argue that important and relevant information should not be acquired simply because it is difficult for agency members (or even agency staff) to manage. The crucial question is: How important, how relevant is the information to making informed policy choices? Based on personal experience and observation, my answer is that most of the information received is not only not useful, but tends to obscure by its sheer volume; by the same token, it also tends to obscure the need for other probative information that is *not* received by the agency. In this respect, the agency's dietary problem is more serious than that faced by a hungry army. If an army is not fed, it may fail to march forward. But an agency that is fed the wrong information may not only fail to march forward; it may actually march vengefully backward. Some think that is exactly what the FCC has done for years.

At least part of the problem of information flow arises from the agency's inordinate reliance on regulated industries for information and analyses. To some extent, of course, this reliance is inevitable. Much of the detailed information needed for policy decisions is unavailable anywhere else. Whether information comes directly from the Commission's files and reports or from the industry in comments or oral discussion, the ultimate source is generally the industry. There is a difference between drawing on information in the agency's files when it is compiled from routine reports and obtaining information intermixed with advocacy in the form of special comments or arguments submitted as part of a particular proceeding. However, in both cases there are serious problems of reliability and accuracy.

The problem is not confined to broadcasting. Even in the case of telephone and telegraph regulation where a uniform system of accounts exists, it has for some time been painfully apparent that the reporting system is inadequate. The problem here is somewhat similar to the problem in broadcasting: The reporting system was not devised to answer problems that arise when the agency is forced to consider the impact of competition on existing firms. However, in the telephone-telegraph field, both the accounting problem and the regulatory problem are made much more difficult by the necessity to regulate prices. If the Commission imposes a more vigorous and more reliable system of reporting, it will be an important step forward. It will not, however, be an easy step to take—particularly in the case of the broadcast industry which has historically benefited from the state of confusion and ignorance about its economic condition, not simply in the context of competitive impact from cable but in other contexts as well.

Why does the Commission allow itself to be swamped by material, a major effect of which is to obscure the absence of probative information? One immediate explanation is to be found in the Commission's historic approach to policy planning. Policy issues are typically framed outside the Commission, mainly by industry interests but sometimes by other interested groups. They are brought before the Commission with a demand that it take action immediately. If the proposal does not have particular interest, or strong political support, it may be shelved for an indefinite period. But if some action seems to be called for, the Commission's first instinct is to invite general public comment without any real effort to study the matter first, to formulate the issues carefully or to focus the resultant public inquiry according to its own independent analysis of the problem. In some cases, as was suggested earlier, the purpose of inviting public comment is simply to remove the issue from the agency's action agenda for a period of time. That is not a principled way of disposing of matters, but it is politically expedient and sometimes even acceptable. Unfortunately, it has become too automatic a response. By thus relinquishing the early initiative to interested parties, the Commission loses an important element of control not only in structuring its general policy agenda, but also in shaping the particular issues in any given policy inquiry.

Ex Parte Contacts

Ex parte contacts are an old problem for the FCC; in the late 1950s the Commission's integrity was seriously compromised by disclosure of *ex parte* contacts with interested persons regarding individual licensing proceedings pending before the Commission.[11] The Commission has still not recovered fully from the damage done to its reputation, despite its effort to correct the problem. As a consequence of these episodes, the Commission (and most of its sister regulatory agencies) adopted a rather stringent set of rules regulating *ex parte* contacts in all adjudicatory cases required by law to be decided exclusively on a formal public record and other nonadjudicatory cases specially designated by the Commission as "restricted."

For the class of cases intended to be affected, the rules have worked well. Communications between "decisionmaking" agency personnel and interested persons (including those within the agency that have investigative or advocacy roles in regard to any particular restricted proceeding) have not been a serious problem since the scandals of the 1950s. Although there are occasionally rumors and allegations about secret influences, most of these are quite groundless. Some are, in fact, deliberate smoke screens created by interested parties seeking to discredit an adverse Commission decision by capitalizing on recollection of some past misconduct.

The real problem lies not in the area covered by its ethical practice rules but in the area beyond them. The recent *Home Box Office* case is illustrative.[12] The case involved a revision of rules restricting and regulating pay exhibition of programs (movies, series, and sports events) on cable systems and broadcast stations. The rulemaking proceeding followed the general form prescribed by the Administrative Procedure Act, with public notice and opportunity to file written comments. No restrictions were imposed on informal, *"ex parte"* contacts between commissioners, staff, and interested industry, and other public groups. In this there was nothing exceptional; such informal contacts have long been thought to be a characteristic of "informal rulemaking"—in much the same sense that lobbying is an accepted characteristic of legislative lawmaking. If there was anything exceptional in the form of the Commission's proceeding, it was the provision for formal oral argument by interested persons (consuming several full days). The provision for oral argument figured prominently in the challenge to the legitimacy of the proceeding; it produced a seeming discrepancy between the informality of allowing individual, off-the-record communications with commissioners and staff on the one hand; and formality of recorded oral argument on the other. The incongruity was more striking because the *ex parte* communications continued after oral arguments on appeal from the Commission's order adopting revised rules, it was argued that the *ex parte* communications were improper at least insofar as they occurred after oral argument when the "record" was officially closed. The court agreed but moved beyond the scope of the oral argument; the court invalidated not only *ex parte* communications after the oral argument, but it proscribed all such communications in rulemaking proceedings any time following issuance of the public notice of proposed rulemaking. In short, all informal communications would be banned from informal rulemaking (which would thus become "formal" to that extent).

The court's opinion admits to making new law on the subject of *ex parte* communications. There is respected precedent for banning *ex parte* communications in informal rulemaking cases that are similar in substance if not in form to adjudication; and there is a more debatable line of recent decisions that have required certain adjudicatory formalities in informal rulemaking to ensure the development of an adequate administrative record. The opinion draws on both lines of precedent, but clearly moves beyond either one.

This is not the place for a legal critique of the *Home Box Office* decision. There is reason to doubt the decision will survive as a precedent with the full dimensions given it in the court's opinion. The sweeping ban of all *ex parte* communications in all rulemaking proceedings was repudiated by one member of the court panel in a separate concurring opinion; more recently the opinion was repudiated in dicta by a different panel in a separate decision involving the FCC. However, whatever the future of the *Home Box Office* precedent, the issue warrants exploration.

Once again, we are confronted with competing models and expectations. Congress has delegated broad legislative power and provided for its exercise through an informal rulemaking process not unlike that used by Congress itself. The courts have ratified this process on several counts—affirming both the delegation and the use of informal rulemaking to implement the delegated powers. However, perceiving that the rulemaking process gives rise to "nonrational" (e.g., political) influences, the courts now seem to be imparting judicial procedures—including a ban on *ex parte* contacts—to the "legislative" arena to bring rulemaking into line with judicial expectations of reasoned decisionmaking.

It is at least a bit curious that the courts, having in the past strongly urged informal rulemaking in preference to more cumbersome adjudicatory procedures, should apparently reverse field and insist on the kind of judicial formality heretofore in disfavor. Inconsistency aside, however, the wisdom of court-enforced judicialization is a difficult issue; arguments on both sides have some credibility.

Against the ban on *ex parte* communications are the considerations of flexibility, speed, and efficiency (in the narrow sense of minimized procedural costs) that would to some extent be sacrificed by requiring all communications with decisionmakers be restricted to recorded requirements. Indeed, more than procedural efficiency could be lost; adding procedural burdens to the flow of information to decisionmakers will undoubtedly reduce the information itself. It is not clear that cutting off informal information flows is the best way to handle that problem. In many cases, indeed, informal external consultation is a means of focusing on the basic material that the parties involved consider essential. It is also an important alternative to reliance on agency staff. In any large bureaucracy, the staff information (and its interpretation) is not always to be trusted. Thus avoiding too great dependence on staff—"staff capture"—may be as important to an effective administration as avoiding "industry capture." Maintaining informal channels to external sources is one means of doing so.

In support of some restraint on *ex parte* communications, it must be acknowledged that a totally unconstrained rulemaking process provides a fertile bed for arbitrary administrative action. For example, a process in which interested persons are able to present their arguments—and facts—to individual commissioners and staff without notice to others imposes no reliable discipline on the kind of facts or arguments presented to decisionmakers. This not only raises obvious fairness concerns, but it raises substantial problems of effectiveness and efficiency.

Whether or not the arguments for and against *ex parte* communications are fully balanced, they are sufficiently stated to commend a solution somewhere between the extremes of laissez faire and total ban. A workable compromise would be a system requiring all decisionmakers to record both the fact and the essential content of all communications with interested persons with respect to any substantive issue within the Commission's concern. Such a record would be available to the public generally—including of course all persons interested in particular proceedings to which such discussions pertain. This would not necessarily improve the low quality of the dialog that *ex parte* communications encourages. However, by making the discussion a matter of public record, it would impose some measure of discipline that is now lacking.

Public Participation

Without doubt, the most important development in administrative law in the past score years has been the demand for increased public participation in the administrative process (and access to the courts to secure or to modify administrative action).[13] During most of this period the emphasis has been on defining the scope of the right to participate. In the case of informal rulemaking, the right of participation—to the extent of filing written comments—was recognized with enactment of the Administrative Procedure Act, but this right in formal agency adjudications was generally limited to persons with a legally protected interest (as was, historically, the right to appeal from an agency decision). The rather complex jurisprudential course by which this right was expanded has been fully explored in the literature of administrative law, and needs no further elaboration here. It is sufficient to state that, although there are still some ambiguities in the scope of the legal right in some cases, the law recognizes a very broad right of members of the public to participate in and to appeal from agency proceedings. The right to participate is particularly well defined so far as the FCC is concerned. Many leading decisions on the subject of public participation have involved the FCC.

The controversy over the general legal right to participate has now shifted to the question of how to make the right more effective, particularly by various forms of public assistance to public participants or to other forms of public interest representation. Again, while it is obviously not the focus of public attention, the FCC has been prominent among the agencies for which greater, more effective public interest representation has been sought.

To this point in its development, the broadened right of public participation does not call for any extended justification but a few observations may be made. Whether or not the philosophical premises of democratic society require direct participation in administrative government, there are sound practical bases for some degree of greater public involvement. One of these is the need for public information from different sources, which was discussed earlier. Public groups can be a valuable information resource, particularly on nontechnical issues such as local operations of radio and television stations or cable systems, etc. However, the value of public participants is less as a source of information than as a source of views and interests distinct from those most prevalently represented before the agency.

It has become a commonplace of administrative government in general and administrative regulation in particular that agencies have a strong bias toward the interests of industry. While often exaggerated, frequently misunderstood, and always incomplete as an explanation of bureaucratic sociology (on which more later), the conventional wisdom about "industry capture" is substantially accurate as far as it goes. Unfortunately, of the remedies that have been prescribed for this malady, few have any significant curative powers. However, one reform that has some practical promise as well as philosophical appeal is active public involvement. Public participation can help offset industry or special interest bias in a number of ways, by exposing agency bureaucrats to perspectives not otherwise represented, by sensitizing them to "public interest" concerns, or by application of legal or political constraints. While this theoretical virtue of public participation as a means of broadening agency perspectives and infusing them with a higher sense of public responsibility seems clear enough, the practical effects are a bit more ambiguous.

First, there is too much presumption in the label "public interest groups" insofar as it suggests a claim to represent *the* public interest. The most that can realistically be claimed is that such groups represent a constituency or a public perspective that is entitled to be heard and respected (particularly since it may be otherwise under-represented at the agency). This is not simply a philosophical caution; the point has enormous practical significance in deciding how to structure and to support public interest representation.

Second, assuming that public involvement is valued not only as a form of participatory democracy but as a means of informing and influencing agency behavior, one must evaluate the practical effectiveness of public participation, particularly compared to other forms of public interest representation. This brings us to the main issue: How can public involvement be made effective?

Not surprising, the key problem is lack of money to support meaningful participation. The public interest groups have been most immediately concerned about financial assistance to cover attorney fees and other legal costs incurred in broadcast licensing adjudications. In this particular area, the need has been partly met by agreements in which the costs of public interest groups are reimbursed by industry parties as part of a voluntary settlement. However, such settlements are a very limited aid because they presume a voluntary settlement. (As the law now stands, neither the FCC nor the courts may order shifting of costs from one litigant to another without specific statutory approval.) More fundamentally, any system where financial support is dependent on cost shifting—voluntary or forced—between litigants tends to skew the character of public interest participation. It tends to channel public interest involvement toward adjudicatory contests where costs can be shifted, and away from involvement in rulemaking proceedings or other informal processes where costs cannot practically be shifted (except to the agency itself). This in turn creates or reinforces a bias in favor of an adversarial role as distinct from an investigative role. And because compensation depends on termination of litigation, cost shifting also tends to discourage sustained, in-depth involvement in a proceeding or an issue.

Proposals have been made to have the Commission itself support public participation from its own funds. In light of a recent court decision, it appears that the FCC would not have authority to expend funds for this purpose without specific congressional authorization.[14] There appears to be substantial political support for general legislation authorizing all regulatory agencies to aid public participation. However, a question remains whether scarce agency resources can be justifiably diverted from internal agency needs to support outside litigants. Given the FCC's limited budget, unless additional funds are appropriated specially for this purpose, the Commission probably should not be expected to allocate significant funds for public interest participation.

Thus far the emphasis here has been on aiding public *participation*, as distinct from public *representation*. In theory, of course, the agency is supposed to represent the public. The substantial clamor in recent years for a special public representative stems from the same disaffection for that theory that has produced the demand for greater public participation. Public representation and public participation are not necessarily matually exclusive; however, insofar as both seek scarce public funds as a major base of support, they are competitors. Therefore, a judgment must be made whether "the public interest" is better served, at the margin, by spending public funds to aid publication participation or to establish a special public representative. This obviously depends on a number of variables—type of representative, expectations as to the kind of contribution to be made, etc.—which cannot be pursued here.

This representative need not be an internal officer; an alternative is an "outside" public counsel—an independent agency that would participate in different agency proceedings to protect public or "consumer" interests. Repeated efforts over more than a decade to establish such an agency have failed, and the failure—in February 1978—of the most recent proposal, supported by President Carter, for an "Agency for Consumer Advocacy" indicates that support for such an agency is waning. (In 1975 bills to create such an agency passed both houses but no further action was taken because of a threatened Presidential veto.)

The failure to establish a consumer advocate agency is not to be regretted. It seems most doubtful that such an agency would yield the benefits promised. The implicit premise on which

The Federal Communications Commission

demand for such an agency rests suggests an underlying infirmity in the concept: The present regulatory bureaucracy has strayed from the path of the public interest, therefore new bureaucracy must be created to set it straight. There seems to be little recognition that part of the present failure of regulation is endemic to the frailties of bureaucratic organization.

The deeper objection does not, however, rest on an ideological distaste for bureaucracy, but on two quite practical objections. First, such an agency adds to the diffusion of institutional resources and responsibilities. Certainly there are many agencies concerned with communications policy in one aspect or another and, among them, they have a wide variety of perspectives. The need is not so much for another agency to duplicate their efforts as for a mechanism to coordinate those agencies, and make them more effective. For example, agencies such as the Justice Department and the National Telecommunications and Information Administration have an important role in making communications policy in a variety of ways—not least of which is participation in FCC proceedings. However, seldom do these agencies in fact participate effectively in FCC proceedings. One major reason is the lack of resources to make a distinctive contribution. This is not to suggest that either of those agencies or both together would satisfy *all* aims of a "public counsel" or "consumer advocate." They can provide invaluable public interest perspectives that would in large degree be replicated by a special agency. A second objection to a separate consumer agency is, superficially at least, somewhat similar to the above: A separate agency would divert federal funds away from funds to support direct participation by public groups themselves. This objection is only partly practical; in part it is also philosophical. As suggested earlier, the central rationale for expanded public participation is the need to expose bureaucracies to new ideas, influences and pressures. While some of this can no doubt be accomplished by a special public advocacy agency, a more effective way of accomplishing this might be to support direct public involvement (citizen groups, etc.) without screening it through another layer of bureaucracy. In a very real sense, the special agency approach is a failure to honor the full logic of participatory democracy.

Internal Organization and Staff

Organization

The Commission's staff is organized essentially along functional lines that broadly correspond to the industries subject to

Commission jurisdiction: broadcasting, cable, common carrier (domestic and international telephone and telegraph services), safety and special services (mobile radio, citizen's band and other radio services), and field operations (monitoring and enforcement). In addition to these main operating bureaus, there are several supporting offices that serve internal functions—the executive director (administrative support), plans and policy, administrative law judges, review board (intermediate appeals in some adjudicatory cases), and opinions and review (opinion writing in adjudicatory cases). There are two exceptions to the functional organization: The chief engineer and general counsel offices are professional in nature.

The basic lines of internal organization probably do not require substantial change at this time, though arguments have been made to the contrary. It has been argued that the present separate organization of operating divisions along lines of industry organization is an anomaly in view of the growing convergence of media that will, it is said, blur if not eliminate distinctions between, say, broadcast, cable, and common carrier services. Another objection to the present organization of the main operational bureaus is that it promotes identification among particular bureaus and the industries within their charge. At the very least, it tends to give such bureaus a parochial perspective in addressing policy issues that affect more than one industry.

With respect to the first objection, a simple answer is that the expected convergence of media has not reached the point where the routine regimen of operating bureaus is affected. As yet we do not have an "electronic highway" (or highways—to preserve the possibility that competitive transmission paths will continue to exist) in which mass communications and point-to-point communications services are technologically, economically, or socially molded into a single operational system. Nor is it plausible to suppose that the time of integration, or convergence, will be greatly affected simply by a merger of the bureaus. On the contrary, the political reality is that a merger of the bureaus might forestall the development of an efficiently integrated system. For example, if the cable bureau were merged with the broadcast bureau, it is naive to assume that cable technology—or broadband mass communications service generally—would continue to command the status it has today either within the agency or outside it. The point of giving a separate bureau status to cable in 1970 was not merely to enhance the status and prerequisites of cable staff personnel (though this clearly was an important element in the dynamics of the cable bureau's evolution); it was to give added status to cable television as an industry, as a technology and as a service. Many critics (myself included) believe that the development of cable television is still being unreasonably restrained to protect the interests of broadcasting and past Commission policies centered on broadcast service. A merger of the cable and broadcast bureaus would, if anything, reinforce the Commission's protectionist instincts.

The second objection—that the bureaus align themselves with industry interests and have a parochial approach to broad policy issues—is undebatable. However, there does not appear to be any simple institutional corrective. If the conflicts among industry interests are unavoidable, in large measure, so are staff organization (for example, along professional lines) is unlikely to be more than cosmetic. It is necessary to have particular staffs deal routinely with particular industries and industry problems as a matter of developing and using experience and expertise, not to mention ensuring reasonable continuity in relations with and actions affecting regulation industries. Where this kind of industry specialization is required, industry identification and parochialism are largely unavoidable. One can perhaps minimize the problem by periodic changes in personnel assignments. In fact, such changes have been fairly common at the FCC in recent years. One should not make too much of this remedy, however; it does not take long for a newly transported bureaucrat to acquire the prevailing bias of his immediate peers.

The appropriate remedy for staff bias and parochialism lies not in changing bureau staff organization but in broadening the professional perspective represented in the bureaus and particularly in the policy staff outside the bureaus.

Composition of Professional Staff

One of the most serious deficiencies in FCC's professional staff is the historic dominance of engineers and lawyers and the underrepresentation of economists. Lawyers and engineers presently comprise over 85 percent of the professional staff; economists and statisticians, less than 5 percent. Particularly noteworthy is the absence of any office of economic analysis comparable to the chief engineer and general counsel offices. For an agency whose regulatory policies influence the economic destiny of a large and growing sector of the American economy, the absence of an office of economic analysis is striking. If the Commission had no other task but to regulate the interstate rates and services of AT&T, one might reasonably think that the magnitude of the task of trying to deal with the world's largest company (which also happens to employ, on a permanent or consulting basis, an uncommonly large share of the country's top economists—some 1600 economists and statisticians in 1970) would warrant an office of economic analysis independent of responsibility to routine bureau functions. When one observes that the agency must deal with the multitude of other firms and multiple, diverse industries, the absence of such a staff—this omission in the FCC's organization—becomes, in Wonderland Alice's words, "curiouser and curiouser."

Although numerous recent studies have noted the need for strengthening the Commission's staff capability for economic analysis, little attention has been given to the need for a separate professional organization for economic analysis, either by critical observers or by the Commission itself. Occasional, casual suggestions in recent years that the Commission establish an office of chief economist, paralleling the offices for engineering and law, have been given short shrift in the past by the Commission despite its recognition of the need for a larger economic staff generally. The reasons for resisting the idea of a separate office for economic analysis are not entirely clear for the Commission has not officially addressed itself to a concrete proposal. However, the view of those with whom this subject has been discussed appears to be that additional economic staff, to the extent that they can be obtained, should be located within the Office of Plans and Policy (OPP) and within the operating bureaus. (An economist was recently appointed to head the OPP; no comparable appointments have been made to senior bureau positions.)

In evaluating this view, we can abstract from the general problem of obtaining additional professional staff or the problem of finding qualified economists. Although both of these are significant problems, they can be deferred for they are essentially irrelevant to the organizational question raised here: Should the Commission establish and seek to staff a separate office of economic analysis as a major priority in organizing and staffing the agency?

We can quickly dismiss any objection that a separate office organized along professional lines would not comport with the functional organization of the Commission generally. The existence of separate professional offices for law and engineering defeats any such objection. In fact, one major reason for maintaining separate legal and engineering offices provides the central rationale for a similar office of economists: to provide independent professional advice that is not bounded by the particular concerns of the different operating bureaus. Even assuming a fully adequate economic staff in the various bureaus (a purely hypothetical assumption, suggested above), there is a strong need for economic assistance and advice to the commissioners that is not tied to the parochial interests and biases of the operating bureaus. For legal and engineering problems, this function is performed by the general counsel and chief engineer (and, of course, to a degree by the commissioners' personal staffs). On economic matters the only staff that can perform this function is the OPP. However, the integration and critical evaluation of bureau analyses and recommendations on individual cases or particular rulemaking proceedings is not properly a function of the OPP, except to the extent that it pertains to major issues of long-term policy. In recent years, one of the chief criticisms of the Commission, and in particular of the chairman, has been that the important resources of the OPP have been misdirected by engaging staff in routine agency affairs, to the neglect of long-range policy analysis and planning. At least part of the explanation for this alliance is the absence of a stronger, high-level staff capability for reviewing the analyses and recommendations put forward by bureau staff. And on economic matters in particular there is virtually no one else to consult outside the bureaus.

If each commissioner's office were staffed with an economist, part of this problem would be alleviated. Although it would perhaps be desirable to strengthen the staff resources of individual commissioners, this is not a solution for the problem at hand. The commissioner's personal staffs serve the special needs of their respective commissioners, not the general needs of the Commission as a whole or of the collective interests of the commissioners as a group. In rendering advice and assistance to the individual commissioners, the staffs simply do not have the time to engage in the kind of independent analysis and study that is necessary to provide the support that a separate office of economic analysis could and should perform.

What is proposed here does not require a large office comparable in size to either the offices of chief engineer or general counsel. Both of those offices perform a number of operational functions that would find no counterpart in an office of economics (for example, the former is responsible for equipment certification; the latter handles all agency litigation). On the premise that such an office would not substitute for hiring additional economics staff elsewhere in the agency, a small staff of perhaps six or seven qualified economists (including the head of the office, the chief economist for the agency) would provide an adequate capability for the function envisioned.

Staff Size and Calibre

This brings us to a more basic complaint concerning agency staffing, i.e., that the agency is understaffed to handle the burgeoning workload. It is easy to demonstrate this by comparisons. One favorite illustration is to compare the agency with the giant monolith AT&T or with the networks. Such comparisons can produce some dramatic statistics; however, such data would produce a sense of futility and hopelessness rather than concrete remedial action, by the agency or by Congress. After all, no one really expects the FCC to match the private resources of AT&T any more than it expects, say, the Justice Department's Antitrust Division to match those of all private firms (including AT&T) that it sues. The goal is not to try to match even remotely the resources of regulated firms; such a goal would presuppose that the agency's function was to manage rather than regulate. Given the agency's broad, and generally sufficient, legal powers to compel the industries it regulates to provide information necessary for its judgments, a relatively small cadre of well-trained professionals should suffice to perform all the supervision that would be needed.

A small staff puts a premium on close scrutiny of work priorities and workforce utilization. This, in turn, necessarily implies that some tasks which Congress and various segments of the public (including the industries) want accomplished cannot be done. That might mean, for example, that less time (if any time at all) would be spent on the task of studying broadcast licensee renewals, cable certificates of compliance, or monitoring of CB radio in favor of more attention to such questions as analysis of general rules governing broadcast, cable and CB industry structure, or more attention to AT&T's interstate rate structure. Such a reordering of priorities is scarcely to be regretted; it is indeed one of the virtues of maintaining a lean administrative structure. Regrettably, for all the recurrent talk of the FCC's thinness, there is little evidence that it has caused the agency significantly to reorder its own priorities.

The Commissioners

Role Models: What is a Commissioner?

Ultimately the strength or the weakness of an agency's staff stems from the direction and support supplied by the agency's appointed leaders, and it becomes impossible to assess the quality of the former without implying something about the latter.

The Federal Communications Commission

Any assessment of regulatory commissioners raises an ambiguity in the role and expectations imposed on such agencies. An evaluation of appointments to regulatory agencies presupposes some relevant benchmark for comparison. A staff professional can be compared, more or less, with professional counterparts elsewhere, in private or public sectors. Comparisons of regulatory commissioners are not so easily made except perhaps with each other, but the interagency comparison does not prove much since the character of appointees does not vary predictably among agencies; and, indeed, it is the quality of all such appointees as a class that is most frequently held up for evaluation. One standard of comparison frequently invoked is that of the federal judiciary. Because agency members perform judicial functions the comparison to federal judges is a natural one. Measured against such a standard, agency members, as a class, probably suffer by comparison.

For the moment we can pass over the question whether merely adopting some of the trappings of federal judgeships would itself significantly alter the character of appointments; the deeper question is whether the courts are the appropriate model for a regulatory agency. While it is certainly true that agency members perform significant judicial functions in deciding individual cases, it is also true that agencies such as the FCC perform more legislative or executive tasks then judicial ones. When the FCC, for example, promulgated regulations barring common ownership of local newspapers and broadcast stations, it was performing a simple legislative task. In reaching a decision, it was neither bound by, nor expected to conform to, the confining procedures or standards of a court. Why then should the decisionmakers be stamped from a judicial cast? Insofar as the agency is given such broad legislative powers and responsibilities, would it not at least be as appropriate to measure the agency against standards pertinent to evaluating legislators? Such a standard merely requires a revision of relative quality of agency members (a revision almost certain to favor agency members), it requires an entirely new frame of reference for judging agency performance. The commitment to carefully reasoned principle—the conventional ideal of judicial decision—necessarily gives way to the dictates of political compromise and practical expediency that are the accepted hallmark of legislative action. Correspondingly, the standard for judging the composition of the agencies stems from an emphasis on trained professionals—presumably lawyers if the judicial tradition is accepted—to an emphasis on representativeness. By this standard we should not seek legal scholars or practitioners for our agencies but rather persons broadly representative of different segments of the public, necessarily including nonlawyers as well as lawyers. Needless to say, measured against this standard, the FCC comes closer to meeting the ideal than it does to meeting the judicial model.

Appointments: Who is a Commissioner?

A recent study of regulatory appointments shows convincingly that selection of appointees is a haphazard process that almost defies description.[15] One characteristic that the study does reveal, however, is that it is not a search for the "best candidates" if by that term is meant those qualified by experience, training, or professional aptitude. If one compares the process of selection with the model that some have imagined to be consonant with the judiciary model for agencies, it does not begin to measure up as rational. Indeed, as was noted, there is not even a requirement that administrative officers be lawyers, and at the FCC currently a majority are not. In itself this is not to be regretted for unless one adheres slavishly to the judicial model, there is no reason for lawyers to dominate the administrative agencies. On the contrary, what is wanted is a mixture of representative talents, skills, and interests.

The appointments process is perhaps no more designed to create a representative body than it is a judicial one. Appointees are not selected to represent political, social, or economic interests. Nor do they represent any particular array of talents and skills that could be objectively identified as relevant to the job. (One surmises that in some cases neither the President nor his chief advisor has any clear notion of what talents would be relevant to the job.) It is often said that appointments are "political," but that is trite. And, standing by itself without further explanation, the "political" label is irrelevant. What one wants to know is not whether political factors enter into the process (one would be astonished if they did not) but whether the politics are constrained by, or informed by, considerations of principle beyond short-range expediency or patronage. It is not usually considered a basis for complaint that a Secretary of State is "political"; the cause for distress arises only if the particular appointment reflects a kind of political choice that is unsuitable or unequal to the demands of the job. So too, *mutatis mutandis*, with regulatory agency appointments. The problem is not that they are "political" but that the politics is patronage politics on a very low order of public responsibility to the regulatory agency functions and public interest purposes.

It is for this reason that the appointments process appears to be so random when viewed from the perspective of talent, experience, and background. The demands of patronage, at least at this level, arise without regard to any of these elements. (A friend, or a friend of a friend, or a friend of the party, etc. can be nearly anyone.) This needs a partial qualification. The process is erratic, but it is predictable in at least one respect: The process is biased against persons with strong views toward important regulated industries. Persons known to hold such views appear to be generally screened out. Appointments do occasionally go to persons who have or develop antagonisms toward the regulated industries but generally this is only because neither the appointments process nor the confirmation process is very efficient in identifying such biases in candidates for such offices, except in rare cases where they have earlier committed themselves to written opinions (in books, articles or speeches) or where they come from one of the regulated industries. Despite that inefficiency, however, the selection process seldom yields candidates who are in fact notably antagonistic to any important industry segment at the time of their appointment, for the simple reason that the typical candidate is not sufficiently knowledgeable about the regulatory issues to have a firm view for or against any particular industry. In a field of regulation such as communications, where there are not only but several different industries and industry interests, the system of selecting candidates for appointments works strongly against the appointment of persons with specific views about regulatory issues because such views are likely to engender opposition from at least one major industry group sufficient to offset any support from another. Even where the political influence of one industry interest group is greater than that of a competitor, if the lesser interest group has a significant political base, it can cause sufficient controversy over a particular candidate to cause the President (i.e., his political advisors) to look elsewhere for a path of lesser, if not least resistance.

The logic behind this is elementary: Neither the patronage value of any individual appointment in particular nor the importance that the President attaches to independent agency appointments in general justifies incurring the substantial political costs of selecting an appointee that is "controversial." These factors work to eliminate from serious contention persons with established track records in the field of communications, or regulation—unless the record is acceptable to all of the affected interests that exercise significant political influence. The appointments process thus typically (though admittedly not always) leads to the selection of persons who generally have no noticeable commitment to an industry viewpoint, or who indeed have no notable views about communications issues—except perhaps some inconsequential views about sex and violence on television, on which all persons are expected to have some views.

It perhaps goes without saying that the relative indifference of the White House and Congress to agency appointments reflects in turn indifference toward, or ignorance about, the agency and its functions. That indifference not only influences the initial selection of appointees, it also affects the environment of the new agency member. The White House and Congress—reflecting general public attitudes—may be relatively indifferent to the agency and its membership; affected industry interests are clearly not. Any vacuum of attention from the former is thus quickly filled by the latter.[16] In this milieu it should scarcely come as a surprise that agencies and their members so often become drawn to the special interests that constantly surround them, thereby giving rise to the now ubiquitous charge of industry capture.

Interest Bias

Conventional wisdom holds that regulatory commissions are captives of those they are supposed to regulate. According to this received learning, the dynamics of administrative regulation go something like this: A social or economic problem arises; to address the issue, a regulatory agency is created by the legislature. Because the legislature is unable or unwilling to give detailed guidance about how to handle the problem—or even to define the problem—the agency is given great but ill-defined power. Initially, the agency attacks the problem with innovative zeal, but eventually the zeal wanes. In time, the agency develops a familiarity with the industry it regulates and with that familiarity develops an identity with industry interests. In short, the agency becomes "captured" by the industry it was established to regulate.

Although the industry bias or "capture" problem is not a new one, it appears to have received increasing emphasis in recent years, and critics have pressed vocally for reforms to eliminate the problem. One prominent "solution" that has gained considerable support in recent years proposes removing one perceived source of industry interest identification—post-term employment of regulators, either in regulated businesses or in law practice that represents regulated firms.[17] This restriction is sometimes coupled with the proposal for a single, long-term tenure (12-15 years) at the agency.[18] The proposal builds on present legal ethics principles that bar government officials from

The Federal Communications Commission

representing private interests with respect to matter in which they had substantial involvement in that official capacity. The ethical practice rules have a twofold thrust: First, to avoid potential conflict of interests (or the appearance of such) that arises when a lawyer is in a position to use for private purposes information obtained in a capacity of public trust; second, to eliminate a potential source of bias on the part of an official charged with requesting (or judging) interests with which that official may later become associated. It is the latter in particular at which the proposed ban is aimed.

The argument against a ban on a post-term industry employment is as simple and straightforward as the case for it: It will impair the ability to recruit qualified people. Even assuming no prior involvement in industry-related activity, the accumulation of experience in an agency is a marketable commodity that has come to be featured as one of the attractions of federal government service. (The FCC is not of course unique; the same "deferred returns" accrue to most holders of top government posts ranging from the head of the Fish and Wildlife Service to the Secretary of State.) Remove that reward, the argument goes, and you diminish the incentives of "good people" to go into federal service. Though plausible this argument must be skeptically received, at least with respect to appointments to the agency, or to policy level staff positions in the agencies. It seems doubtful that good people have to be "lured" to such positions by the promise of post-term employment with industry. There may be more substance to the argument with staff specialists: A bright young lawyer who wants to develop a specialty in communications law may not want to serve with the FCC if it means a choice between permanent civil service and giving up his or her specialty because conflict-of-interest rules effectively forbid him from representing communications interests.

The more important objection to restrictions on post-term employment, for appointees to policy level positions at least, is that it does not touch the core problem. The conflict-of-interest rules focus solely on the question of how to maintain regulators' integrity; they do nothing to assure that regulators with integrity are selected in the first place. The conflict-of-interest rules may ensure that regulators do not "go bad," but it does nothing to ensure that they are any "good" in the first place. It should be noted, however, that the references to "good" and "integrity" perhaps suggest too great emphasis on ethical considerations when the more fundamental problem is simply one of caliber and experience. Given the widespread dissatisfaction with the regulatory appointments, it is indeed curious that critics have been so preoccupied with conflict-of-interest rules that do not address the question of the quality of appointments or the nature of the appointments process.

Important as it may be to offset the influences that may cause bias among agency members after their appointments, it is even more important to influence the initial selection of appointees. It is, of course, extremely difficult to alter substantially the political environment that influences the appointments process. The appointments process might operate somewhat differently if the FCC's activities (not merely its role in renewing broadcast licenses or reviewing television programming) were better understood—if communications policy were more widely perceived to be important to social and economic welfare. Greater public and political attention would then be given to the process, and this would result in greater pressures to improve the standards of selection.

Improving the Selection Process

Given present circumstances, it is unlikely that we will see any radical transformation in public or political attitudes toward regulatory agencies in general or the FCC in particular within the immediate future. However, there is one quite simple measure that could make some improvements in the appointments process without putting overwhelming demands on political systems—this is to establish some form of public selection and screening process for agency appointments.

Precedent for such a mechanism can be found in the review of candidates for federal judicial appointments by the American Bar Association's Committee on the Judiciary. Although there is no legal basis for it, review of candidates by the Committee has become an established means of providing public evaluation of nominees. It would be naive to pretend that the ABA's review process has been responsible for producing the highest qualified persons for the bench; the selection of judicial appointees, like the selection of agency appointees, has been influenced by patronage considerations. But if the ABA review process has not eliminated this much-criticized aspect of judicial appointments, it has almost certainly had a good effect in exposing some of the poor products of the selection process to public scrutiny and thereby acting as a modest deterrent to the selection of unqualified appointees.

Basic Agency Structure

In critical discussions of basic regulatory organization, two issues have dominated all others: independence versus executive control and multi-member commissions versus the single administrator as a form of organization. Although the two questions, independence and commission form, are typically linked, there is no necessary connection between them. The independent regulatory agencies have invariably taken the form of multi-membered commissions, while regulatory agencies within the executive branch have been headed by a single administrator—or commissioner (for example, the head of the Food and Drug Administration is a "commissioner," even though he does not head a commission in the traditional sense of a multi-member body). But, history aside, there is no reason to presume that "independence," whatever it means, should require a commission form or executive control should preclude it. It is appropriate therefore to treat the two elements of organization separately.

Independence

Although the independent regulatory commission has become such a familiar and seemingly permanent feature of administrative government, the concept of a regulatory agency exercising nonjudicial policymaking authority beyond the control of the executive continues to draw attack from regulatory reformers. As recently as 1971, the Ash Council appointed by President Nixon urged elimination of the independent commission form in the fields of transportation, power, securities regulation, and consumer protection, in favor of executive control of these functions.[19] Somewhat surprisingly, it proposed that the FCC retain its independent nature because of the supposed need to insulate its functions from political (executive) control. In light of the Nixon Administration's hostile relationships with the communications media, this omission of the FCC may have been simply a matter of political expediency: Congress would not have accepted such a reorganization, and there was no reason to generate needless controversy in such a specially sensitive field. However, other critical studies have suggested that communications policymaking and administrative functions be assigned to an executive agency with "judicial" functions transferred to a special administrative court.

While the creation of a special administrative court for judicial matters might avoid some of the objections that would otherwise be raised against presidential control of communications matters, it probably would not substantially alter the general political objection to a shift of "independent" policymaking functions to the executive. Despite repeated attacks on the independent commissions, they continue to be created with fair regularity—a reliable indication of their secure position in administrative government. While it is conceivable that, in particular instances, Congress might permit some transfer of functions from independent commission to the executive branch, it is difficult to imagine that Congress would assent to this reorganization of communications functions. A number of rationales have nevertheless been advanced for such a reorganization. Most, but not all, of them center on the perceived need for greater public and political accountability, stronger leadership, and unified policy direction for functions, which are now divided between several executive agencies and the FCC.

The matter of political accountability needs no development here. It is sufficient to note that the absence of direct political accountability does not seem to be a special flaw of organizational independence. Even casual observation casts doubt on the assumption that this accountability is, as a practical matter, materially greater for executive agencies performing functions comparable to those performed by the FCC. (Is the FCC less "politically accountable" than, say, the Forest Service is within the Department of Agriculture or the Food and Drug Administration within the Department of HEW?) It should be emphasized that the regulatory functions that concern us are, with a few exceptions, the responsibility of subordinate echelons of the executive bureaucracy that are unlikely to be within the close scan of presidential attention, or his direct control.

Similar considerations apply to the argument for high-level political leadership. Communications policy needs and warrants greater presidential attention than it now receives. The appropriateness of direct presidential involvement in FCC policy deliberations seems to me beyond question. In this regard, the commonplace observation (as embraced, for example, by the Ash Council) that communications is too sensitive to allow presidential influence is both naive and quixotic. Given the President's powers with respect to every other vital interest that touches our lives, and our liberties, it is difficult to understand why basic communications policy should be beyond his influence. Of course, his activities, no less than those of the FCC, should be subject to legal and constitutional constraints. But presidential influence is no more inherently suspect on legal or constitutional grounds than the activities of the FCC or of congressional leaders. Subject to the constraints that limit the activities and influence of all government officials in this area, presidential involvement in communications is desirable. It could provide a broad perspective to policymaking that is all too often missing at the present time. However, the question is not whether executive organizational control is a necessary or sufficient basis for presidential leadership. It is perhaps neither. There is little evidence to indicate that presidential supervision of an executive communications agency of stature comparable to the FCC would be much greater than it is now. The means of executive leadership and influence on major policy issues already exist but they have not been effectively used, not for any reasons of organizational theory but simply for want of sustained interest.

It has been argued that executive organization would lead to greater presidential care in agency appointments than is the case with independent agency appointments. The premise is that Presidents are careless about independent agency appointments insofar as the performance of the agency is not the President's direct responsibility; by contrast, Presidents are said to be more reluctant to appoint incompetent commissioners to executive agencies, for "their failure will be his failure and continuing to keep them in office, despite poor performance, will be his responsibility."[20] Again the question is, at bottom, an empirical one: Is the President in fact more personally involved, more careful about the quality of executive agency appointments—at the level with which we are concerned—than he is with independent regulatory agencies? Casual inspection of agency appointments and more studied empirical observation suggest that the assumption of greater care and attention to executive appointments is, other things being equal, dubious. However, we will return to consider a related issue: whether the single administrator form of organization typical of executive agencies does not have some advantages over multimembered commissions in this respect.

Plural Membership

Besides the question of independence, a number of criticisms have been made of the commission form of organization. The principal criticism is the one made by the Ash Council: It creates inefficiency and diffusion of responsibility.

In terms of internal management (scheduling, personnel hiring and assignments, etc.), a multimembered commission is an awkward form of management. But, as applied to the FCC at least, such a criticism is mooted by the fact that these matters are typically either delegated to staff (the individual bureaus or the executive director) or are handled by the chairman's office with only episodic involvement by other commissioners. Thus, while it is possible that a plural membership that actively sought to run the internal affairs of the agency as a "committee of the whole" could create confusion and inefficiency, this simply has not been a significant problem at the FCC.

The more plausible argument against the commission form is addressed not to the intrinsic inefficiency of plural decision-making but to the nature of the commissioners themselves. It has been supposed, for example, that a single-administrator agency would attract better appointees. On the premise that it is better to be a chief than a "co-leader," the point seems largely irrelevant for, as noted earlier, finding good people who are willing to serve on the FCC is a negligible problem; the real problem is to create a political environment in which the best persons can and will be selected, and that environmental variable would not presumably be altered by the form of organization. It is nevertheless probable that a single-administrator organization would, on the average, yield better appointments insofar as the enhanced responsibility and power of a single administrator would cause greater attention to be given to that single appointment than to merely one of seven. It is undoubtedly true today, for example, that the position of chairman—of the FCC and of the other regulatory commissions as well—receives far greater presidential attention than the position of commissioner. It would be too much to argue that appointments to positions of commission chairmen are invariably better than those to commissioner positions. But it does not seem exaggerated to claim that, at the very least, the former receive more careful consideration of a kind designed to yield more highly qualified persons. Each additional position thus dilutes the average care and attention given to regulatory appointments.

From this it would seem to follow that the Commission ought for this reason alone to be reorganized as a single-headed agency. (It could, presumably, remain an independent agency.) However, we have looked only at the negative side of commission organization. There are some affirmative things to be said of commissions. One of the prominent arguments for the commission form is that it diffuses power. Insofar as this may create inefficiencies and may diminish the importance of the agency and its appointees, it has been considered a liability. But there are some offsetting advantages. The "collegial form" of the agency can provide different perspectives at the decisionmaking level in the agency.

Yet, whatever the advantages of collegial control and diversity, there are surely limits. Assuming that the advantages of collegial form are at least sufficient to withstand the pressures for a single-headed agency, somewhere a line must be drawn between a single administrator and a multitude of commissioners. Almost any number is arbitrary; however, it is noteworthy that, except for the Interstate Commerce Commission, which few people regard as a model to be emulated, all of the major commissions formed both before and after the FCC began have been limited to five members. The selection of the number seven for the FCC is a reflection of a historical idiosyncrasy that has long since ceased to have any significance. (Commissioners were once responsible for different geographic regions—initially five, then seven; the regions were long ago abolished.) However, even the number five is probably too high; if the number of members is reduced as a means of inducing greater attention to be given to appointments, the number probably need not be greater than three—a number which would allow some diversity of perspective while offering at least some of the advantages ascribed above to single-headed agencies.

Separation of Functions

Related to, but logically distinct from, the proposals for executive control of communications functions and a single administrator are proposals to separate different agency functions into two, perhaps three, separate agencies.

The most well-known proposal is that of former CAB Member Louis Hector and former FCC Chairman Newton Minow, who argued that the combination of adjudication, policymaking/administration and investigative/prosecutorial functions produced unnecessary confusion of roles, inefficiency in procedure and—in the case of the combination of prosecutorial and adjudicatory functions—unfairness.[21] To correct these deficiencies, it was proposed to place the adjudicatory functions in a special administrative court, the policymaking/administrative functions in an executive agency concerned with transportation or commerce, and the investigative/prosecutorial functions in the Department of Justice.

These proposals have generally evoked a negative response from administrative law commentators, and for good reason. The assumption that the combination of functions is inefficient is simply incorrect. In fact, on pure efficiency grounds an argument can be made that the institutional integration of these functions increases efficiency and effectiveness by making each task more immediately responsible to the other. The problem of confusion in roles—for example the supposed distraction of policymakers who must adjudicate and adjudicators who must make policy—is more theoretical than real. This complaint presupposes a sharper definition of FCC tasks into judicial, legislative, administrative, and enforcement categories than exists. Most of the daily work that occupies a commissioner's time (or the time of top-level staff) would be difficult to sort into such neat compartments. As far as fairness to litigants is concerned, the supposed conflict between prosecutorial and adjudicatory function is, for the most part, mooted by the fact that the commissioners seldom participate in investigative or prosecutorial functions that they subsequently adjudicate.

The real conflict is not competition among different procedural functions (for example, adjudication and policymaking). Rather, the problem is the competition between different substantive regulatory responsibilities. It is often said, for example, that commissioners are preoccupied with broadcast and cable regulation to the neglect of policy issues in other areas, most notably in the field of telephone and telegraph services. As a consequence, major policy issues in the latter area are decided at the subordinate bureau level where they are less publicly visible or accountable. In the past few years, this imbalance in commission attention has slightly altered as common carrier issues of major public importance have literally forced themselves on the commissioners for decision. There continues to be a substantial imbalance, however, in top-level management's consideration given to major policy issues in the mass communications are on the one hand, and telephone/telegraph area on the other.

It remains to be seen whether there is any simple institutional solution to this problem. Former FCC Commissioner Robert Bartley proposed that the FCC's responsibility should be reconstituted into four different agencies. Broadcast (and presumably cable) matters would be addressed by an independent, five-member "broadcast commission." Domestic and international telephone and telegraph services would be regulated by a similar independent common carrier commission. Safety and special radio services would be the domain of the Department of Transportation (on the rationale that most of these services relate to mobile radio). Spectrum allocation and treaty responsibilities would be assigned to an office in the Congress. A similar proposal has been advanced more recently by Dean Burch, who called for a division of responsibility within the FCC between mass communications and telecommunications.[22]

These proposals have some appeal; unlike earlier proposals they are at least addressed to a practical, not a conceptual, problem. Unfortunately, these proposals would institutionally separate policy issues at a time when technological advances and service developments are combining them. The introduction of broadband technology into both mass communications and point-to-point communications requires a comprehensive and integrated policy focus. How else will it be possible to deal with such issues as satellite broadcasting and integrated telephone-broadcast services?

Substantive Improvement of Regulatory Performance

Implicit in the search for ways of doing things better is the assumption that they are worth doing at all. It is on that premise that I have identified various faults in the FCC and its processes and suggested ways in which they might be improved. It should, however, be recognized that procedural or reorganization change alone offers only modest opportunity for substantial improvement in regulatory performance. At some point in the analysis of reforms, one is drawn inevitably into questions of substantive policy. Earlier instances were cited of licensing policy (comparative licensing, licensing of CB, certification of cable) as illustrations of this point. In each case, the FCC's processes have been widely criticized as inefficient, but few critics seem to have recognized that the larger inefficiency may reside in the nature of the task itself. In the case of cable, for example, the question whether it is necessary to have a certification process is but a preface to a more basic issue—is it necessary to have regulation at all? (Or at least, is it necessary to have the elaborate scheme of regulation that has been constructed?)

It is necessary to rethink our regulatory objectives. In addressing the issue of agency bias it was noted that the industry capture explanation of regulatory behavior is supported by substantial evidence. However, the conventional criticism of "industry capture" is a somewhat superficial explanation for a more complex phenomenon. The chief weakness of the industry capture thesis, and procedural reforms that are derived from it (such as conflict-of-interest rules), is that it personalizes what is all too often an institutional bias of regulation itself. It is true that the regulator tends to develop a familiarity and an empathy with regulated interests. What is not true is the inference that this familiarity and empathy are necessarily contrary to the political intent and purpose of regulation. In communications, as in many other regulated sectors of the economy, regulation has generally been instituted or extended on the initiative (or at least with the consent) of the regulated interests—typically as a means of "stabilizing" the industry, i.e., restricting open competition.[23]

Thus, it is not simply that regulators have become tools of private interests, but that *regulation* has become a tool of private interests. Industry interests may complain about regulation, but given a choice between regulation and competition, most opt for regulation. Unfortunately, so too do many others, in the belief that regulation is necessary to improve the marketplace. It is a belief shared by most regulatory bureaucrats. If there is one predominant bias of regulators that stands out, it is a preference for regulation over open competition as a vehicle for social and economic control. Regulation is, or quickly becomes, a lifestyle of the bureaucrat. As such, it often takes on a force all its own, one quite independent of original purpose albeit one still connected in some general way with the concept of "the public interest."

There are, no doubt, cases where regulation serves legitimate general public interest objectives. The classic case is that of the natural monopoly where the industry structure is not conducive to competition; and there are doubtless other instances where some overriding public objectives dictate regulation, even at the expense of competition.[24] This is not the place to try to identify which particular regulatory tasks fall on one side of the public interest fence and which on the other. It is enough simply to note that the legitimacy of regulation should not be uncritically assumed and that institutional or procedural reforms in the regulatory system may be insufficient, even counter-productive, if what is really needed is an entirely different policy regime. Increasingly it has become apparent that the regulatory scheme for communications requires a fundamental reevaluation. Until that is done, reforming the processes and institutions of regulation can only be incomplete.

Notes

1. A more extended treatment of all the matters treated here is given in Glen Robinson, "The Federal Communications Commission: An Essay on Regulatory Watchdogs," *Virginia Law Review*, Vol. 64, 1978.

2. On the topic of delegation, *see* Glen Robinson, and Ernest Gellhorn, *The Administrative Process*, Chapters 2 and 11, West, St. Paul, 1974. On the scope of FCC powers in particular *see*, e.g., *National Broadcasting Company v. U.S.*, 319 U.S. 190, 1974; *United States v. Southwestern Cable Company*, 392 U.S. 157, 1968; *United States v. Midwest Video Corporation*, 406 U.S. 649, 1972.

3. The seminal FCC decisions are: First Report on CATV, 38 F.C.C. 683, 1965; Second Report on CATV, 2 F.C.C. 725, 1966. The present scheme derives essentially from Cable Television Report,

36 F.C.C. 2d 143, 1972. The history of cable regulation and how it grew is recounted in detail in Don LeDuc, *Cable Television and the FCC*, Temple University Press, Philadelphia, 1973.

4. *See* Paul Berman, "Computer of Communications? Allocation of Functions and the Role of the Federal Communications Commission," *Federal Communications Bar Journal*, Vol. 27, 1974, p. 161. A recent illustration is provided by AT&T's Dataspeed 40 service offering, AT&T 62 F.C.C. 2d 21, 1977.

5. *See* Senate Committee on Government Operations, 95th Cong. 1st Sess., *Study on Federal Regulation*, Vol. 4, Committee Print, 1977.

6. The measures taken to expedite applications in the broadcast and in other areas, and the results in backlog reduction, are reviewed in detail in the testimony of FCC Chairman Richard Wiley in Hearings before the Oversight and Investigations Subcommittee of the House Committee on Interstate and Foreign Commerce, 94th Cong., 2d Sess., Regulatory Reform, Vol. 5, 1976, pp. 201, 253-77.

7. For an exhaustive treatment of the subject (which cites most of the earlier commentary) see Robert Anthony, "Towards Simplicity and Rationality in Comparative Broadcast Licensing Proceedings," *Stanford Law Review*, Vol. 24, 1971, p. 1.

8. *See* the dissenting opinion of Commissioner Robinson in Cowles Florida Broadcasting, Inc., 20 F.C.C. 2d 372, 435, 1976. The seminal exposition of the idea is that of economist Ronald Coase, "The Federal Communications Commission," *Journal of Law and Economics*, Vol. 2, 1959, p. 1. Subsequent discussions have extended the same idea to allocation of frequency rights more generally. *See*, e.g., Ronald Coase, "The Interdepartment Radio Advisory Committee," *Journal of Law and Economics*, Vol. 4, 1962, p. 175. A DeVany, R. Eckert, C. Meyers, D. O'Hara and R. Scott, "A Property System for Market Allocation of the Electromagnetic Spectrum: A Legal Economic-Engineering Study," *Stanford Law Review*, Vol. 21, 1969, p. 1499; Charles H. Jackson, "Technology for Spectrum Markets," unpublished Ph.D. dissertation, Massachusetts Institute of Technology, 1976. *See generally*, Harvey Levin, *The Invisible Resource*, Johns Hopkins Press, Baltimore, 1971.

It should be noted that the basic idea of support for this auction proposal has been endorsed by such certified liberals as Senator William Proxmire, *Broadcasting*, May 7, 1973, p. 47 (userfee proposal) and Congressman Henry Reuss, *see* House Rept. 11893, 85th Cong., 2d Sess., 1958 (proposal to require the Commission to establish basic threshold qualifications of competing applicants and, in the event of a tie, to award the license to the highest bidder).

9. No single reference can do justice to this sweeping statement. However, a 1976 critique by the staff of the House Interstate and Foreign Commerce Committee presents an excellent statement of the case against the FCC's cable policies, as well as a short bibliography of critical studies and writings. Communications Subcommittee of the House Committee on Interstate and Foreign Commerce, 94th Cong., 2d Sess., *Cable Television: Promise Versus Regulatory Performance*, Subcommittee Print, 1976.

10. James Q. Wilson, "The Dead Hand of Regulation," *The Public Interest*, Vol. 25, 1971, pp. 39, 48.

11. Henry Friendly, *The Federal Administrative Agencies*, Harvard University Press, Cambridge, 1972, p. 54.

12. *Home Box Office v. FCC*, 567 F. 2d 9, Dist. of Columbia Cir., 1977, cert. denied, 98 S.Ct. 111, 1977.

13. On public participation, Richard Stewart, "The Reformation of American Administrative Law" *Harvard Law Review*, Vol. 88, 1975, p. 1671, reviews the development of the legal doctrine. On the practical aspects that emphasize the need for financial assistance to participants, *see* Ernest Gellhorn, "Public Participation in Administrative Proceedings," *Yale Law Journal*, Vol. 81, 1972, p. 359; and Note, "Federal Agency Assistance to Impecunious Intervenors," *Harvard Law Review*, Vol. 88, 1975, p. 1815; Senate Committee on Government Operations, 95th Cong., 1st Sess., *Study on Federal Regulation*, Vol. 3, Committee Print, 1977.

14. *Greene County Planning Board v. FPC*, 565 F. 2d 807, 2d Cir., 1977.

15. *See* Senate Committee on Commerce, 94th Cong., 2d Sess., *Appointments to the Regulatory Agencies*, Committee Print, 1976. *See* also Senate Committee on Government Operations, 95th Cong., 1st Sess., *Study on Federal Regulation*, Vol. 1, Committee Print, 1977. It is difficult to generalize about FCC commissioners; they have varied backgrounds. A few of them have come from the agency staff but this is not a pattern. Some have come from congressional staffs, which are a common source of regulatory appointees generally, but this has not been a dominant source of FCC commissioners. Still others come from industry or industry-associated positions, but the number of such appointments to the FCC has not been such as to justify the degree of concern often expressed. A considerable number have come from other government posts, but the variety of such positions and the different circumstances of the transfer defy easy generalization. The random sources of appointees befits the randomness of the selection process.

16. The process by which Washington lobbyists follow Aristotle's law of nature (which abhors vacuums) was succinctly described several years ago by former FPC Chairman Lee White in a Brooking's Institution conference on regulatory reform:

> A successful lawyer in Keokuk is appointed by the President to serve on an independent regulatory agency or as an assistant secretary of an executive department that exercise regulatory functions. A round of parties and neighborly acclaim surround the new appointee's departure from Keokuk. After the goodbyes, he arrives in Washington and assumes his role as a regulator, believing that he is really a pretty important guy. After all, he almost got elected to Congress from Iowa. But after a few weeks in Washington, he realizes that nobody has ever heard of him or cares much what he does—except one group of very personable, reasonable, knowledgeable, delightful human beings who recognize his true worth. These friendly fellows—all lawyers and officials of the special interest that the agency deals with—provide him with information, views, and more important, love and affection. Except they bite hard when one regulator doesn't follow the light of their wisdom. The cumulative effort is to turn his head a bit.

Quoted in Roger Cramton, "The Why, Where and How of Broadened Public Participation in the Administrative Process," *Georgetown Law Journal*, Vol. 60, 1972, pp. 525-530, n. 14.

17. *See*, e.g., Henry Geller, *A Modest Proposal for Reform of the Federal Communications Commission*, P-5209, The Rand Corporation, Washington, D.C., April 1974, pp. 47-52.

18. The rationale of extending tenure is not entirely clear. While there has been some recurrent concern that many commissioners do not serve for full terms, it has yet to be demonstrated that this is a serious problem. If it were, extending the length of the term would be a curious way to solve it. This extension only increases the likelihood that members will leave prior to the expiration of the full term (under present law this would necessitate other short-term tenures, inasmuch as the terms run from fixed calendar dates rather than from the beginning of each appointment). Whether or not a current problem exists with respect to turnover, the present seven years is, in my opinion, too long. A shorter term of four to five years would not only better conform to actual experience at the FCC (the median tenure is between four and five years), it would conform to the traditional expectation of appointees to high-level government posts generally.

19. The President's Advisory Council on Executive Reorganization, *A New Regulatory Framework: Report on Selected Regulatory Agencies*, 1971. Most of the criticisms of the commission form of organization that are discussed here are found in the

Ash Council report, though most can also be found in numerous other works. The subject of reorganization is reviewed in Glen Robinson, "On Reorganizing the Independent Agencies," *Virginia Law Review*, Vol. 57, 1971, p. 947, part of a symposium critique of the Ash Council report. Most points that are made below are taken from that 1971 article. *See also* Roger Noll, *Reforming Regulation*, Brookings Institution, 1971, Washington, D.C., for an excellent critique of economic regulation in general and of the Ash Council report in particular.

20. Phillip Elman, "The Regulatory Process, A Personal View," BNA Antitrust and Trade Reg., Report No. 475, 1970, p. D5.

21. Louis Hector, "Problems of the CAB and the Independent Regulatory Commissions," *Yale Law Journal*, Vol. 69, 1960, and Newton Minow, "Suggestions for Improvement of the Administrative Process," *Administrative Law Review*, Vol. 15, 1963, p. 146.

22. Bartley's proposal is outlined in his Address before the Illinois Broadcaster's Association, May 23, 1968, FCC Release No. 17280. Burch's proposal is briefly discussed in Geller, *supra.*

23. *See*, e.g., George Stigler, "The Theory of Economic Regulation," *Bell Journal of Economics and Management Science*, Vol. 2, 1971; Richard Posner, "Theories of Economic Regulation," *Bell Journal of Economics*, Vol. 5, 1974, p. 335.

24. Even in such cases, there is reason to be skeptical about what regulation can achieve. There is a significant economic literature on the efficacy of regulation. Most of the studies draw negative conclusions even where the regulatory purpose seems justifiable in general principle. *See*, e.g., George Stigler, and Claire Friedland, "What Can Regulators Regulate: The Case of Electricity," *Journal of Law and Economics*, Vol. 5, 1962, p. 1; Paul MacAvoy and Roger Noll, "Relative Prices on Regulated Transactions of the Natural Gas Pipelines," *Bell Journal of Economics*, Vol. 4, 1973, p. 212. Most of the economic studies of this genre have dealt with the efficacy of utility rate regulation—the type of regulation most relevant to evaluation of communications regulation. However, recent studies of safety regulation, which similarly conclude that regulation is ineffective and in some cases counterproductive, are also pertinent. *See*, e.g., Sam Peltzman, "An Evaluation of Consumer Protection Legislation: The 1962 Drug Amendments," *Journal of Political Economics*, Vol. 81, 1975, p. 1049; Albert Nichols and Richard Zeckhauser, "Government Comes to the Workplace on Assessment of OSHA," *The Public Interest*, Vol. 49, 1977, p. 39.

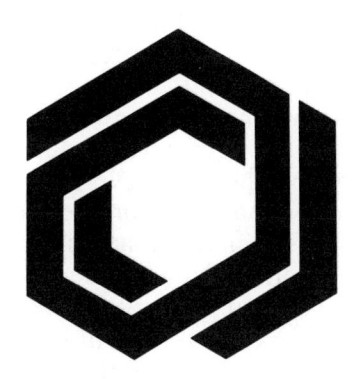

11.
The Executive Branch

Forrest P. Chisman

This chapter considers the extent to which Executive Branch policy and operational functions in the communications field are and should be under the jurisdiction of one or more government agencies, and it evaluates five potential structures for integration: a small policy unit within the White House, something like the former Office of Telecommunications Policy (OTP); a communications deputy secretariat in the Department of Transportation or Department of Commerce (as this was written the OTP has been reorganized into a National Telecommunications and Information Administration—NTIA—in the Commerce Department[1]); an independent executive communications agency like the Environmental Protection Agency; and a separate, cabinet-level Department of Communications. For purposes of the analysis, Executive Branch communications functions can be divided into policy functions and operational functions. These will be considered separately and then together.

Some definitions and a brief preamble are in order. "Communications" (as opposed to "telecommunications") here means policy and operational issues involving both electronic and nonelectronic forms of information distribution, whether through mass media or point-to-point exchanges. "Information" policy functions denote pervasive issues extending throughout all forms of dissemination and beyond them to other activities in our society: privacy protection, for example, and the equitable spread of social services, and competitive or concentrated industry structures. To avoid the specter of "big brother," however, two areas are conspicuously excluded from our candidates for executive branch integration: 1) the President's personal, political, parochial interest in the use of communications to advance his own program or popularity or both (this would apply to aides and cabinet officials as well); and 2) coordination of government collection, processing, and distribution or failure to distribute official information from and to the populace—all the issues that go under the names of "sunshine," "freedom of information," "national security," "executive privilege," and the like. The first I would leave to the standard public relations apparatus of the executive, while the second—which essentially defines the relationship of the governors to the governed—is best left centralized in the Office of Legal Counsel in the Department of Justice, subject to oversight by the House and Senate Committees on Government Operations.

We are talking, then, of policy and perhaps operational leadership in communications and information for society as a whole. An introductory question is, why seek any coherent executive leadership? Apart from Canada, no other advanced Western nation has a cabinet-level Department of Communications, and even in Canada, experienced hands are beginning to mutter about how that ten-year-old agency is "a ministry in search of a mission." The answer takes the following three forms:

(1) The United States is the world's first "information society," whose occupations and earnings now derive more from information processing than from any other pursuit. This has given rise to a host of new socioeconomic risks and opportunities—accelerated by the fast pace of introduction of new information distribution technologies and services—demanding comprehensive and advanced planning of the type described in Alvin Toffler's *Future Shock*.

(2) Neither the FCC nor the courts, both tribunals that receive and act on pleadings and petitions, can be expected to break free of their reactive modes to engage in systematic advanced planning. Moreover, the committee structure of Congress (even after recent reforms), when combined with the multiple demands on members' time and attention, makes it very difficult for that body to

sustain a comprehensive view of future challenges. There is therefore a need for some form of integrative and forward-looking leadership; by default if not by immutable constitutional plan, that requirement falls on the executive.

(3) As we shall see, the executive branch operates a massive communications system that is crucial to the accomplishment of agency missions such as national defense and social service delivery. In addition, it engages in extensive research, development, and systems design, and it represents the United States in international telecommunications negotiations. All of these are widely conceded to be executive functions, and it is, therefore, vital for the executive to develop institutions and processes to see that they are carried out effectively.

In short, communications is, and should be, an executive function, and this paper considers how far it should be carried.

Policy Functions

There are five major executive policy functions: 1) addressing long-range social and political issues (such as the revision of the 1934 Communications Act, privacy, and the future of the Postal Service); 2) addressing current "hot" issues (such as FCC dockets on pay cable and congressional concerns over sports blackouts); 3) spectrum policy development (as opposed to spectrum allocation, which is a technical function); 4) international policy development; and 5) national security policy (which includes emergency preparedness).

There is currently very little integration among executive agencies within any of these policy functions. For example, on the long-range issue of electronic funds transfer, the FCC, the Federal Reserve System, and NTIA are all involved; as in an *ad hoc* EFTS (Electronic Funds Transfer System) Commission, but the interaction among these bodies is at best sporadic. International policy development is carried out by a four-headed monster: NTIA, State (including the new International Communications Agency), FCC and the telecommunications industry. This means that in each policy area, no one agency or consortium of agencies can develop and implement an overall view of the nation's interests. The result has been that initiatives are weak and fragmented; often as in the international area, they create bad relations within and among governments. It seems necessary for the Executive Branch to take on the responsibility of pulling together government's activities in these policy areas.

There is also very little integration among the executive functions. Long-range policy development has not received adequate attention. When there is work on long-range policy, it is seldom coordinated with short-range responses within the same agencies. For example, over the last few years, the (now-defunct) White House Office of Telecommunications Policy issued a series of pronouncements on issues of competition within the telephone industry that, in sum, favor "full and fair" competition. But since June of 1976, the agency several times changed its stand on the merits of the "Bell bill," which frontally challenged the carefully developed prior OTP position. The Defense Department vocally disagreed with that position.

Spectrum policy is an underdeveloped area, which seldom if ever is separated from spectrum management and also seldom is integrated with other policy issues. For example, one might wish to consider fundamental social and economic tradeoffs between use of the spectrum for broadcasting and for land mobile (including CB) radio—but at present there is no mechanism for doing this.

International issues would seen to merit some organic relationship to long- and short-term domestic policy development, but in practice they are conducted in isolation by a small group of technical experts. Finally, national security concerns in communications are primarily the responsibility of the Defense Department, which has shown little desire to cooperate with other executive agencies in its spectrum or hardware demands. Security restrictions prevent most "outsiders" from gaining any real, working appreciation of the Defense Department operation, but its most visible part, the National Communications System, is a small operation that does little to integrate executive functions beyond preparation for natural disasters.

The organizational implications of these observations about policy functions seem clear. All the functions interact with each other and it is necessary for some agency to take charge of creating an overall view to insure the orderly development of federal policy toward communications. The complexity of these functions indicates that no small White House unit by itself can do the job. Even short-range policy could not be handled adequately by such a unit because that policy not only must be enunciated, it also must be sustained through FCC pleadings and other means. Looking strictly at policy functions, it is not clear, however, that a large executive entity is necessary. In other areas, organizations such as the Council of Economic Advisers and the

The Executive Branch

Council on Environmental Quality, with staffs of roughly the same size as the former Office of Telecommunications Policy, have been able to conduct effective policy work. In large part this has been because they have chosen their targets selectively, and because their White House location has given them the visibility and the stature necessary to enforce policy decisions throughout the government.

One could argue that, although an executive telecommunications entity on this scale might not be able to address all significant issues, comprehensiveness is not absolutely necessary. But one also could say that there are enough issues of sufficiently great policy importance to demand that the Executive Branch not neglect them, and that therefore a small organization would be overwhelmed. To choose between these two views of size and adequacy, one must move beyond the strict policy functions of an executive communications entity and look at the operational needs of the executive branch.

Operational Functions

There are seven major operational executive telecommunications functions as follows:

(1) Operation of telecommunications systems, such as the Defense Department and General Services Administration telephone and record systems.
(2) Design of systems and procurement, such as executive policy on competitive bidding.
(3) Technology tracking, that is keeping abreast of the most recent developments to make sure that government operations, along with civilian spin-off applications, are up to date.
(4) Spectrum management conducted by a small branch of experts in the Interagency Radio Advisory Committee (IRAC).
(5) Scientific hardware and social-applications research and development carried out by a large number of agencies such as the Defense Department, National Aeronautics and Space Administration (NASA), Commerce, Health, Education and Welfare (HEW), Housing and Urban Development (HUD), Agriculture, Justice, and its Law Enforcement Assistance Administration (LEAA), etc.
(6) Social service activities such as NASA's efforts to commercialize satellites and the Rural Electrification Administration's support of rural telephone systems.

(7) Negotiations on international issues which, as contrasted with international policy development, are conducted largely by the Department of State.

Looking at operational functions strictly by themselves, the need for organizational integration within each function is probably not as great as for public functions. The agencies that operate separate communications systems seem to have achieved the necessary economies of scale and specialization, although it is probably desirable to upgrade the National Communications Systems. Centralized procurement and operational planning would be of dubious value given specialized requirements, although there is probably some need for supervision in light of past discovery of inefficiencies at Defense and the General Services Administration. There should certainly be a centralized government effort at technology tracking, but this can be accomplished without further reorganization. Present spectrum assignment procedures seem to work well through cooperative arrangements among the various executive departments.[2]

There is probably no need for a single, centralized government agency to manage all scientific research and development on telecommunications hardware and social applications—although it would be highly valuable for experiments and demonstrations to be coordinated by a central mechanism. Similarly, with regard to social services activities, fragmentation is a genuine problem. There is considerable agreement that someone outside the mission-oriented executive agencies must coordinate the necessary inter-agency efforts to deploy social service applications of advanced communications technology, such as applications in rural areas. Greater integration in international negotiations is probably not necessary, although the State Department must upgrade its capacity to understand the telecommunications intricacies it is negotiating.

To summarize, a mechanism is needed to coordinate communications R&D and social service applications that are being or should be pursued by a variety of federal agencies. No further integration is needed within other operational functions.

Integration between operational functions, however, presents a more ambiguous picture. Those who operate communications systems should have a great deal to say about their design and procurement but, as discussed earlier, their operations and procurement probably should be supervised for possible duplication and overlap. With regard to technology tracking and research and development, however, the need is primarily for good working relations between agencies rather than for integration. This is probably true as well for spectrum and international negotiations. Social service activities, if properly coordinated, will entail such working relations.

In sum, requirements for integration of operations do not mandate a large executive agency, because procurement functions might be supervised by the Office of Management and Budget (OMB), and the social services could be coordinated by an agency the size of the present NTIA (or former OTP).

Integration of Policy and Operational Functions

The case for a larger executive communications agency becomes more persuasive when one looks at integration between policy and operational functions. This is most clearly seen in the relationship between policy and, respectively, procurement, spectrum, R&D, social service activities, and technology tracking. In these areas, there is little integration between policy and operational activities. OTP issued integration between policy and operational activities. OTP also issued several circulars, encouraging competition in procurement and careful reviews of system design, but overall, these efforts have been extremely limited and ineffective. Much more could be done to use operational activities of all sorts as policy tools.

Consider mobile communications, for example. The future of this technology in the United States will depend heavily upon governmental policies in spectrum and upon judgments about technical and economic feasibility. An executive communications unit might decide that, as a matter of policy, mobile communications of a certain sort should be promoted; working through IRAC, it might make the necessary spectrum available to fleets of government vehicles. In addition, however, it might bring down the cost and advance the technology in this area quickly by encouraging government agencies to procure mobile technology of a certain sort.

There are other examples of perhaps a clearer sort. Consider the issues raised when GSA proposed "FEDNET" which would have integrated all federal-agency data retrieval; or when the FBI pushed "CRIMENET" which would filter law-enforcement information from all the states through a central gathering point; or when the Federal Reserve proposed to own and operate an EFT system for all member banks. In each of these cases, major policy interests come into play around the issues of privacy, the maintenance of a healthily decentralized federal system, and delimiting government incursions into the domain of private enterprise. In short, operational activities in these areas can be seen as policy activities, or at least as their natural outgrowth. This means that to take an active role in developing policy, an executive agency must have an operational arm capable of giving that policy a push.

This point can be generalized briefly to underscore the case for flexible and effective executive leadership. Shorn of operational capability, a communications policy office would have to rely entirely on the responsiveness of the FCC and of congressional committees to put its policies into force. In many cases, this is just as it should be because the process provides a check against overzealousness and an opportunity for affected interests to be heard. In other cases, however, the asymmetry between the immediate business of the FCC and of Congress and the long-range perspective of an executive communications policy office makes for muffled or deferred responses. Furthermore, the role of commenting upon FCC docket proceedings confines the executive to considering remedies within that agency's statutory purview: opening up or narrowing barriers to entry into the intercity, private-line communications market, for example, with uncertain effects upon residential and rural rates for ordinary telephone service. An integrated and coordinating executive office could turn a whole armory of potential remedies and select those most calculated to fill a specific social need: tax incentives, for example, or targeted revenue sharing. It would be no barrier to such an office, as it would be to a congressional committee or the FCC, that such policy tools are outside its immediate jurisdictions.

An opponent of this point of view might argue that operational activities do not combine to form a "critical mass" for policy in the sense that, contrary to the notions of some observers, policy-makers are not more clear about the problems involved in the communications field and no more psychologically prepared to face them realistically in a strictly "policy shop" that in an agency that has operational responsibilities. In addition, critics might argue that it is overly paternalistic for government to guide the direction of developments in the communications industry, and that all operational functions are essentially "mission support" functions of federal agencies, which as a result should not be subject to overall coordination.

There is no critical mass for policy functions. Policy shops can and often do work well. With regard to the argument about paternalism, the best answer is that the federal government in its role as purchaser, researcher, designer, and vendor already has a massive effect on the communications industries and it is

The Executive Branch

irresponsible to argue against rationalizing that effect. Of course, there must be a vigorous role for the free market, but that is only to say that policies should not be executed operationally at the federal level without careful consideration of need—a question industry itself can be counted on to address vigorously. This is obviously a matter of careful analysis in each case, but it seems contradictory—if not foolhardy—for the government, for example, to decide that the market price for communications equipment should be competitive but for it not to press government agencies to take competitive procurement seriously.

With regard to the basic preference for mission orientation, one can say that communications functions of the federal government should be strictly in accordance with the perceived needs of the respective agencies and that agencies in the executive branch should have no further concerns. Citizens deserve performance on communications issues from their government just as they deserve performance on defense or environmental issues. No one need say (or ever could say) that the Defense Department could not have a particular missile system because the spectrum required by the system was needed elsewhere. Rather the choice would be between a more or less expensive system for more or less use of spectrum. To the extent that there are fundamental disagreements between any executive communications entity and a mission agency, these would be and should be raised to the level of the President. That is, after all, one of the reasons he is there. The nation has realized a need to depart from a strict mission orientation in many infrastructures, notably including energy and the environment. With half our work force now working in the information sector, there is an equally strong argument for recognizing the need to depart from a strict mission orientation in the use of the communications infrastructure. There obviously will be a process of balancing, but that is inherent in all government operations, and the final arbiter should be the President.

The strongest case for integrating policy and nonpolicy activities is in the areas of spectrum and international negotiations. Spectrum assignment currently proceeds on an *ad hoc* basis with very little relation to general communications policies. Clearly, such assignments always should keep in view long-term spectrum planning as well as social service, system design, procurement and other operational policies. For example, Defense utilizes approximately one-third of the electromagnetic spectrum. Someone should hold an authoritative inquiry into whether the Defense Department really needs it and is using it well. In international affairs, it is unsatisfactory for the State

Department to conduct negotiations while policy is formulated by a consortium, as at present. International negotiations cannot proceed from hard and fast policy positions. Rather, positions must be developed through negotiation, and someone must have the authority and expertise to engage in that process. The State Department is inadequately staffed and inadequately prepared for this mission; given its career structure and political orientation, it may well never be an appropriate body. It therefore seems to make sense for an executive communications agency to lead in both the formulation and negotiation while accepting the need to coordinate these functions with other agencies.

Overall Considerations

The conclusion of this discussion is, therefore, that operational functions should be considered as extensions of policy functions and integrated with them to the greatest extent possible. Because the policy net is so wide, there should be greater integration among operational functions, too. If this is the case, the organizational implications are clear: A small organization in the White House is not sufficient to conduct executive functions in the telecommunications field.

The strongest argument against taking the functions outside the Office of the President is that the Department of Defense would object to having spectrum policy set by any more subordinate office. This, however, probably could be overcome if there was a final appeal to the President, perhaps with his science advisor or a special communications policy advisor acting as an assistant.

The arguments in favor of taking fully integrated executive functions out of the White House Office are more numerous and stronger. To begin with, keeping those functions that would fit inside the executive office would lead to fragmentation or to too large a communications establishment within that office, or both. Or, alternatively, it would mean that the executive communications agency could not grow to keep pace with the information sector. Second, on the basis of general administrative policy, it would seem desirable for the White House office to engage only in those functions which necessarily are attached to the President as an individual or to the presidency as an institution. Communications, while an executive function, is not one of these any more than is transportation, housing, or energy. Third, the experience of the Office of Telecommunications

Policy was that the White House location is of little advantage. A legislative mandate in another location on balance might strengthen an executive telecommunications entity. Fourth, there are many restrictions on what a communications entity within the White House can do. Staff members constantly have to consider that they are "speaking for the President" and as a result, are restricted in many of their pronouncements. Fifth, the job of adequate formulation and implementation of communications policy is probably too large for even a group of White House office agencies. Adequate supervision of executive branch agencies' procurement policies by itself would require a sizeable bureaucracy, as has been demonstrated in the computer field. If we agree that this should be integrated with other telecommunications functions, the capacity of the White House quickly is exhausted. Finally, and most importantly, many observers report that the White House is a bad place to formulate long-term policy. Short-term political pressures quickly divert energies to firefighting.

In the author's opinion, the executive communications policy and operational functions ultimately can be performed either by a deputy secretariat in Commerce or Transportation or by a freestanding executive unit similar to the Environmental Protection Agency. In no case should an executive communications entity be of lower stature than an agency like the EPA.

Whether we chose a deputy secretariat or an independent agency, the staff required is on the order of 200 to 300 individuals. The choice between these two alternatives is rather difficult to make. Putting communications in an existing agency would draw upon a present body of expertise and prestige. But communications easily might be lost within a large agency that has other functions and would have to deal from a position of relative bureaucratic weakness. An independent executive agency would deal from the strength of a new legislative mandate and would not have the past perspectives or predilections of an existing department to hinder it. Nevertheless, its creation would entail a seeming growth of bureaucracy, contrary to the principles of the present administration, which would have to be countered by a statement or reorganization stressing improved coherence and the absence of staff increases.

The choice ultimately depends upon one's theory of the degree to which executive agencies should be consolidated. If we believe in consolidation and are prepared to accept the consequences in terms of realistic responsibility of lower level bureaucrats, the location in an existing department is probably desirable. On balance, however, telecommunications is a field that has great importance, but not great visibility within the federal government, and an independent executive agency should be created, if for no other reason than to upgrade the perceived importance of this vital function in our society. The mission of such an agency would be precisely to make and enforce telecommunications policy for the government and industry to the extent that this is possible given executive powers.

One day there may be the support and substance needed for creation of a full-scale Department of Communications. That day is not now in sight, and the route to it—if we ever reach for it—would lie through a smaller independent agency. Given only the functions discussed here, a Department of Communications probably never will make sense, however. As noted, a staff on the order of 200 persons probably could perform those functions and this is clearly too small for a cabinet department. A Department of Communications might make sense, however, if one or more of three decisions were made: 1) that the Postal Service should be returned to the government and that the growing importance of electronic mail as well as the general interrelatedness of all communications functions requires that policy for the Service be integrated with other communications policy; 2) that management of governmental telecommunications functions should be taken away from the Department of Defense and GSA and integrated into a single unit; and 3) that the "quasi-legislative" functions of the FCC (such as setting rules for cable operations or establishing basic principles for calculating rates of return) should be removed from that agency and placed in the executive branch. Many thoughtful people support each of these steps and argue that in each case the function indicated should be combined with the executive functions mentioned earlier to form a cabinet department. Indeed, some experts argue that unless the FCC is stripped of its quasi-legislative functions, no executive communications agency can succeed. In the author's view, however, a conclusive case has not been made for these arguments.

Two final issues remain to be considered: 1) how an independent agency can gain leverage and a proper hearing of the policy councils of government once shorn of its White House moorings; and 2) how much political capital Congress should be asked to invest in its creation. The two issues, as can be seen, are closely related.

Despite the lack of personal interest by the President, OTP on occasion commanded special attention because it was presumed by its very location to be expressing an Administration view. That leverage is lost in the shift to an independent location, and

The Executive Branch

the question arises what deference thereafter would be paid to it by State, Defense, the Federal Reserve Board, the Antitrust Division of the Department of Justice, or—for that matter—the Congress or the FCC. Part of the answer would have to come from its mandate, which should in clear and unmistakable terms make the new agency the Administration policy formulator in the communications field. A great deal could be gained by making the administrator of that office also a special advisor to the President. Finally, as a practical matter, linkages would have to be worked out with OMB guaranteeing that the interagency supervisory and coordinating functions of the new agency are backed up by budgetary enforcement. Clearly, much detailed thought and drafting must precede these arrangements.

Notes

1. The Executive Order delegating responsibilities to NTIA authorized the Secretary of Commerce, acting through an Assistant Secretary for Communication and Information, to "serve as the President's principal adviser on telecommunications policies pertaining to the Nation's economic and technological advancement and to the regulation of the telecommunications industry; develop and set forth these policies and perform whatever studies might be necessary; coordinate Federal telecommunications assistance to State and local governments; assign Federal Government frequencies, establish Federal spectrum use policy, and with the Federal Communications Commission, compose a long-range plan for management of all electromagnetic spectrum resources; with the State Department and other agencies, develop policies and programs relating to international telecommunications issues and negotiations; serve as the chief point of liaison between the President and Comsat; coordinate Executive Branch telecommunications activities and help formulate policies and standards to meet issues—such as privacy—arising from these activities; advise the Office of Management and Budget on the procurement and management of federal telecommunications systems and conduct whatever studies this function might necessitate; and participate with the National Security Council and the Office of Science and Technology with respect to the use of telecommunications for emergency and national security purposes." (Executive Order No. 12046)

2. Some economists argue that the present IRAC structure should be replaced by a system of economic allocation. In such a system, agencies would have to pay for spectrum and, as a result, would be forced to trade off their demands against the possible economies to be gained from, for example, the use of spectrum-conserving technology. In this case, spectrum assignment might be transferred to the Office of Management and Budget (OMB) or a specialized executive communications entity. This indeed would require more integration, but more work is needed to prove its value.

The National Telecommunications and Information Administration is divided into four major program areas, each directed by an Associate Administrator: policy analysis and development, telecommunications applications, Federal systems and spectrum management, and telecommunications sciences.

12.
The Judicial Role

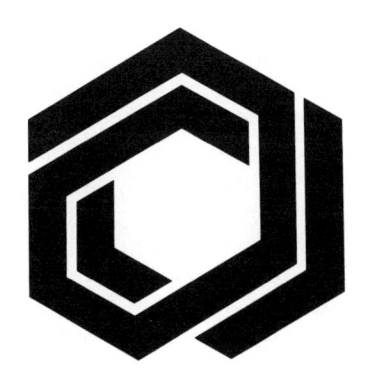

Glen O. Robinson

The role of the judiciary in reviewing administration action has always been controversial among lawyers, who quarrel over legal arcana, and political scientists, who debate the relationship of the judiciary to administrative government. This discussion of the judicial review of FCC decisions is a mixture of both legal and philosophical perspectives on the judicial role. It attempts to measure the results of judicial review to evaluate the judicial role in light of these results. It concludes with some brief comments on changes in judicial review that have been proposed or informally suggested.

While this discussion is directly concerned with the role of the courts in reviewing and influencing *communications* policy, this concern cannot be isolated from the broader question of the role of the courts generally. Some people have expressed anxiety about the "new judicial activism" in communications policy, reflecting a view that the courts have begun to intervene more aggressively in matters previously left largely to executive or legislative discretion. For example, judicial review of FCC actions in the area of broadcast law seems to have become more rigorous than it was two decades ago or so. One striking instance of this is a series of decisions by the U.S. Court of Appeals for the District of Columbia Circuit, insisting that the FCC hold hearings into cases involving changes of "unique" broadcast formulas.[1] The tone of several other opinions in cases involving the FCC also suggests more intensive judicial scrutiny.[2] But this apparent increased judicial activism in reviewing administrative actions can be seen in all areas of law.[3] We obviously cannot here evaluate the extent to which such a trend toward activism is real or merely apparent.[4] It is enough here to note simply that insofar as there is a general trend of increased judicial activism, it becomes difficult to identify its special effects in the communications field.

To begin, we must identify two distinct questions. The first is a question of measurement: What are the courts doing to influence communications policy? The second is a question of normative judgment: What is the proper role of the courts? Of course, both questions require some standard and, given the absence of any clear objective benchmarks for assessing the judicial impact or the judicial function, this standard must be somewhat subjective.

The Legal Framework: A Primer on Judicial Review

Before assessing the judicial role, it is necessary to outline the legal framework of judicial intervention.

With a few minor exceptions, FCC actions are reviewable only in the U.S. courts of appeals. For various reasons, appeals involving all aspects of communications policy have been heavily concentrated in the Court of Appeals for the District of Columbia. One reason is that all cases involving broadcast licenses must by law go to that court and there are more of these cases than any other kind. Another reason is that most communications lawyers practice in Washington, which makes the court convenient. Still another possible reason is the assumption common among communications lawyers that the District of Columbia Circuit is stricter in its review of administrative agencies than are other courts of appeals. As will be shown below, the assumption is probably incorrect in communications cases. However, it is the perception, not the reality, that dictates the choice of venue where choice is legally possible. (Conversely, the greater familiarity of the District of Columbia Circuit with a broad range of cases and the absence of any special geographic bias could induce review in another circuit where a lawyer seeks a more parochial judicial perspective.)

The concentration of cases in a single court has several consequences for an evaluation of judicial review. First, that court has greater familiarity with communications policy issues than any other single federal court. (The same is true for most administrative regulation.) Thus, while some observers have suggested that an "expert" tribunal be established for communications (and other administrative agency) decisions, such a court already exists, after a fashion at least. The idea of an "expert" court is not universally appealing, however, and this suggests a second implication of the concentration of cases in the District of Columbia Circuit: that concentration not only enhances the court's knowledge of communication policy issues but also enhances its influence and power, which have caused concern in some circles because of the court's reputation for judicial activism. We will discuss later to what extent the court is more "activist" than other courts in communications cases. For now it is sufficient to note that an evaluation of judicial review must consider especially the performance of the District of Columbia Circuit.

The law allows specified classes of applicants and holders of licenses and permits to seek court review of FCC "decisions" and "orders." In addition, any other person who is "aggrieved" or whose interests are "adversely affected" by an order of the Commission can seek court review. In recent years, courts have interpreted this provision broadly, allowing cases to be brought not only by persons having personal, legally protected interests but also by those seeking to vindicate a more general public interest affected by the action. Indeed, almost any person who is sufficiently concerned to seek review is typically allowed to do so, as long as there is an "order" sufficiently final to permit review—a requirement that has also been given a liberal interpretation. This liberal access to the courts must intensify concerns about judicial activism, for it increases the exposure of agency policy to judicial scrutiny. The more liberal the access to the courts, the greater the range of policy questions that can be brought before them, hence the greater their potential impact on overall performance.

However, the central issue of the judicial role here is less a question of access to the courts than it is what the judges do with the agency's decisions when cases reach them, what administrative lawyers call the "scope of review" question. Trying to define the proper scope of review of agency action has always been an exercise in confusion, darkened by terms and formulas with no definite or accepted meaning. The main statutory guidelines are those of the Administrative Procedure Act (APA). In pertinent part the Act reads as follows:

> To the extent necessary to decision and when presented, the reviewing court shall decide all relevant questions of law, interpret constitutional and statutory provisions, and determine the meaning or applicability of the terms of an agency action. The reviewing court shall:
> (1) compel agency action unlawfully withheld or unreasonably delayed; and
> (2) hold unlawful and set aside agency action, findings, and conclusions found to be—
>   (a) arbitrary, capricious, an abuse of discretion, or otherwise not in accordance with law;
>   (b) contrary to constitutional right, power, privilege, or immunity;
>   (c) in excess of statutory jurisdiction, authority, or limitations, or short of statutory right;
>   (d) without observance of procedure required by law;
>   (e) unsupported by substantial evidence in a case subject to sections 556 or 557 of this title or otherwise reviewed on the record of an agency hearing provided by statute; or
>   (f) unwarranted by the facts to the extent that the facts are subject to trial de novo by the reviewing court.
> In making the foregoing determination, the court shall review the whole record or those parts of it cited by the party, and due account shall be taken of the rule of prejudicial error.

The words of the Act do not help us much in determining the proper scope of review. The phrase "arbitrary and capricious," for example, does not mean to a lawyer what it means according to Webster (caprice: a sudden whim or fancy), and agency actions that would not merit such an epithet from literate laymen might well receive it from a reviewing court.

If Webster is not a useful guide to the standards of review, neither are most judicial opinions. The use of APA formulas tends to be ritualistic; some opinions do not use the Act's formulas at all, or use them along with other formulas that require clarification themselves. To add to the confusion, in some cases the Act's formulas are used incorrectly—contrary to statute as well as conventional use.[5] Obviously, the absence of clear, fixed standards for review presents great difficulty in evaluating judicial performance.

However, we can pass over this difficulty for the moment since it is appropriate to examine the factual question what that performance has been, before passing on to the normative question, what it ought to be. One cannot, of course, derive values from facts. But knowing a few facts does at least help to clear the mind about the nature and significance of normative judgments. Such a factual inquiry illuminates the normative question of what courts should do in several ways. First, it illustrates the kinds of policy issues that are exposed to review and the frequency with which they are exposed. Second, within flexible limits of interpretation, it shows the results of judicial action, and suggests which factors do or do not appear to influence the decisions. Third, it focuses the normative issue and the range of relevant debate over it. In discussion of the subject one often encounters opinions that the courts should be more or less "active"—relative terms that refer to some perceived state of affairs. Unfortunately, many people base such opinions on selective and impressionistic perceptions, relying on a case here, some judicial obiter dicta there, and so on. What is needed is a more complete factual background on judicial performance.

Measuring Judicial Performance

There are various ways of measuring judicial review. The simplest is, of course, to sample judicial opinions to see what they say on the subject. As was already observed, however, the language of opinions is a highly uncertain guide. For example, in recent years some courts have gone so far as to suggest that reviewing courts are in "partnership" with the agencies "in furtherance of public interest."[6] At first glance, this partnership concept seems at odds with the premise on which the legitimacy of judicial review in a democracy has been thought to rest. The grant of power to a politically unaccountable judiciary has been reconciled with democratic norms by supposing that the judicial power serves legal principles distinct from other public-interest considerations that influence the agency or public officer whose actions are subject to review. This is not necessarily a narrow role. Besides insuring that agencies conform to basic constitutional and statutory standards—substantive and procedural—the courts may legitimately require agencies to articulate rational explanations for their decisions. However, it is important to the legitimacy of these functions that the court should not perform as merely another gear in the machinery of government policy, but should maintain some separateness, and distance, from the particular policy concerns that prompt agency action. To describe the court as a "partner" of the agency connotes an alliance in purpose and outlook that undermines this vital separation. Such an alliance implies both too much and too little judicial intervention, depending on whether in its role as a "partner" the court in a particular case disagrees or agrees with agency policy. But in general it implies a judicial role in policymaking that seems out of keeping with the traditional concept of judicial review.

Nevertheless, one must be cautious about drawing conclusions from phrases alone. Phrases like "partnership" are roughly akin to such expressions of judicial piety as statements that courts are the "guardians of liberty," "defenders of justice," and so on. Plainly one cannot rest simply on verbal filigree; it is necessary to consider the objective results as well. Unfortunately, measuring such results is extremely difficult in the absence of a clear, objective function to measure. One promising approach would be to compare cases reviewing FCC decisions with those involving other agencies. However, a detailed comparison was too ambitious to present here. Thus, except for a couple of observations about other cases, this chapter is confined to FCC decisions.

As a first effort to deal with the problem, an examination was made of the rates at which appeals courts or the Supreme Court either affirmed or reversed FCC decisions during two seven-year time periods: 1950-1956 and 1970-1976. The decisions are tabulated in the Appendix to this chapter. Before examining some of the results, the approach and its limitations require explanation. The periods selected for study are somewhat arbitrary but they at least provide a point of departure for testing judicial attitudes. The first period covers part of an era before the modern judicial "activism" that has been so frequently cited began to emerge. It also represents a period of considerable FCC activity (the early television licensing period), and, correspondingly, of considerable judicial activity (not necessarily activism) as well. The second period is a sample of the modern era.

In addition to measuring the overall affirmance/reversal rates in these two periods, the cases were broken down in two ways. First, court of appeals cases were classified by whether they were heard in the District of Columbia Circuit or in some other circuit. Supreme Court decisions were separately identified. The purpose was to test the perception of communications lawyers that the District of Columbia Circuit is more "activist" than other circuits. The numbers involved, as well as the absence of any clear hypothesis with respect to other circuits, made it pointless to break this latter category down by individual circuits. Second, cases were classified by substantive subject matter, using three major functional classes of communications law. This was intended to test the hypothesis that the results of judicial review might be distinctively different among such categories. Additionally, distinctive types of cases within each class were identified. Unfortunately, even with such a further breakdown, this attempt at classification does not reveal much about the nature of the issues or their importance. However, because limitations on sample size made it impractical to attempt more detailed breakdown, more refined analysis must be left to a commentary on the individual cases.

Describing results presents problems. One problem is how to classify cases that involve partial affirmance and partial reversal. Another is how to characterize disposition on procedural grounds. The first problem was resolved by judging the relative importance of the issues affirmed and reversed in each case. If, for example, a case was remanded for clarification or modification of a relatively minor issue but affirmed on all other points, it was treated as an affirmation, and vice versa. The second problem was more vexing. Most cases can be characterized as procedural in some sense. For example, a reversal for failure to articulate the grounds for decision, or to hold a hearing on a particular issue—both common cases for reversal—is in a sense procedural. Indeed, the typical case is simply remanded to the agency for further clarification or for a hearing. However, to classify such cases as procedural would make it impossible to test for distinctive results in different substantive areas. More important, to classify cases in this way would imply that the procedural mechanism for reviewing agency action is independent of substantive issues. Such an assumption is naive. Procedural classifications have therefore been ignored except in cases where the procedural issue involved no issue of substantive communications policy (e.g., a Freedom of Information Act claim) in which case it was placed in a miscellaneous category. Obviously this may bias the results by suggesting the courts are more likely to intervene on the merits of a case than they actually do. However, it seemed best to correct for this problem in subsequent commentary rather than in the tabulation of the results.

A simple tabulation of affirmances and reversals in the various categories lacks the scientific precision one might bring to bear on the problem with rigorous statistical analysis. One might, for example, employ a statistical technique such as discriminant analysis to identify and assess the relevance of different independent variables such as venue, type of case (broadcast, cable, etc.), type of issue (programming, tariff rates, etc.), composition of review panel, and the like. Such a test might at least allow us to identify the most important variables with greater confidence. However, such statistical rigor would put too fine a point on the analysis given the inherent ambiguity in the dependent variable (affirmance/reversal). And it is at least doubtful whether, on the data available, more refined statistical techniques would be more meaningful than simple tabulation.

More fundamental than the absence of statistical rigor in the tabulation, however, is the absence of any measure of the relative importance of individual cases. Some have suggested that courts may be more likely to affirm decisions in minor cases than in cases involving major policy issues. The breakdown of cases by broad subject matter might be taken as a partial surrogate for such a classification if one assumes, for example, that common carrier cases are, as a class, likely to be more important than cable certification cases. This would be at best an unreliable guide, however, for even if one could reliably rank the categories of cases, one would still know nothing about the relationship between the importance of policy issues and the likelihood that a court will affirm or reverse. There is no clear objective criterion for classifying the cases according to the importance of the issues, and I did not attempt to employ a subjective evaluation of all the cases for this purpose. However, review of the individual cases indicates there is not a disproportionate number of important cases in either the affirmance or reversal category.[7]

Subject to all the qualifications mentioned, the results of this examination reveal several things that are both interesting and seemingly at odds with conventional wisdom. First, the reversal rate was lower in the 1970-1976 period of court-agency "partnership" than in the earlier period, supposedly before the emergence of modern judicial activism. The reversal rate in the current period, standing by itself, is difficult to interpret. One could not say that it is low or high. But the fact that the reversal rate is lower than that of an earlier period brings into question the widespread supposition that courts have become more active in the past decade than before. One possible explanation—which is masked by the method of classification—is that the sample from the earlier period contains an unusually large number of reversals on questions of procedure in a formative stage of legal development. (The sample is dominated by broadcast licensing cases, most of which involved hearing rights of applicants, protestants, or both; two of the courts of appeals reversals on these grounds were in turn reversed by the Supreme Court.) However, since the current period also contains a number of cases in which the same explanation could be made for it, it is difficult to know how much weight the explanation can carry without a case-by-case, issue-by-issue comparison. Unless some other explanation presents itself,[8] one cannot discount the possibility that the greater apparent activism of courts today is more a matter of style than substance. This discussed in greater detail below.

Perhaps even more striking than the time period comparison is the comparison between the District of Columbia Circuit and all other circuits. Contrary to the conventional supposition that the former is more "activist" than the latter, the data show little difference overall. A reading of the individual opinions themselves suggests no explanation for this, except the conclusion that the conventional wisdom is incorrect.

The data on affirmance and reversals for the different classifications are interesting in several respects. Decisions on common carrier cases were affirmed more frequently than other kinds of cases between 1970 and 1976, suggesting that courts were more likely to defer to the FCC in this class of cases than in others (in contrast to the earlier period where, however, the limited number of cases makes interpretation unreliable). This result is to be expected because in these complex, frequently technical cases the claim of administrative expertise has the greatest credibility. But it is difficult to know how much weight to attach to this judgment. Although the substantive issues involved in these and similar cases may lie outside the range of common experience, the particular issue presented to the court may be quite elementary and familiar—whether a hearing should be held, whether the Commission has statutory power to take a particular action, and so on. The crude classification used here does not indicate how many of these decisions actually turned on review of the complex issues and how many were resolved on related but simpler grounds. A reading of the individual opinions is inconclusive on this score; however, it does not suggest any special judicial deference in these cases as a general rule. In some cases, one gains the impression that the greater the claim to deference, the greater was the court's determination not to extend it. Outside the communications field there are corroborating indications of just such an attitude—the notion being apparently that the court can and, to be faithful to its review responsibility, must rise to the challenge of whatever task is presented to it.[9]

The highest reversal rate in the 1970-1976 period is in individual cable television cases. Again, it is difficult to reach reliable general conclusions about the scope of review in these cases. All the cases that were reversed involved highly individualized controversies—chiefly disputes over signal carriage or exclusivity protection in an individual situation. One case touched upon the somewhat more general question of FCC authority to pass on the character of individual franchises, but even this case was an unusual one, and was so regarded by the Commission.[10] While the cable systems "won" the majority of the cases, not much significance can be attached to this in view of the small number of them. From the opinions themselves it is evident that no clear bias is evident on behalf of cable; nor is there any evident hostility toward the rules or basic policies.

The next highest number of reversals occurs in the area of program regulation. However, the significance of this is not quite what one might suppose. Contrary to any notion that the courts necessarily intervene to protect free speech from FCC regulation, several decisions reversed the FCC for refusal to regulate, and several others affirmed significant FCC regulation of programming.[11]

Evaluating the Judicial Role

From a glance at the tabulated results it is apparent that judicial activity has substantially increased. The number of reported opinions in the 1970-1976 period is more than double that of the 1950-1956 period. This reflects a number of changes: the expansion of the Commission's jurisdiction in the cable area; an increase in the Commission's regulatory activity in traditional areas, most notably in the common carrier area; and, perhaps most important, increased litigation by public interest groups, who were virtually unrepresented in the 1950-1956 sample.

This expansion of judicial activity is not, of course, unique to communications. It is part of a well-publicized trend everywhere. Obviously this trend makes it more important to define the appropriate role of the courts in reviewing administration action. However, it also suggests that the phenomenon of judicial "activism" cannot be viewed simply as a question of the scope of review. In fact what is popularly seen as judicial activism may indicate more extensive rather than more intensive oversight. To that extent, it is equally important to examine what courts are asked to do as it is to look at how they are doing it. On this issue my personal judgement is ambivalent. On the one hand, it is discomforting that we seem increasingly to be developing, as the old cliche puts it, a government of lawyers rather than of law. On the other hand, such a concern seems scarcely relevant to the question of the role of the courts in reviewing agency action. Even a vigorous judicial review adds relatively little legalistic formality compared to what is mandated by legislative mandate—reflecting a tradition of legalistic formality deeply embedded in American attitudes of governance.[12]

Such legalism as is contributed by judicial review is both necessary and important. First, judicial review adds—it *should* add—a general dimension to policy which, particularly in an age of increased bureaucratic specialization, is essential. Second, review can add principled rationality to the political pragmatism and expediency of administrative decisionmaking. Finally, review can make administrative decisionmaking visible and provide a means for challenging it. Granted that courts themselves are no more politically accountable than bureaucratic agencies (in fact they are probably less so); by providing a means of public challenge to agency action, judicial review nevertheless helps to promote a greater degree of democratic responsibility than would otherwise be the case in bureaucratic decisionmaking. These are general aims of judicial review that are appropriate for the courts to serve. It remains to be asked whether the courts have adequately done so in the field of communications policy. We will look at the above aims in reverse order—the order in which I would rate the judicial performance.

In providing an effective means for public challenge to agency decisionmaking, it is difficult to find fault in the performance of the courts. Liberal interpretations of jurisdiction, standing, and reviewability have made it possible for the public to be involved in shaping public policy to an extraordinary degree. Decisions in the communications field have, in fact, become a model for other areas of administrative government.[13]

On the point of public visibility, the only notable deficiency is the scarcity of Supreme Court pronouncements on issues of national importance. Supreme Court review serves purposes beyond the mere oversight of lower courts. It serves also as a focus for public scrutiny of legal issues of national importance beyond that which can be obtained with review by a single court of appeals. Unfortunately, this "showcase" function seems to have received little recognition as a basis in itself for Supreme Court review. Emphasis has instead been given to supervision of the lower federal courts—in particular the elimination of conflicts in circuit court decisions—and the correction of constitutional error, particularly in state tribunals. Most of these cases will also involve issues of important national policy so that the showcase function is incidentally served. However, in the communications field there are few circuit conflicts to force review of ordinary communications policy on that basis. No doubt this is partly because of the dominance of the District of Columbia Circuit in reviewing communications policy. In addition, there are relatively few constitutional issues that can lay compelling claim on Supreme Court attention. Review by the court of appeals is a fairly reliable safeguard in this respect, particularly when it is seldom determined by circuit conflicts. As a consequence, many important policy issues do not receive high-level review and the greater national attention that ordinarily attends such review. In effect, the courts of appeals—which in this case practically means the District of Columbia Circuit—have virtually become the Supreme Court.

This fact obviously gives added importance to the question of how well the appeals courts have achieved the other aims of judicial review identified above—principled rationality and generality of perspective. Opinions are deeply divided on this matter. Some believe the courts have reasonably fulfilled the reasoned principle function; others believe that they have overstepped their proper bounds. As noted earlier, many believe the Court of Appeals for the District of Columbia in particular has arrogated to itself responsibility for policy judgments that properly reside in the Commission.

There is no doubt that the current rhetoric of judicial opinions suggests a more active interventionist role than that of a decade or two ago. Reading today's lengthy and often discursive appellate opinions, one senses that many judges are attempting to influence policy as much by dicta as by their opinions. Whether or not the dicta influence the agency directly, they can have an indirect influence by helping to shape the intellectual framework in which dialog about policy takes place. (It is a common criticism that increasingly judges are "writing for the law reviews and casebooks.") But this effort to extend the realm of judicial influence by going beyond the immediate controversy before the court is not a product of the "modern era," and it is by no means confined to communication cases. Apart from the fact that the opinion of today frequently reads more like a law review article than it did in past years (suggesting perhaps less regard for the limited offices of a judicial opinion), the only obvious difference between present day and earlier judicial activism appears to be that the latter was predominantly associated with politically conservative bias against government action while today it tends to be identified with a liberal bias in favor of government involvement in social and economic affairs. The shift in attitude, in short, seems more one of political outlook than of distinctly judicial behavior. However, if this is the case, the field of communications is, again, not unique.[14]

Despite an undeniable activism in style and apparent attitude, most opinions that were examined individually (all of these in the 1970-1976 period) seem nevertheless defensible by traditional conceptions of the role of a reviewing court.[15] The record of affirmances and reversals itself does not suggest that the courts are usurping agency prerogatives in making basic policy choices. Indeed, many of the reversals were remands for additional agency findings or minor corrections; which, when made, were subsequently affirmed. Nor does a reading of the opinions, with some exceptions, suggest that the courts are embarked on a policymaking "frolic" of their own.

I draw this conclusion cautiously on the strength of the 1970-1976 opinions, mindful of the fact that in 1977 the FCC suffered an unprecedented number of reversals—some on major policy issues. Indeed through November 1977, more than half of all appeals of FCC decisions to the court of appeals resulted in a reversal of some significant part of the FCC's decision. Such a high rate of reversal—especially on major issues—would appear to call for reconsidering the above judgment based on 1970-1976 decisions. The sudden spate of reversals seems to suggest factors at work beyond a mere "hard look" at the agency's decisions—perhaps an increasing loss of confidence in the FCC's regulatory attitudes.

There is, however, reason to be cautious about the weight to be placed on the 1977 decisions. First, in a number of these decisions the course of review is not yet final. In some cases, either a rehearing before the court of appeals en banc or a further appeal to the Supreme Court was pending as of Spring 1978. (I have made no attempt to pursue the survey into 1978.) Second, in several cases it is difficult to judge the scope and impact of the reversal until the FCC responds to the judicial remand. Third, and most important, a survey of the individual decisions suggests that the courts were not quite as interventionist as might first appear from the results.[16]

Of those cases in which the FCC was reversed in major part in 1977, about one-third seem to me to be difficult to justify under a concept of limited judicial review. While the final result of these cases may indeed be sound, the court's imposition of that result on the agency clashes with the traditional notion that judicial review is confined to ascertaining whether the agency's action was "rational" or, in the case of adjudicatory decisions, supported by "substantial evidence." Nevertheless the substantial majority of the 1977 decisions seem to me to comport with the traditional conception of judicial review. Obviously, individual judgment about this will vary greatly. It is difficult for an outside observer to separate the merits of a particular controversy from his or her views of the proper role of the courts in reviewing it. And since it must be equally difficult for a reviewing judge to make such distinctions, one is inclined to be more tolerant of judicial excursions beyond the narrow confines of reviewing agency decisions for mere rationality.

This is not to say that judges cannot recognize the proper limits of their function, or that outside observers cannot discern whether they have followed it. But, allowing for close calls, it is my judgment that the courts have generally stayed within the proper bounds in reviewing FCC decisions.

If at times the courts (most notably, but not exclusively, the Court of Appeals for the District of Columbia) have seemed to impose their own policy views, at other times they have been too acquiescent. Indeed, although some judges characteristically show negligible deference to agency judgment, many courts appear to be quite uncritical in accepting the agency's frame of reference in judging issues of general policy. In doing so, they fail to provide the generalist perspective. In most routine review cases where the issues are fairly narrow (e.g., review of a minor license or tariff decision, review of minor procedural questions), it is enough to see that the agency has sufficiently "crossed its t's and dotted its i's." But in cases where the agency embarks upon a major new venture and calls into question such matters as its regulatory powers or the character of its regulatory mission, one might expect the courts to take a broader perspective and to question the basic assumptions underlying the agency's action. Too often they have not done so. Witness, for example, the almost uncritical attitude of the courts toward the Commission's policies in the field of cable regulation. (An important exception is the recent decision invalidating the FCC's rules restricting subscription television on cable.) Whether or not the Commission's policies have support in the statute or in the public-interest purposes of regulation, the scope of the agency's undertaking and its tenuous connection to the traditional purposes of radio and television regulation called for more skeptical and demanding judicial scrutiny than was given. Similarly, in the case of broadcast licensing the courts have often been remarkably tolerant in accepting the Commission's bias in favor of the status quo and stability.

The one area where the courts seem to have been most aggressive in reversing the Commission is radio and television program and related regulations; however, even here there are relatively few instances when the courts have been willing to set aside the agency's basic premises. For example, the radio format decisions mentioned earlier seem to me to be wrong, but not because they interfered with agency discretion. They are wrong because they entered a domain where neither the court nor the agency belongs. To illustrate the point suppose the FCC had issued the same order as the court in the format cases. The court would owe little or no deference to the agency on the First Amendment issue raised by such action—the FCC itself would probably not seriously argue the contrary—and I would argue further that the court should reverse. If that conclusion is sound or even plausible, it follows that the error of the court was not giving insufficient deference to the FCC but giving insufficient deference to the First Amendment. The point bears emphasis: The remedy, if there is one, for judicial "activism" of this type certainly does not lie in rearranging the terms of the agency-court "partnership" agreement. It lies in cultivating in both partners a greater sense of restraint in the use of power. (It is worth noting in this regard that the FCC has vigorously resisted the court's mandate in the program format cases in the expectation of securing ultimate Supreme Court review; this resistance is perhaps a constructive step toward establishing that a "partnership"—in contrast to a master-servant relationship—is not entirely hierarchical.)

Reforms

Even if one were to conclude that the courts have not followed their proper role in reviewing FCC decisions, it is not obvious there is any institutional reform that would solve the "problem." Three possible reforms have been proposed: changes in the standards governing the scope of review to restrict judicial "activism"; creation of a specialized reviewing court for communications and other regulatory decisions to bring greater "expertise" to bear on these decisions; and, changes in venue to distribute the caseload outside of the District of Columbia. Obviously, these three proposals do not all look in the same direction. The first proceeds on the explicit premise that present judicial review is too intrusive. The second at least implicitly rests on the opposite premise—that review ought to be more demanding by being more "expert." The third has been advanced, by different groups, on both premises.

None of the proposed reforms has been justified. The first suggestion seems useless even if one accepts the underlying premise. As mentioned earlier, the formulations of scope of review (e.g., "arbitrary and capricious," and "substantial evidence" tests) serve more as labels than as guidelines for the reviewing function, and it is doubtful that an attempt to adjust or refine the labels will significantly alter judicial practice. One could, of course, eliminate judicial review altogether for certain types of decisions or issues, but no one has yet proposed such a drastic reform. Another alternative would be to curtail standing of litigants seeking to vindicate only a general "public interest" as distinct from a personal, legally protected right. Such a limitation, however, would cut only a handful of the decisions that have been seen as evidence of "activist" intervention by the courts. If one examines the decisions that give rise to the charge of undue "activism," he will not find more than a bare handful in which appellant did not have standing on the most limited and traditional criteria. There is, therefore, no reason to think that, short of cutting off review altogether, a change in the standing rule would materially affect the scope of review. Moreover, on grounds of principle it would be difficult to justify denying access to the courts by, say, public interest groups while retaining the conventional review right of, say, licensees protesting the grant of a license to competitors.

Even if it were possible to develop useful formulations for limiting the scope of review through redefined standards, the case for doing so has not been made. Granted, there are cases where the courts have been too aggressive in interfering with administrative prerogatives, but there are probably as many where they have not been aggressive enough in forcing the agency to take a "hard look" at the problems before it. What is needed then is a formula for review that will produce neither uniformly greater timidity nor aggressiveness, but a discriminating and sensitive application of toughness with restraint—a review that, in the words of Goldilocks is "just right." I know of no recipe that will produce such a broth.

The second proposed reform, creation of a special tribunal for review of communications and other administrative law decision, requires only brief mention. Insofar as the proposal aims to provide more expert surveillance of administrative decisions on communications policy, it is not needed, conscientious application or ordinary intelligence is fully adequate to perform the task of review. Indeed, this holds for administrative regulation in general if the role of judicial review is to provide not another level of expert decisionmaking but rather a generalist surveillance of the "experts."

The third reform, altering the venue for reviewing communications decisions, has been proposed by different advocates for opposite reasons. The proposed reform has focused on the provisions of the Communications Act that give the job of reviewing broadcast licensing decisions (most notably license renewals) solely to the Court of Appeals for the District of Columbia. Some public-interest advocates have urged elimination of the exclusive venue to permit actions to be brought in the communities where the station is located. The underlying motive is mixed. In part it rests on grounds of convenience: Where review is sought by local groups it is supposed to be less costly to bring action in a court near to the community. However, this rationale seems insubstantial because review would still be in the respective circuit court of appeals which will often be no more conveniently located (in terms of travel time and cost) than the court in Washington, D.C. Probably the deeper rationale is the fact that some citizen groups who advocated this change were dissatisfied with their recent lack of success in licensing cases brought in the District of Columbia Circuit and have imagined that they might receive a more sympathetic reaction elsewhere. Recently, after winning a number of major decisions in that circuit, these groups appear to have lost interest in changing venue. However, the same reform has been suggested by broadcasters on precisely the opposite premise: that courts outside the District of Columbia will be more sympathetic to the broadcaster.

In fact, there is a little tangible evidence of a significant difference among the circuits in regard to communications decisions generally, and although no comparison can be made on license cases, there is no reason to think this class of cases is unique. For this reason, a change in venue is unlikely to be meaningful. Perhaps one might argue that distributing the review function more widely would diffuse responsibility and lessen the concentration of influence now given to one court, even if the content of decisions were not changed. On the premise that specialized courts are not desirable, one might suppose that spreading responsibility would enhance the generalist perspective. This reasoning, however, overlooks the fact that the exclusive venue provisions apply only to a limited class of cases—cases that frequently raise no important issues of general policy even within the limited field of broadcasting. Thus, the theoretical advantage of redistributing responsibility is of limited practical significance at best.

Conclusion

For those uneasy about the judicial role in communications law—because it is either too active or not active enough—it is difficult to find reliable, objective evidence to justify such uneasiness. That does not mean that the pattern of review is consistently satisfying; it means that the extremes of judicial activity seem fairly well balanced. For every occasion in which a court strays quite far in one direction, there is another decision that strays in the opposite path. Obviously if the mean were entirely a product of opposing extremes, there would be reason for concern and for reform, for it would indicate an unacceptable lack of legal stability. But the vast majority of decisions fall quite close to the center of what one would predict for the judicial pendulum. No large reforms would seem therefore to be in order (or in any event practical) beyond a more vigorous public scrutiny of individual decisions and a reproof of those few that represent significant deviations from the norm.

Notes

1. See *Citizens Committee to Save WEFM v. F.C.C.*, 506 F.2d 245 (District of Columbia Cir.), 1974, (the latest court of appeals decision in the series); Entertainment Formats, 57 FCC F.2d 580, 1976.

2. See, for example, *Home Box Office, Inc., v. F.C.C.*, No. 751280 (District of Columbia Cir.), 1977, (pay television rules), cert. denied, 98 S.Ct. 111, 1977; *National Citizens Committee for Broadcasting v. F.C.C.*, 555 F.2d 938 (District of Columbia Cir.), 1977, (newspaper-broadcast ownership rules), cert. granted, 98 S.Ct. 628, 1977.

3. A recent, and most interesting, review of judicial activity by Warner Gardner, "Federal Courts and Agencies: An Audit of the Partnership Books," *Columbia Law Review*, 1975, p. 800, is suggestive of the scope and degree of activism in reviewing federal agency action. It also shows, however, the difficulty of reaching firm conclusions on this point since the judicial performance is anything but clear, single-minded, or consistent. On judicial activism, see N. Glazer, "Towards an Imperial Judiciary," *The Public Interest*, Vol. 41, 1974, p. 104.

4. So far as general activity is concerned, it is undoubted that the number of court actions has steadily increased, but so too have the underlying activities of both government and private enterprise which give rise to legal disputes. For example, the Environmental Protection Agency and the Consumer Product Safety Commission are both "new" agencies, much of whose work is subject to judicial review in one way or another. Statutes establishing such agencies increase the amount of litigation that

comes to court, but one could hardly explain this increase by reference to the judicial review philosophy of members of the judiciary. Similarly, the Civil Rights Act of 1964 and a variety of other recent legislation creating and implementing executive orders provide rights of action for discriminatory practices in housing, education, and public accommodations, and federal contract compliance rules seeking greater fairness in employment practices, have increased the volume of litigation to a large, if undetermined, degree. Much of this judicial activity is irrelevant to the question of judicial "activism," a term which implies a disproportionate interest in judicial involvement relative to the increase in legal conflict. And one must be wary of inferences of increased "activism" from the mere fact of increased activity: If Congress, for example, commands judicial review of all EPA decisions, it scarcely denotes an increase in judicial activism for the courts to obey.

5. See K.C. Davis, *Administrative Law Text*, 3rd edition, 1972, p. 351. An example of formulas that do not rely on the wording of the Administrative Procedure Act is the "hard look" formula (to borrow a phrase currently used in the Court of Appeals for the District of Columbia Circuit) which has now become popular in the District of Columbia. See, for example, *WAIT Radio v. F.C.C.*, 418 F.2d 1153 (District of Columbia Cir.), 1969. Examples of incorrect usage include *United States v. Midwest Video Corp.*, 406 U.S. 649, 671, 1972; and *Citizens to Preserve Overton Park, Inc., v. Volpe*, 401 U.S. 402, 414, 1971, mistakenly suggesting the applicability of the substantial evidence rule to rulemaking proceedings.

6. *Greater Boston Television Corp. v. F.C.C.*, 444 F.2d 841, 851 (District of columbia Cir., Leventhal, J.), 1970, *cert. denied*, 403 U.S. 923, 1971:

> The process (of judicial review) combines judicial supervision with a salutary principle of judicial restraint, an awareness that agencies and courts together constitute a 'partnership' in furtherance of public interest, and are 'collaborative instrumentalities of justice.' The court is in a real sense part of the total administrative process, and not a hostile stranger to the office of first instance. This collaborative split does not undercut, it rather underlies the court's rigorous insistence on the need for conjunction of articulated standards and reflective findings, in furtherance of evenhanded application of law, rather than impermissible whim, improper influence, or misplaced zeal.

This partnership is not, of course, limited to the FCC. See, for example, *Associated Industries v. Department of Labor*, 487 F.2d 342, 354 (2d Cir.), 1973. Whether or not this "partnership" connotes a new, more active judicial role than heretofore is debatable in the absence of any precise study of the question. Some judges have perceived the modern role to be more active than heretofore—see *Environmental Defense Fund v. Ruckleshaus*, 439 F.2d 584, 497 (District of Columbia Cir.), 1971, Bazelon, J.—but that is hardly dispositive. (It can hardly escape notice that the author of the opinion in *Environmental Defense Fund* is known to be among the more "activist" judges in the federal judiciary.)

7. In the 1970-1976 period the courts affirmed in whole or in major part FCC decisions on such undeniably major issues as entry into telephone service markets, *Washington Utilities and Trans. Comm. v. F.C.C.*, 513 F.2d 1142 (9th Cir.), 1974, *cert. denied*, 423 U.S. 836, 1976; *American Telephone and Telegraph v. F.C.C.*, 539 F.2d 767 (District of Columbia Cir.), 1976; interconnection of telephone terminal equipment, *North Carolina Utilities Commission v. F.C.C.*, 537 F.2d 788 (4th Cir.), 1976; affirmed on rehearing, 552 F.2d 1036, 1977, *cert. denied*, 98 S.Ct. 222, 1977; allocation of frequencies to land mobile and structure of land mobile services, *National Association of Regulatory Utility Commission vs. F.C.C.*, 525 F.2d 630 (District of Columbia Cir.), 1976; major reinterpretation of equal time rules, *Chisholm v. F.C.C.*, 538 F.2d 349 (District of Columbia Cir.), *cert. denied*, 429 U.S. 890, 1976; rules limiting network program time, *National Association of Independent Television Producers and Distributors v. F.C.C.*, 516 F.2d (2d Cir.), 1975; two major broadcast comparative renewal cases, *Fidelity Television Inc. v. F.C.C.*, 515 F.2d 182 (District of Columbia Cir.), 1975; rules governing computers and communications, *General Telephone and Electronics Service Organization Inc. v. F.C.C.*, 474 F.2d 724 (2d Cir.), 1973. The above list by no means exhausts the list of important policy decisions affirmed; it simply illustrates the range and type of important cases.

By comparison, the reversals of important policy issues in this same period are quite limited in scope and in number, they include the following: *Citizens Communications Center v. F.C.C.*, 447 F.2d 201 (District of Columbia Cir.), 1971, (license renewal hearing policy); *Business Executives Move for Vietnam Peace v. F.C.C.*, 450 F.2d 642 (District of Columbia Cir.), 1971, *reversed sub nom.*; *Columbia Broadcasting System v. Democratic National Committee*, 412 U.S. 94, 1973, (FCC denial of right of access reversed by Court of Appeals, which in turn was reversed by the Supreme Court); *Citizens Committee to Save WEFM v. F.C.C.*, 506 F.2d 256 (District of Columbia Cir.), 1974; *National*

The Judicial Role

Broadcasting Corporation v. F.C.C., 516 F.2d 1101 (District of Columbia Cir.), 1974, *cert. denied*, 424 U.S. 910, 1974, (fairness doctrine); *Midwest Video Corp. v. U.S.*, 441 F.2d 1322 (8th Cir.), 1971, *reversed* 406 U.S. 649, 1972, (FCC rules requiring cable program origination reversed by court of appeals, which in turn was reversed by Supreme Court); *National Cable Television Association v. U.S.*, 415 U.S. 336, 1974, (authority to collect fees from regulated firms).

A number of important decisions were handed down in 1977; these are discussed in Note 16 *supra*.

8. Commenting on my choice of 1950-1956 period, one knowledgeable communications lawyer noted that this was a period that saw considerable scandal over such matters as *ex parte* contracts and that the court must have been influenced by this fact "to examine each and every Commission decision carefully" resulting in a "rash of remands." The apparent inference to be drawn from this is that the rate of reversals in this period is higher than it would otherwise be. However, plausible as that might seem at first blush, I am skeptical. In fact the *ex parte* scandal was not publicly discovered until late in the period under review—around 1956. Even supposing that the judges heard of the scandal as soon as it was revealed, it could not have influenced their decisions before the last year—or at most two—of the 1950-1956 period. Indeed the legacy of the 1950s scandal probably affects present judicial attitudes more than those of 1950. See *Home Box Office v. F.C.C.*, Note 2, *supra*.

9. Especially noteworthy is the recent *Ethyl Corp. v. E.P.A.*, 541 F.2d 1 (District of Columbia Cir.), 1976, where the majority opinion (by Judge Wright) examines the technical record in uncommon detail in the course of affirming EPA lead standards for gasoline. Responding to a concern voiced by Judges Bazelon and McGowan about the dangers of judicial presumption in reviewing highly technical issues (which are contrasted to "FCC judgments concerning diversification of media ownership" 541 F.2d at 66), Judge Leventhal observes: "Our present system of review assumes judges will acquire whatever technical knowledge is necessary as background for decision of the legal questions.... Our obligation is not to be jettisoned because our initial technical understanding may be meager when compared to our initial grasp of FCC or freedom of speech questions." 541 F.2d at 68-69.

10. *Teleprompter Cable Systems v. F.C.C.*, 543 F.2d 1379 (District of Columbia Cir.), 1976 (reversal of FCC decision denying authorization to local cable system which had bribed

local officials in obtaining operating franchise).

11. With respect to the first category, reversal for failure to regulate, see the *WEFM* and *Business Executives Movement* cases, Note 7 *supra*. Judicial affirmances of FCC program regulation include *Illinois Citizens Committee for Broadcasting v. F.C.C.*, 515 F.2d 397 (District of Columbia Cir.), 1975, (affirming FCC control of obscenity); *Brandywine-Main Line Inc. v. F.C.C.*, 473 F.2d 16 (District of Columbia Cir.), 1972, (affirming FCC denial of license for violation of fairness doctrine).

12. No doubt if judicial review were removed to such an extent as to remove any effective enforcement of "due process," formal legal procedures would disappear. However, the alternative of no judicial enforcement is not a credible or pertinent speculation. The pertinent inquiry is whether, presupposing at least a modicum of judicial review to enforce minimal standards of procedural regularity in agency proceedings, how much *additional* "legalism" is added by extensive, and intensive, judicial review of agency decisions. In my view it is very little.

13. See, for example, *Office of Communications of the United Church of Christ v. F.C.C.*, 359 F.2d 994 (District of Columbia Cir.), 1966.

14. See Glazer, Note 3 *supra*, at 109: "In the past the role of the activist courts was to restrict the executive and the legislature in what they could do. The distinctive character of more recent activist courts has been to extend the role of what the government could do, even when the government did not want to do it." Corroborating evidence is provided by the program format decisions, the recent cross-ownership decision and the short-lived *Business Executives Movement*, Note 7 *supra*.

15. While documentation is difficult for such a subjective judgment, the following are offered as illustrations of the kind of cases on which the judgment was based; the interested reader can judge for himself whether they support my judgment. *Meadville Antenna Inc. v. F.C.C.*, 443 F.2d 282 (3rd Cir.), 1971, (FCC denial of request for waiver of cable exclusivity rule; inadequate findings); *American Telephone and Telegraph v. F.C.C.*, 499 F.2d 439 (2nd Cir.), 1971, (partial Telpak tariff; FCC prescription of unlimited sharing of Telpak services held to be in conflict with earlier decision; FCC also failed to make findings that a total ban on sharing would be unlawful as a remedy for discrimination); *Friends of the Earth v. F.C.C.*, 499 F.2d 1164 (District of Columbia Cir.), 1971, (FCC refused to apply fairness doctrine to advertising of gasoline and automobiles; reversed as inconsistent with prior decision applying doctrine to cigarette ads); *American Telephone and Telegraph v. F.C.C.*, 487 F.2d 865 (2d Cir.), 1973, (FCC order to AT&T not to file tariff for particular private-line service without

The Judicial Role

prior FCC approval reversed for lack of statutory authority); *Columbia Broadcasting System v. F.C.C.*, 454 F.2d 1013 (District of Columbia Cir.), 1971, (FCC order requiring CBS to give time to Republican Party after CBS had extended time to Democrats to respond to President Nixon's messages, reversed for inconsistency and arbitrariness); *Midwest Video Corp. v. U.S.*, 441 F.2d 1322 (8th Cir.), 1971, (reversed by court of appeals for lack of statutory authority; court of appeals reversed in turn by Supreme Court 406 U.S 649).

16. The cases represent a broad range of policy issues. It is not possible to review them in detail. I will merely outline the principal cases and offer a few comments on them. The comments apply only to reversals of major agency policy decisions since it is the reversal rate that is the basis of controversy, and it is major policy issues that are of concern to this paper. By way of disclosure it should be noted that, as a commissioner, I participated in many of these cases, sometimes voting for and sometimes against the agency decision.

Broadcasting. One of the most controversial of the 1977 decisions was the *National Citizens Committee for Broadcasting v. F.C.C.*, Note 2, *supra*. The court affirmed an FCC rule barring common ownership of newspapers and broadcast stations in the same community, but reversed its remedy which was to apply the ban prospectively only—except in a handful of cases. The court ruled that divestiture was presumptively the appropriate remedy in all cases. The decision is extraordinary in its refusal to accord the agency a wide latitude in its choice of a remedy. An agency's choice of remedies is not unbounded, but normally the courts intervene in this area to limit a remedy that is too broad, rather than one that is too narrow. Nevertheless, the Commission contributed considerably to its defeat in this case by relying on a number of rather infirm grounds for refusing to order divestiture. Once these infirm supports were removed from the decision, as they easily could be by careful review, the only thing that the Commission had to fall back on was an unarticulated administrative/legislative discretion. ("There is no reason for it, it is just our policy.") It is understandable that the court would not accept the mere fact of agency power as sufficient basis for agency decision. However, it is troublesome that the court did not permit the agency on remand to develop a rationale for its exercise of discretion or to reconsider the entire issue of cross ownership in light of the court's mandate on the remedy. In ordering divestiture as presumptive remedy, the court went beyond its role and imposed its own conception of regulatory policy. (Before this book went to press, the Supreme Court reversed the court of appeals and affirmed the F.C.C.s decision in all respect.)

After the *NCCB* case, probably the most noteworthy court cases in the broadcast field are a series of decisions involving FCC enforcement of equal employment opportunity. The Commission was successful in *National Organization for Women v. F.C.C.*, 555 F.2d 1002 (District of Columbia Cir.), 1977, where the court sustained the Commission's decision that the licensee had not discriminated against women in programming and hiring practices. However, the court reversed agency actions in three additional employment discrimination cases and demanded greater agency sensitivity to discrimination charges. In *Black Coalition of Richmond v. F.C.C.*, 566 F.2d 59 (District of Columbia Cir.), 1977, the court rebuked the FCC for refusing to hold a hearing on serious discrimination charges solely on the basis of the licensee's post-term corrective measures. The court found inadequate licensee efforts to attract and train black employees and directed the Commission to demand a more meaningful accounting for conduct during the contested licensing period as well as more exacting standards for future employment practices. In *Bilingual-Bicultural Coalition on Mass Media*, No. 75-1855 (District of Columbia Cir.), 1977, *vacated in order to be reheard en banc*, the court held that where renewal challenges are filed based upon unemployment discrimination charges, the agency must afford challengers the right to prehearing interrogatories. In *Office of Communications of the United Church of Christ v. F.C.C.*, 560 F.2d 529 (2d Cir.), 1977, the court reversed as arbitrary and capricious the FCC's decision to extend its exception from equal employment filings from stations with five or fewer employees to those with ten or fewer. Although conceding that the former exception level was arbitrary, the court held that a change could not be made unless the agency could show reasonable grounds for doing so—grounds which the court could not locate in the record. While none of the court's decisions in the three EEO cases seem to me compelling, the last two EEO opinions seem particularly dubious—even though, on the merits, I would not be troubled if the Commission had taken such a position. Contrary to *Bilingual*, there is, in my view, no legal basis for a court directing the Commmission as to the use of agency procedures for validating claims of this kind. This is not a case where constitutional due process rights of the appellants are at stake; it is not a case where a prima facie case for a hearing under the statute has been made out. So far as I can detect, the only basis for the court's commandment is its own view that the Commission was not, in general, sufficiently sympathetic to EEO complaints. In the *Church of Christ* case, the court attempted to determine that the Commission did not sufficiently articulate the reasons for changing its policy. The fact is that the initial exception level was, as the court is prepared to concede, arbitrary. Thus, there is no a priori reason to prefer the old exception level to the new except on the somewhat artificial notion that once an arbitrary line has been drawn it cannot be changed unless the agency finds a new one that can be specially justified by well-reasoned principles—a test that, by assumption, the earlier standard would fail.

In other broadcast cases, the Commission's record on appeal was mixed. In *Pacifica Foundation v. F.C.C.*, 556 F.2d 9 (District of Columbia Cir.), 1977, *cert. granted*, 46 U.S.L.W. 3436, 1978, the court (Judge Leventhal dissenting) reversed an FCC decision proscribing the broadcast of "indecent" language during certain times of the day. Though I think the FCC's decision defensible, I consider the court's decision as clearly not beyond the pale of appropriate behavior; the case involved an issue of law (First Amendment claims in particular have always been scrutinized closely by the courts) which is squarely within the province of the courts. Moreover, the Commission's action was at the very boundary of an area of speech regulation which the Supreme Court had defined as the permissible area of government regulation. The Commission itself recognized as much and openly invited judicial "guidance." (After the above was written, the Supreme Court reversed the court of appeals.)

In four comparative broadcast licensing cases, appeals were filed by the losing applicant. In three, the Commission was summarily affirmed. In the fourth, *Pasadena Broadcasting v. F.C.C.*, 555 F.2d 1046 (District of Columbia Cir.), 1977, the court remanded the Commission's decision for deviating without an evidentiary basis from its policy of favoring applicants offering original services to a community over applicants in areas already served by existing licensees. As discussed in Chapter 10, comparative hearings seem to me a waste of time. On that view, the court's rather mechanical insistence that such a hearing be held is cause for distress. However, evaluating the opinion in terms of legal precedent—which favors comparative hearings—it is difficult to say that it exceeds the proper standard for agency review. In another broadcast licensing case, the court vacated and remanded an agency decision denying a license renewal on several grounds, including the licensee's action in filing a "strike" petition against a prospective competitor. The court's reversal is on a very narrow and largely unimportant question of law involving the weight accorded witnesses in the hearing. However, on a more substantial issue the court directed the Commission to clarify its policy towards strike petitions and how it distinguishes these from legitimate petitions filed by licensees to deny applications of prospective competitors. *Faulkner Radio v. F.C.C.*, 557 F.2d 866 (District of Columbia Cir.), 1977. Again, it is difficult to say that the decision is beyond the pale of proper judicial review. I voted with the majority of the Commission and still believe the decision is correct, but the court's concern about an over-broad policy on strike applications is a legitimate reason for remand (the court's point about the weight accorded to certain testimony is trivial but also legitimate).

The Commission was affirmed in major part in two appeals involving the fairness doctrine. In *Georgia Power Project v. F.C.C.*, 559 F.2d 237 (5th Cir.), 1977, the court affirmed an FCC finding that the fairness doctrine was not violated by several power company advertisements. In a more important decision, the court affirmed the FCC's basic fairness doctrine of enforcement policy including its policy determination not to apply the fairness doctrine to routine commercial product advertising. See *National Citizens Committee for Broadcasting v. F.C.C.*, No. 74-1700, (District of Columbia Cir.), 1977, (NCCB II). However, the court remanded the case to the Commission to consider further two proposals it had rejected: a proposal to allow broadcasters to provide free access periods in lieu of fairness, and a proposal to require licensees to list in their renewal ten major controversies of public importance. The reversal and remand on these issues seems justifiable given the Commission's summary treatment of the proposals. (I confess to partial bias in as much as I urged the Commission to adopt the access proposal. However, this bias is offset by an even stronger bias against the proposed listing of controversial public issues.)

Finally, in *Community Service Broadcasting of Mid-America v. F.C.C.*, No. 75-2057 (District of Columbia Cir.), 1977, the court (Judge Leventhal dissenting) invalidated statutory and agency requirements that non-commercial broadcasters record all programming in which issues of public importances are discussed. The court viewed the requirement as an apparent device for censorship; however, it rested its decision on the fact that, because similar requirements were not imposed on commercial broadcasters, the statute and rules violated equal protection. The decision seems to me to be a rather wooden application of equal protection principles. Sensitivity to the character of government-supported media ought to counsel some deference to the intent of Congress in singling out public television for special scrutiny. However, it must be noted that, in any event, the decision is directed primarily at the statute, not the agency's rules, which were designed only to implement the clear requirement of the statute. Thus, the decision bears only tangentially on the scope of review of *administrative* discretion.

Cable. The courts reviewed three cable decisions in this period, affirming only one—involving revision of network program non-duplication rules. *Columbia Broadcast System Affiliates v. F.C.C.*, 555 F.2d 985 (District of Columbia Cir.), 1977. In *Teleprompter Cable Communications Corp. v. F.C.C.*, 565 F.2d 736 (District of Columbia Cir.), 1977, the court examined the narrow application of a grandfather clause in the Commission's signal carriage revision and reversed the agency's denial of permission to carry certain previously approved signals. The court's decision seems unequivocally correct; the FCC literally played fast and loose with its own rules and disregarded precedent in ruling its own prior grant of authorization to the system.

The most important and most controversial cable decision of the period reversed Commission rules restricting pay television on cable. *Home Box Office v. F.C.C.*, Note 2, *supra*. The court's opinion both in language and in approach suggests a lack of restraint in pursuing agency discretion. Nevertheless, the court did not overstep the proper boundaries of judicial review in ruling that the pay cable rules were arbitrary and beyond the agency's authority, unless one is prepared to read prior Supreme Court opinions as giving the Commission carte blanche authority. (As a commissioner, I dissented to most of the pay cable rules.) The court's judgment that the rules also violated the First Amendment seems to me strained, however. Given the court's own prior rulings on regulation, directed specifically at radio and television programming and particularly given the now extensive jurisprudence on the scope of economic regulation which incidentally affects speech, the First Amendment objection is dubious. Perhaps most dubious, and certainly most widely criticized among members of the bar, is a ruling that the Commission acted improperly in engaging in informal "ex parte" communications with interested parties in the course of the rulemaking proceedings. As discussed more fully in Chapter 10, this ruling marks a bold departure from accepted law on the subject of rulemaking procedures. However, a subsequent decision by a different panel of the court has rejected this broad ban on all informal contacts and it seems very likely that the second decision will ultimately prevail as the decision of the court. See *ACT v. F.C.C.*, 564 F.2d 458 (District of Columbia Cir.), 1977.

Common Carrier. The courts heard appeals in nine common carrier cases during this period and overturned the Commission in three. In *MCI Telecommunications Corp. v. F.C.C.*, 561 F.2d 356 (District of Columbia Cir.), 1977, *cert. denied*, the court reversed a Commission decision that a specialized common carrier was limited to providing private-line telephone services. The court held that the Commission had made no finding that such a limitation in services was required by the public interest. In *American Telephone and Telegraph v. F.C.C.*, 551 F.2d 1287 (District of Columbia Cir.), 1977, the court reversed an agency refusal to give AT&T an evidentiary hearing on the company's mobile telephone dialing operations. In *RCA Global Communications, Inc. v. F.C.C.*, 559 F.2d 881 (2nd Cir.), 1977, the court vacated a Commission formula for distributing international telegraph traffic for lack of an adequate evidentiary record. Of the three only the MCI case is sufficiently important to warrant discussion here. Essentially the decision holds that the Commission had not made a reasoned finding, in this or in earlier decisions, that the public interest requires that competitive entry be limited to private-line services. On the merits, I think the Commission's decision was probably correct in concluding that competitive entry should not at this time be permitted beyond private-line service—reflecting a judgment that the public-switched services are properly monopolistic (because of "natural monopoly" characteristics). However, I cannot dispute the court's holding that such a judgment had not been explicitly articulated and explained by the Commission in this or other decisions. The Commission in fact has never squarely confronted the problem of defining the boundaries of competition and monopoly in this context. I doubt that this decision will in fact change the Commission's policy regarding the limits of competitive entry, but it should force the Commission to explain the basis for that policy.

The six common carrier cases affirmed in the courts were contests over tariff schedules and over agency decisions permitting the introduction of new carrier services. Three are particularly noteworthy: *North Carolina Utilities Commission v. F.C.C.*, 552 F.2d 1036 (4th Cir.), 1977, *cert. denied*, 98 S. Ct. 222, 1977 reaffirming a 1976 decision (tabulated above) which affirmed the FCC's decision allowing interconnection of customer-supplied equipment; *ITT World Communications Inc. v. F.C.C.*, 555 F.2d 1125 (2nd Cir.), 1977, affirming the Commission's grant of authority to AT&T for international dictaphone service; *Communications Satellite Corp. v. F.C.C.*, No. 75-2193 (District of Columbia Cir.), 1977, affirming in major part the FCC's specification of Comsat's rate of return (the court remanded for consideration of several narrow components of the rate determination).

Appendix Table 12-1
1970-1976 Decisions[a]

Type of Case	Total Affirm	Total Reverse	Percent Affirm	D.C. Circuit Affirm	D.C. Circuit Reverse	Percent Affirm	Other Circuits Affirm	Other Circuits Reverse	Percent Affirm	Supreme Court Affirm	Supreme Court Reverse	Percent Affirm
Radio and Television												
1. Broadcast: licensing (initial license and renewal adjudications; general standards, including equal employment opportunity; multiple ownership and network rules)[b]	32	9	78	29	8	78	3	1	75	--	--	--
2. Broadcast: technical (interference standards, technical coverage, etc.)	6	1	86	4	0	100	2	1	67	--	--	--
3. Broadcast: programming (fairness; equal time; obscenity, etc.; format including rules)[b]	20	11	65	12	9	57	7	2	78	1	0	100
4. Cable: certification (all cases of individual application of rules)	10	8	59	3	4	57	7	3	70	--	--	--
5. Cable: general rules	5	2	71	48	22	69	23	9	72	2	0	100
Total	73	31	71	48	22	69	23	9	72	2	0	100
Common Carrier												
1. Rates (domestic and international rates)	6	1	86	4	0	100	2	1	67	--	--	--
2. New entry (specialized common carrier cases and rules; new services, domestic and international, including mobile radio)	13	1	93	8	1	89	5	0	100	--	--	--
Total	20	2	90	12	1	92	7	1	88	--	--	--
Miscellaneous (fees, practice cases and rules, and procedural issues unrelated to above categories)	9	7	57	5	5	50	2	0	100	0	1	0
Total, all cases	**102**	**40**	**73**	**65**	**28**	**70**	**32**	**10**	**76**	**4**	**1**	**80**

Notes: (a) All court of appeals and Supreme Court decisions from January 1, 1970 through December 31, 1976. District court opinions are not included. (b) Some cases in broadcasting (1) and (3) overlap. For example, the network prime-time access rules involved issues that could be classified as licensing. Other cases involve issues that could be classified as procedural; they are included here where the procedural issue is related to substantive policy.

Appendix Table 12-2
Major Court Decisions, 1950-1956 [a]

Type of Case	Total Affirm	Total Reverse	Total Percent Affirm	D.C. Circuit Affirm	D.C. Circuit Reverse	D.C. Circuit Percent Affirm	Other Circuits Affirm	Other Circuits Reverse	Other Circuits Percent Affirm	Supreme Court Affirm	Supreme Court Reverse	Supreme Court Percent Affirm
Radio and Television												
1. Broadcast: licensing (initial license and renewal adjudications; general standards, including equal employment opportunity; multiple ownership and network rules)[b]	16	12	56	14	12	52	--	--	--	2	0	100
2. Broadcast: technical (interference standards, technical coverage, etc.)	3	4	42	2	4	33	--	--	--	1	0	100
3. Broadcast: programming (fairness; equal time; obscenity, etc.; format including rules)[b]	0	1	0	--	--	--	--	--	--	0	1	0
4. Cable: certification (all cases of individual application of rules)	--	--	--	--	--	--	--	--	--	--	--	--
5. Cable: general rules	--	--	--	--	--	--	--	--	--	--	--	--
Total	19	16	53	16	16	50	--	--	--	3	1	75
Common Carrier												
1. Rates (domestic and international rates)	0	1	0	0	1	0	--	--	--	--	--	--
2. New entry (specialized common carrier cases and rules; new services, domestic and international, including mobile radio)	3	3	50	2	2	50	1	1	100	0	1	0
Total	3	4	42	2	3	40	1	0	100	0	1	0
Miscellaneous (fees, practice cases and rules, and procedural issues unrelated to above categories)	11	7	61	11	7	61	--	--	--	--	--	--
Total, all cases	60	33	55	29	26	53	2	0	100	3	2	60

Notes: (a) All court of appeals and Supreme Court decisions from January 1, 1950 through December 31, 1956. (b) See note b, Table 1.

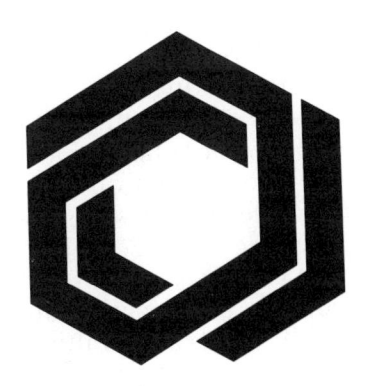

13.
The Role of Congress

Ernest Gellhorn

Congress has been criticized frequently for lack of organization and for inability to perform its primary task of legislating basic policy. This criticism often is extended to the way Congress performs its next most important function, oversight of the executive branch and the independent agencies. Some observers have even concluded that few in Congress understand the mechanics much less the aims of legislation and oversight.

However, change is in the air. In recent years Congress has made a serious effort to develop a sophisticated and effective role in the budgetary process. Committees have been restructured, new assignments have been made, and even the seniority system has been modified. Legislative policymaking is more frequent, and Congress is reexamining its oversight role.

These changes are affecting the role of Congress in making communications policy. For the first time in more than forty years, a major effort is underway in Congress to rewrite the basic communications law. Congress also has accepted a reshuffling of the executive structure for communications by creating the position of Assistant Secretary for National Telecommunications and Information Administration in the Department of

Commerce. Given the new impetus, this would seem to be an appropriate time to develop a deeper understanding of the congressional role in communications policymaking.

Until recently, Congress has not exercised either its authority or its power to legislate policy. The first basis for congressional authority is found in the Constitution which confers on Congress specific and exclusive powers to make all laws, to appropriate all government funds, and, in the case of the Senate, to confirm the appointment of executive officers.

Clearly, if Congress is to legislate effectively or to appropriate funds wisely, it must be able to determine the needs and evaluate the effectiveness of its programs. Although the Constitution does not specify any ancillary or supporting authority, Congress generally is acknowledged to have almost plenary powers for overseeing the implementation of the laws it adopts—a second major basis for congressional authority. In addition, and perhaps the most significant source of power available to Congress, is its position at all "crossroads" in Washington, and consequently its ability to command public attention. All Presidents, even the most imperial ones, ultimately have had to reckon with the fact that Congress is a powerful player in the exercise of governmental power.

It seems odd that for so long, despite this wide-ranging authority and power, Congress has played such a limited and often mechanical role in establishing communications policy. To understand this anomaly requires closer examination of how Congress has used its authority and exercised its power.

The Power to Make All Laws—Regulating without Direction

Although Congress is theoretically all-powerful in lawmaking, it is not disturbing and should not be surprising that Congress' actual exercise of its authority is more limited. Constitutional powers are best stated broadly, and legislative discretion often results in the use of only some of those powers. Communications issues are complex and evolving technology often makes specific solutions quickly obsolete. However, the congressional role in communications policy has been far more restricted than these points might suggest. Generally, Congress has used its power to establish a regulatory agency, the Federal Communications Commission (FCC), with broad policymaking authority, and to check on the FCC's actions by various formal and informal methods.[1]

Congress has exercised its legislative authority to make communications policy only in the most general sense. In creating the FCC, Congress gave that agency a vague mandate to serve "the public interest, convenience and necessity." Beyond these undefined terms, the enabling legislation has added little of substance. For example, Congress has not directed the FCC about its preference for either over-the-air or cable television. In fact, Congress has yet to acknowledge the existence of cable television in communications legislation.[2] One consequence of this neglect is that congressional intervention in communications policy has been almost entirely reactive rather than creative.

Perhaps this responsive posture by Congress is as it should be because few in Congress have the expertise, time, or ability to formulate a comprehensive communications policy. Also, Congress, operating either as a body or by committee, cannot be as informed or expert on policy needs as the FCC or other executive departments. However, many communications policy issues involve basic values, and it is Congress which is "expert" on normative questions, moreover, one frequent result seems to have been a political policy vacuum that provides an opportunity for those with particular interests to veto changes that would threaten the status quo.

Major FCC regulatory policies reflect this pattern of limited and reactive congressional participation. For example, it was the FCC, not Congress, that struggled with the problem of allocating frequencies between VHF and UHF television licensees (the "deintermixture" controversy). Only after a compromise was reached under the Commission's auspices did Congress respond by adopting the prearranged solution of a new law requiring that all television sets be equipped to receive both VHF and UHF signals. The role of Congress in regulating common carriers—primarily, interstate telephone service and charges—has been even more passive, allowing the FCC to shift its power in some cases from comprehensive regulation to limited competition with virtually no legislative mandate, guidance or reaction.

Yet policy requirements involving 1) equal time for political broadcasts, 2) fairness in airing all sides of controversial issues, and 3) limits on sex and violence during family viewing hours, reveal a varying mixture of legislative command, pressure, and oversight. Because of the possible effect on the political process—and every Congressman's chance for reelection—Congress has closely regulated access to station time for political broadcasts (Section 315 of the 1934 Communications Act). In this instance the FCC has played primarily an interpretive or reactive role.

A related concept, the fairness doctrine, provides access to broadcast time for controversial views or those opposed to a licensee's position. Its development reflects a still different approach by Congress and the FCC. This rule, which originated with the Commission, is now not only accepted by the Supreme Court but also is considered an essential element of the FCC's mandate by congressional committees. Congress ratified a Commission-developed doctrine and that rule now has the status of unwritten legislation. Congressional hearings and pressure from individual members of Congress forced the FCC to "initiate" the family viewing hour policy which limits sex and violence in network television programming. Given this pressure but uncertain of its constitutional role in program regulation, the Commission's Chairman encouraged networks to monitor the programs they aired during family viewing hours.[3]

As this review suggests, it seems unclear whether Congress actually has abdicated its role or merely has used its resources less directly. To the extent that Congress has left a policy vacuum, that void has been partly filled by the FCC with increasing help in recent years by other offices in the Executive Branch. A full understanding of policymaking in this area also should recognize the often significant and usually negative role of the judiciary (see Chapter 12). However, the evidence is clear that Congress until now has operated at only a secondary level in communications policy.

This limited congressional role and the absence of strong executive leadership in communications are costly. The leadership gap has slowed the development of new technology and at times has prevented its introduction. There is no evidence, however, that a specific grant of authority to one executive agency or more vigorous congressional leadership would result in a wise, comprehensive, and forward-looking communications program. For example, the question of government regulation of television program content in terms of sex and violence is a crucial current issue. Whether government can or should set limits on broadcast entertainment in a country whose Constitution prohibits abridgment of the freedom to speak involves extensive debate that will not be resolved by urging stronger congressional leadership. Communications policy is basic to the kind of society we are and will become. In a country that prides itself on openness to new ideas and people and which does not have a tradition of elitist standards, it seems highly unlikely that the legislative branch will mandate limits on broadcast entertainment. Without a national norm, it is neither possible nor desirable for Congress to force a consensus on this and similar content-oriented communications issues.[4] This illustration is not atypical. Similar debates have occurred and will continue to occur regarding such issues as a community-needs policy, the fairness doctrine, the role of cable and pay television, and the role of competition in a regulated telephone industry.

The failure of Congress to act is not merely a general legislative lack of will or insight. There are many areas in which Congress writes specific laws covering broad subjects, as reflected by the several hundred page Agriculture Act of 1977, the tax code, and the current proposals for an energy law. In communications, however, it is still unclear whether Congress can generate the necessary consensus for writing a broad-ranging communications plan. First, Congress is not a unified organization; it is a representative body reflecting numerous constituencies. Unless a consensus exists because of an immediate crisis or, alternatively, an agreement can be forged by various "tradeoffs," Congress seldom can be moved to adopt broad-ranging and far-reaching legislation.[5] Neither situation exists in communications today. Second, even specific or noncontroversial legislation requires substantial and undivided industry and agency support. With the exception of occasional successes developed from prearranged compromises, such as the all-channel receiver law requiring that television sets receive UHF as well as VHF signals, forces contending for control of communications policy have not found it in their self-interest to resolve differences.

In delegating legislative power to the FCC, Congress has neither diminished its obligation to formulate communications policy nor has it relinquished all authority to direct that policy. However, the absence of substantive legislative guidance may have made the FCC more susceptible to informal congressional control as well as more vulnerable to excessive interference by congressional investigation and oversight. The Commission, for example, is not buttressed by a specific statutory directive. Ironically, one consequence is that the FCC, which is often criticized for regulation by raised eyebrow, has itself become subject to congressional facial movements.[6] In addition, Congress' power over appropriations and confirmations can have a profound impact on communications policy.

Informal controls are often less visible and only infrequently noticed—except, of course, by the regulators and the regulated industry. Also, since there is no clear mandate in the 1934 Act, informal controls do not meet Congress' oversight responsibilities to insure that the FCC adheres to congressional intent in administering the law. As a result, congressional oversight focuses on influencing specific policy and investigating inefficient or corrupt practices.

Authorization (legislation), appropriations, and government operations committees, particularly their chairmen, exercise indirect controls on communications policy. Authorization committees are responsible for legislative control of the agencies, appropriations committees for fiscal monitoring, and government operations for investigative inquiries. Although this division of responsibility is not neatly observed in practice, it does express the general orientation of most committees. Of course, only a few senators and representatives can participate in supervising communications policy and administration. However, all are intimately affected by the role of communications policy in the electoral process and few would seriously challenge broadcast industry interests.

The many techniques available to control communications policy indirectly create numerous opportunities for congressional influence. It is clear that Congress' indirect power is both real and substantial—despite its somewhat indeterminate impact. As one commentator has described the role of Senator Warren Magnuson, the former chairman of the Senate Commerce Committee and ranking member of the Appropriations Committee, "He doesn't have to ask for anything. The commission [FCC] does what it thinks he wants it to do."[7] Former Senator John Pastore is said to have exercised similar power as chairman of the communications subcommittees of both the Senate Commerce and Senate Appropriations committees. In this process, however, the legislative power of Congress is more a threat than anything else.

The Power of the Purse—Paying to Look

Since most of the work of Congress concerns spending issues, the greatest potential for congressional control of administration would seem to be in the appropriations process. Studies of Congress, therefore, commonly conclude that the appropriations committees are "the outstanding instruments for congressional oversight."[8] In this connection, a recent study found that appropriations hearings are "currently the only routine comprehensive review of the agencies by Congress."[9]

As a process for systematic review, appropriations oversight of the FCC would seem to conform to this pattern, especially since Congress rarely has sought to influence FCC policy through specific legislation. Yet the facts of congressional review of FCC appropriations hardly fit this model. Few changes are made; little inspection, analysis, or oversight occurs.

The Role of Congress

Two characteristics of appropriations hearings deserve notice, however. First, they are only a process for review. Questions asked at the hearing usually focus on how the agency has performed in the past and whether the current request for funds is justified. While basic policy issues and priorities sometimes surface, they do not receive the committees' primary attention. The merits of specific policies or the consideration of alternative goals seldom are at issue. Second, appropriations hearings focus on budget increments. This element takes on additional significance because the initial budgetary consideration is often cursory—especially if, as customary, few resources had been committed originally. Therefore, once a program has begun and has been reviewed by the committee previously, the only question usually considered by the committee in later years is whether the increment is justified. The appropriations hearing is oriented toward the FCC's self-defined needs, rather than whether its programs are appropriate or effective. In fact, the largest determining factor for approval of the current year's budget is the budget of the previous year.

This limited, step-like process toward appropriations review is supported, and perhaps cause, by several pressures. Foremost among them is the fact that the absence of a policy consensus is likely to be even more marked at the appropriations committee level than at legislative hearings. During appropriations hearings there is no force pressing for policy consideration or agreement. In addition, both House and Senate committees with responsibility for FCC appropriations have other responsibilities.[10] One consequence is that the actual number of days devoted by these committees to FCC budgets since 1965 has declined substantially in the Senate (from 27 to 6 days) and less significantly in the House (from 73 to 67).[11] Other factors include the press of other business on Congress, the relatively small FCC budgets and the limited political rewards for vigorous appropriations review.

Selectivity, then, is the key to the appropriations oversight role. The hearings examine only new expenditures, with the committees looking for "soft spots" in the agency's estimates. Led by politically sensitive and astute FCC chairmen, congressional investigators have spotted few weaknesses in the Commission's budgets in recent years. Rather, attention has been given to other matters. For example, both House and Senate appropriations committee reports for 1976 called attention to the problem of license fees created by the Supreme Court's 1974 ruling that regulatory charges must be limited by the "value to the recipient." Having obtained no assistance from its legislative committees, the FCC apparently used the appropriations process to urge the legislative committees to provide guidance in this area.[12]

Many other agencies including the Departments of Commerce, State, and Defense, and the Federal Aviation Administration, spend substantial sums on communications—particularly on research. Yet there is little exchange of information or coordination of appropriations decisions among the various revenue committees. There has been no significant effort to define the communications role of these agencies. Similarly, the Budget and Impoundment Act of 1974, which requires legislation for "new budget authority, new spending authority, or changes in revenues or public debt limits," seems to have had no appreciable impact on communications appropriations.

Although occasionally specific actions result from appropriations oversight of communications policy, such as the pressure imposed on the FCC to settle disputed land mobile allocations, the primary impact cannot be pinpointed easily. However, Congress' limited oversight has had at least four observable effects. First, the fact of the annual review, with its potential for a comprehensive inquiry provides some assurance that the FCC will order its priorities and performance to obtain congressional approval. Second, this annual review fosters informal contact and exchanges between the agency's leadership and Congress. As a result, congressmen are not surprised by what is presented during the hearings and the Commission informally obtains congressional support. Third, consequently the views and sometimes the requests of individual congressmen carry disproportionate authority. Finally, the FCC leadership realizes the agency will encounter the least congressional criticism by avoiding controversial policies and by acting cautiously.

The Power to Confirm—Ignoring and Pushing, Yawning and Applauding (Politely)

Occasionally statements are made that policy oversight is a major facet of the confirmation process through which the Senate exercises its constitutional advice and consent power regarding major executive appointments.[13] Whether this observation conforms to practice generally is beyond the scope of this discussion, but the record of FCC appointments and confirmations contradicts the assertion that the advice and consent power constitutes a powerful congressional weapon. More compelling is the study by Graham and Kramer,[14] that the selection process is, in reality, a contest among those with powerful connections and sufficient

The Role of Congress

support to avoid political inquiry. They note that, as a result, the confirmation process is actually a perfunctory process with little attention given to the candidate's qualifications, aptitude, or regulatory philosophy. Even acknowledged mediocrity is seldom a bar to FCC or other appointments.[15]

Before considering why this process contributes so little to communications policy and whether steps might be urged to improve regulatory appointments, it should be noted that not all appointments to the regulatory agencies conform to this pattern. There have been several, though not enough, truly superb appointments that were supported by prior achievement and borne out by subsequent performance. Robinson at the FCC, Cary at the Securities and Exchange Commission, Elman at the Federal Trade Commission, and Kahn at the Civil Aeronautics Board, are examples of high-quality appointments, although most of these had no major political opposition.

The selection or appointment process by which the President determines whom to nominate cannot be separated from the procedures used by Congress to examine and confirm these appointees. For example, knowing that the members of Congress are less likely to question their former colleagues or senior assistants closely, Presidents can avoid tough questioning for their appointees and embarrassing political battles by suggesting defeated legislators or aides of senior members as regulators. This tactic also serves both to enhance political allegiances and to establish future obligations. Since a President's major programs often face detailed congressional scrutiny and have a higher priority than regulatory programs, it is not hard to understand why presidential regulatory appointments often have been both political and disappointing. Every political campaign creates debts and regulatory appointments make decent consolation prizes for the losers or rewards to helpers denied more favorable opportunities. The fact that these appointees are likely to work effectively with Congress is a bonus.

Therefore, no administration has developed a systematic process for selecting commissioners for the FCC. When a vacancy is expected or occurs, campaigns on behalf of candidates begin. The appointment process is handled, in the first instance, by a presidential assistant who reviews the names submitted and who sometimes looks at the roster of names on an inventory of available talent. Pressures will build; the trade journals, *Broadcasting* and *Telecommunications Reports*, will list the viable candidates, and the campaigns accelerate. As in most political battles, few rules have emerged and most contests are unique. If there is one common denominator, it is that the "winner" is usually determined by the political necessities existing at the time of the nomination. The candidate's regulatory philosophy is often secondary. It is not uncommon that the top contenders will have opposing or very different political and regulatory positions. The most important question is who is for the candidate and the next significant is whether the nomination will arouse serious opposition. The nomination process is a series of ad hoc decisions not designed to assure competence or qualification—and the FCC and public have suffered as a result.

Another result of legislative inattention to agency appointments is that many, probably most, confirmation hearings on FCC nominees are perfunctory at best. During one morning in the early summer of 1974, the Senate Commerce Committee considered three nominees. The hearings "stretched" into the afternoon only because one nominee persisted in reading a statement of his philosophy for the record, despite the Senators' seeming lack of interest. All were routinely approved. Recent hearings have proceeded even more quickly when the President selected the Senate majority's former counsel as one nominee and a black lawyer from a distinguished tax law firm as another. Occasional battles develop, of course. Searching questions may be asked and serious debates examining the nominee's possible conflicts of interest, regulatory views, and competence occur. These examples of serious purpose, however, are the exception rather than the rule.

A review of past battles reveals that confirmation "difficulties" usually develop only when attention has not been paid to detail or when political debts have not been honored. In most instances, probing inquiries are made and persistent questions asked not to establish a nominee's qualification, but rather to remind the President of the rules of the game and to reestablish former conditions of political payoff and compromise. Even the Senate Commerce Committee's time-honored catalog of candidate questions is readily avoided by the nominee's assurance that he or she is open-minded and will decide the issues later on their merits. It seems almost a disqualification for the nominee to have thought about the serious issues facing the agency, as if knowledge were an insuperable barrier to impartiality and success.

One is forced at some point to ask whether things are likely to improve or whether regulatory commissions must be consigned to permanent mediocrity. Although recent appointees, on the whole, seem somewhat better in ability and background, that is the usual pattern at the beginning of almost every administration. Some progress was suggested by President Carter's campaign theme of merit appointments, but that remains an unfulfilled promise. This suggests that some systematic response would be a desirable possibility.

One idea currently being considered by the American Bar Association is the establishment of a committee to evaluate appointees to the regulatory agencies.[16] If the ABA is joined by other groups, the President and Congress would be provided with independent and outside evaluations. They would focus attention on the nominees and would encourage better appointments and more rigorous confirmation hearings.

Others have suggested more formal structures. Senators Ribicoff, Percy, and Javits have proposed the creation of an Office of Presidential Nominations to assist the Senate in checking and evaluating nominees.[17] Still others, such as Common Cause, have recommended that formal requirements be established including vacancies in the *Federal Register* and developing a more open and potentially quality-related screening process. Congress currently is studying techniques for bringing public attention and pressure to bear on the appointment and confirmation process.

These formal suggestions seem likely to create still more costly bureaucratic procedures and therefore seem inadvisable. However, until some changes are made, it is not unfair to describe the process as one in which Congress ignores the FCC nominees with the exception of its own favorite candidates. Once the political compromise is reached, Congress quietly yawns and politely applauds the new commissioner.

The Power to Investigate and Oversee— Regulating by Indirection

If there is one respect in which Congress actually has used its authority to make communications policy, it is in the area of legislative oversight of agency performance. Indeed, two authors go as far as to report, "no other Federal agency has been the object of as much vilification and prolonged investigation by Congress as the FCC."[18] Almost nothing has escaped scrutiny or criticism, whether it was the FCC's organization and personnel in the 1940s, the corrupt sale of valuable rights and licenses by Commissioner Mack and others in the 1950s, the payola scandal of the 1960s, or the problems of network domination, or sex and violence in the 1970s.

This is not to suggest that these investigations or their criticism of the FCC was undeserved, or that it is inappropriate for Congress to examine the agency's personnel or performance closely. However, many investigations have gone much farther than basic policy and probed into minute details and personnel hirings. Such after-the-fact inquiries aimed at narrow programs are a poor substitute for careful examination of agency appointments or for the failure to articulate communications policy by legislation. Appropriations review does not have to be limited to a few questions about budget changes and oversight proceedings do not have to be narrowly focused, uncoordinated, episodic attempts to garner personal publicity for the investigators.

As this discussion suggests, congressional investigations and oversight hearings into FCC performance and policy—as well as informal individual pressures—are often substitutes for more direct and general policymaking by Congress in the communications field. Too frequently these indirect methods appear to involve substantial intrusions into administration decisionmaking. In spite of this indirect influence, however, Congress in fact plays only a limited role in the communications policy field.

A limited congressional role is, of course, not inconsistent with the independent authority of the FCC to make policy within the broad legislative mandate given the agency. Under the theory of agency independence and expertise, Congress should wisely limit its activities to a few vigorous investigations to insure honesty, and to oversight hearings when the agency strays beyond the political consensus.

This approach is not without its difficulties, however. It assumes the correctness of the expert agency model, an idea whose time may have come and gone.[19] As previously noted, even this reliance on administrative government requires some congressional guidance—beyond the "public interest, convenience and necessity" standard of the Communications Act of 1934. It also depends on effective congressional oversight, which in this case means periodic and comprehensive hearings to check on FCC policy. However, Congress has many other and often more pressing responsibilities. To maximize time and resources, congressional oversight theoretically would benefit from coordination and planning. Yet oversight of the FCC by Congress tends to respond to specific pressures, immediate demands, and the rewards of media publicity. Seldom are investigations coordinated. The pressure for publicity and reelection often means that hearings are repetitive and that committees with different jurisdictions and responsibilities vie for the same spotlight.

Effective oversight is not likely to attract media attention or provide political rewards. It requires extensive homework and study, wide access to information, a knowledgeable staff, and attention to detail that is not always exciting. Also there will not

The Role of Congress

always be a substantial or immediate payoff. Therefore, despite increasing concern within Congress and thoughtful studies, it seems unlikely that Congress will ever do more than respond to obvious crises or glaring weaknesses.

One can argue that this is as it should be. If the agency is to regulate and make communications policy, the FCC's leaders must be free to discuss, debate, and decide these questions. Congressional hearings are not only distracting but also may consume the agency's time and energies. Moreover, comprehensive and constant oversight is likely to lead to interference by Congress into administrative decisions as well as to result in a focus on less significant issues. Clearly, investigations and other oversight activities cannot be substituted for affirmative policymaking. Even in the ideal world where Congress would focus on policy issues and leave administration to the agency, oversight hearings act as a negative constraint. They are not designed for the development and enforcement of policy. Their purpose is to check abuses or control administrative power so that it is responsive to the electorate. Given the hazard that the hearings could be used to develop policy, the episodic and limited nature of most investigation and formal oversight may be a strength or at least is not a serious drawback. Only when the oversight function is forced to serve a purpose for which it was neither authorized nor designed, does the strain become too great.

It seems unlikely that suggested remedies to correct the so-called institutional weaknesses of oversight—inadequate staffing, the lack of independent sources of information, and failure to develop program review standards[20]—will have much effect on communications policy. Program review cannot be meaningful unless a means exists for determining legislative intent, since there is nothing with which the program in question can be compared. Nonetheless, the more general purpose of oversight, that of insuring that administrative programs are efficient and responsive to the public interest, arguably can be accomplished in the absence of a clear statement of legislative intent or constant review. Similarly, legislative hearings publicizing national problems by developing the facts and outlining the dimensions of the program, are undertaken frequently in communications. One illustration of both generalized oversight and particularized investigation was the series of hearings led by former Senator Pastore into the effects of television violence on children.[21] They educated the nation, led to the Surgeon General's report, and focused FCC attention on the question.

Reform of Legislative Control—Sound and Fury

Since it seems improbable that Congress will reverse itself and for the first time in more than half a century of communications regulation adopt a specific communications policy by legislation, additional methods of oversight potentially could develop.

One effort to reassert legislative control over the FCC and similar agencies is the proposed "sunset law."[22] Essentially a statute of limitations on government programs, the sunset law would reverse the usual presumption of continuing agency authority by terminating every program automatically unless Congress voted to continue it. The central feature of a sunset law is what its sponsors fondly call its "action-forcing mechanism."

Applied to communications policy, however, the proposals seem as useful as the now discarded PPBS (planning, programming, budget system) technique of the 1960s or the discredited management-by-objectives process of the 1970s. Even adherents of widespread deregulation do not propose elimination of all broadcast or common carrier regulation. Enactment of the sunset proposal, therefore, would not result in an abolition of the FCC. It is also unlikely that a meaningful program evaluation would occur without any prospect of termination or broad-ranging revision of the communications law. When statutes have terminated or specifically limited programs—such as the draft or the national debt limit—renewal or extension has been readily forthcoming. The only likely result of a sunset law would be some delay and a bogus congressional debate on whether to extend the Communications Act without seriously intending to let it expire. As Behn has observed, "To terminate a program is to deny some people long-held, and they believe, honestly-earned government benefits. They will fight back."[23] Members of Congress do not like to make "real enemies"—and certainly not for lost causes.

This is not to suggest that there is no utility in any sunset-type proposal. For example, if the law is targeted narrowly and is combined with limited zero-based budgeting requiring individual programs to be explained, justified, and approved, the review mechanism of a limited sunset proposal might invigorate the process of legislative oversight. However, any broad sunset proposal would promise much but offer only an opportunity to negate an entire regulatory program, an action which seems unlikely to lead to general regulatory reform.

A somewhat more sophisticated approach is that suggested by the "legislative veto," that reserves for Congress the right to review and to veto potential agency rules and policy before they are implemented. The purpose of this proposal, already incorporated into numerous statutes, is to give Congress an opportunity to void agency policies which, in its judgment, exceed statutory authority or implement unsound policy.[24] The legislative veto proposal is, in part, a response to the inability of Congress or the courts to control the delegation of lawmaking authority to the agencies. Instead of controlling agency power in advance within the statute creating the agency, Congress would control policy—except adjudication which is constitutionally protected—after the agency's expert staff and members of the public had an opportunity to contribute.

Aside from possible constitutional objections, the legislative veto approach raises serious practical questions to the extent it would apply to agency rules adopted after notice-and-comment proceedings. First, it is only a negative check on policies proposed by the agencies; it is not a substitute for legislating policy directly. Second, a study of the uses of this veto power in five separate contexts[25] shows that it spurs negotiation and compromise and, occasionally, results in serious impasse between the agency and Congress. This in turn creates opportunities for circumvention of public participation and for influence by interest groups that use Congress to pressure agencies. Serious problems exist also in that the veto process is inconsistent with effective judicial review and has a major impact on agency policymaking. The last point is most important. The FCC often has been criticized for pursuing an ad hoc, "rudderless" course. The FCC uses rulemaking frequently but usually for technical and noncontroversial questions. Vigorous policymaking, especially by rulemaking, would stir interest group pressures and congressional opposition. Numerous examples of FCC efforts to avoid congressional displeasure suggest, therefore, that the availability of automatic review of FCC rules by Congress is likely to encourage less rather than greater agency vigor. Agencies are also likely to rely on constitutionally protected adjudications rather than on rulemaking to establish policy. This would represent a substantial loss because the public would be deprived of a full opportunity to participate and the agency would be confined by less flexible procedures. In addition, oversight would not necessarily be improved, since review would be limited to the subject matter of the proposals. The legislative veto, therefore, is unlikely to rescue Congress from itself.

Conclusion

The role of Congress in making and supervising communications policy is currently in a state of flux. Significant steps are being considered to provide positive direction including the enactment of legislation outlining basic policy, the establishment of tighter controls over the budget process, the provision of closer scrutiny of Presidential appointees, and the development of additional procedures for oversight of agency programs and performance. One suspects, however, that a return to these questions a decade or two from now would not show great improvement or even many changes. However, some proposals for change may result in incremental improvements in the legislative direction of communications policy. Yet the fundamental pressures that have led to a limited congressional role—the absence of a crisis or available compromise, the numerous demands on Congress' attention, and the political nature of Congress—are not likely to change. The influence of these pressures on Congress is inherent in the tripartite structure and in the system of checks and balances designed to preserve a democratic government. This influence is increased by the constitutional requirement of frequent congressional elections. Nonetheless, Congress has failed to write basic communications policy or to demand more qualified FCC appointments. Supporting democratic values should not mean that inadequate congressional performance should be accepted. Certainly, the public has a right to expect more from its elected representatives.

Notes

1. Other executive departments, such as Commerce, Defense and State, have contributed to communications policy, but the FCC has played the major role. The authorities and resources of the Office of Telecommunications Policy (OTP), established in the White House in 1970, were transferred in April 1978 to the National Telecommunications and Information Administration (NTIA)—newly created within the Department of Commerce." 553, 90 Stat. 2541 (1976), 17 U.S.C. §§101-810.

3. See Writers Guild of America, West, Inc., v. F.C.C., 423 F. Suppl. 1064 (Central Dist. Calif.), 1976, appeal pending, No. 77-1602 (9th Cir.).

4. For a careful analysis of this point *see* L. Jaffe, "The Illusion of the Ideal Administration," *Harvard Law Review*, Vol. 86, 1973, p. 1183.

5. Congressman John Brademas, the House Democratic Whip, aptly described the difficulties of adopting far-reaching legislation when, in discussing the impasse on energy legislation, he commented:

> It is tough enough with separation of powers and the absence of disciplined parties to enact legislation when there is some national consensus. It is really difficult when half the people don't even believe there is a problem.

6. *See* E. Krasnow and L. Longley, *The Politics of Broadcast Regulation*, New York, St. Martins Press, 1973, p. 57.

7. L. Kohmeier, *The Regulators*, New York, Harper and Row, 1969, p. 67. *See* also J. Landis, *Report of Regulatory Agencies to the President-Elect*, 86th Cong., 2d Sess., Committee Print, 1960, p. 53: The FCC "has been subservient, far too subservient, to the subcommittees on communications of the Congress and their members."

8. Macmahon, "Congressional Oversight of Administration: The Power of the Purse," *Political Science Quarterly*, Vol. 58, 1943, pp. 160, 380. *See* also, Richard Fenno, *The Power of the Purse; Appropriations Politics in Congress*, Boston, Little Brown, 1966, p. 17: "The [Appropriations] Committee is expected to keep itself and the rest of the House informed on the way in which executive agencies use the money granted to them and, indeed, to influence the use of public funds in ways prescribed by law."

9. *Study on Federal Regulation*, Vol. II, "Congressional Oversight and Regulatory Agencies," 95th Cong., 1st Sess., Committee Print, 1977, p. 42.

10. These subcommittees are also responsible for several other agencies including the Federal Trade Commission and the Securities and Exchange Commission, as well as for large executive departments such as State, Justice and Commerce. With a small staff and divided responsibilities, most of their time is consumed by the departmental budgets. *Id.* p. 24.

11. *Id.* p. 22.

12. *See* House Rep. No. 94-1126, 94th Cong., 2d Sess, 1976, p. 34; Senate Rep. No. 94-864, *id.* p. 49.

13. *See*, for example, G. Mackenzie, *The Appointment Process: The Selection and Confirmation of Federal Political Executives*, unpublished thesis, Harvard University, Department of Government, 1975, cited in *Study on Federal Regulation, supra* note 9, p. 59; J. Harris, *The Advice and Consent of the Senate*, Berkeley, University of California Press, 1953.

14. J. Graham and V. Kramer, *Appointments to the Regulatory Agencies: The FCC and FTC (1949-74)*, 94th Cong., 2d Sess., Committee Print, 1976.

15. As Graham and Kramer carefully demonstrated, "[p]artisan political considerations dominate the selection of regulators to an alarming extent. Alarming, in that other factors—such as competence, experience, and even, on occasion, regulatory philosophy—are only secondary considerations. Most commission appointments are the result of well-stoked campaigns conducted at the right time with the right sponsors, and many selections can be explained in terms of powerful political connections and little else...." *Id.* p. 391.

16. *See* E. Gellhorn and R. Freer, *Pre-Appointment Review of Appointments to the Regulatory Commissions*, unpublished draft prepared for the Section on Antitrust Law of the American Bar Association, June 15, 1977.

17. See Remarks of Senator Ribicoff, "(Proposed) Senate Resolution 258," *Congressional Record*, S14415 (daily ed. September 8, 1977), p. 123.

18. Krasnow and Longley, *supra* note 6, p. 58. *See* also W. Emery, *Broadcasting and Government*, 1961, pp. 395-96.

19. *See* generally R. Stewart, "The Reformation of American Administrative Law," *Harvard Law Review*, Vol. 88, 1975, pp. 1669.

20. *See,* for example, J. Pearson, "Oversight: A Vital Yet Neglected Congressional Function," *Kansas Law Review*, Vol. 23, 1975, p. 277.

21. *See* E. Krasnow and H. Shooshan, "Congressional Oversight: The Ninety-Second Congress and the Federal Communications Commission," *Federal Communications Bar Journal*, Vol. 26, 1973, pp. 81, 105-06.

22. *See,* for example, Colorado Rev. Stat. art. 24-34-104 (Supp. 1977); Senate Bill 2925, 94th Cong., 2d Sess., 1976; Senate Rep. No. 94-1137, *Id.* p. 22. For a cogent critique from which I have borrowed, *see* Behn, "The False Dawn of the Sunset Laws," *The Public Interest*, Vol. 49, Fall 1977, p. 103.

23. *Id.* p. 116.

24. For an examination of the practical problems created by legislative vetoes of agency rules, *see* H. Bruff and E. Gellhorn, "Congressional Control of Administrative Regulation: A Study of Legislative Vetoes," *Harvard Law Review*, Vol. 90, 1977, p. 1369.

25. *Id.* (Reviewing the use of vetoes in programs administered by the Office of Education; Health, Education and Welfare; Federal Energy Administration; General Services Administration; and Federal Election Commission.)

Part 5.
Communications Issues, Institutions, and Processes: An Overview

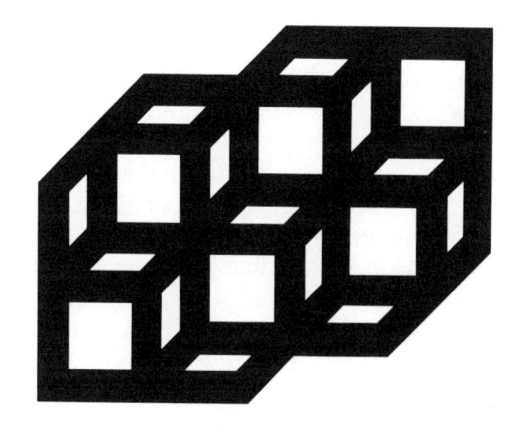

14.
Communications for the Future: An Overview of the Policy Agenda

Glen O. Robinson

It is a convention of historians to label historical periods according to those events or social attributes which seem most distinctively characteristic of the age. For the careful student of history, such labels can be somewhat misleading—obscuring relevant events and characteristics beneath the superficial gloss of words like "renaissance," "age of rationalism," "industrial age," etc. But for the most part, the commonly used labels of history serve a useful purpose as convenient generalizations about the past. When it comes to describing both the present and the future, however, the use of labels becomes a bit more problematic. Lacking the perspective of historical distance from the age described, we run the danger of confusing purely epiphenominal events with the deeper, more enduring, attributes of the period we are describing.

Contemporary sociology and political science abound in labels for our time and the time to come—each label having its distinctive emphasis on those social characteristics that are perceived to predominate now or will predominate in years to come. Among those characteristics, one that has moved to the forefront in discussions of economic and technological trends is the increasingly central role of information—its production, manipulation and exchange—in our economy and our social organization. The importance of this role has been recognized in qualitative terms by scholars such as Daniel Bell for some time. But only recently has the magnitude of information-based activity been given quantitative dimensions through the seminal work of Marc Porat.

In a study of national income accounts whose results are described in Chapter 1 of this volume, Porat has estimated that information activities (including, but not limited, to those directly involving communications) now account for nearly one-half of the Gross National Product and involve almost one-half of the labor force. We have become, he concludes, an "information society." The accuracy of Porat's data may be debatable insofar as it rests on numerous changes in conventional economic classifications. (For example, physicians' services are partly reclassified as an information activity based on time budget studies showing the portion of time spent in such "information activities" as performing diagnoses, reviewing patient histories, etc.) However, even allowing for debate over such reclassification of accounts, Porat's central conclusion seems unexceptionable. One need only casually examine our modern economy to find that it rests to a growing degree on the production, processing and dissemination of information. While one would not, of course, ascribe the characteristics of the modern American society to the entire world, all indications are that technical and economic advancement world-wide is in the same direction—with ever greater reliance on information technologies and services.

If the trend seems clear, the public policy implications that attend it are not. Exploration of those implications to date has been rather limited. However, I shall not attempt to deal with the macroeconomic issues here that Porat's discussion suggest. Here, I merely review some of the particular communications policy issues—issues which have especial public importance when set in the economic and social framework that Porat presents.

The pace of technological innovation in this area is rapidly and relentlessly pressing for a change in our ways of thinking about communications and information. Nowhere is this more evident than in the field of electronic communications, which is the

Communications for the Future

central concern of this book. The 1934 Communications Act was premised on technologies of limitation. Since radio waves easily nullify each other radio stations were required to be licensed. Because one network of wires easily connects all telephones, point-to-point communications markets were regulated to discourage duplication. But the philosophy behind these restrictive provisions was not conceived or designed for such diverse technologies as microwave, satellite, microprocessor, coaxial cable, or optical fiber. The explosive growth in data communications generated by the computer was not considered at that time. Services such as broadband cable, citizens band radio, electronic funds transfers, video discs, facsimile transmission, teletext, or telemedicine were not anticipated. This mix of old regulatory concepts with new technologies has yielded the current anomaly of rationed abundance.

It would seem, therefore, that the first question for policy analysts is how to cope with the new environment of abundance in communications and information services described by Walter Baer in Chapter 2. Many now doubt that the full promise of new communications technology ever will become a reality. This skepticism is understandable given the recent history of cable television, once widely perceived as the vehicle for delivering this abundance to every American office and household.

The development of cable communications in the 1960's brought with it glowing projections of a "wired nation" in which nearly everyone would be linked to a broadband transmission system carrying communications and information services. The first flush of rosy enthusiasm has faded somewhat in the 1970's as cable has not expanded as fast as some of its early boosters have predicted. Nor has it brought forth the innovative new services that were at first promised, as its critics have been quick to note. The reason for cable's "halted growth"—if we can so describe its failure to meet original expectations—are doubtless multiple. Many critics have been quick to place the responsibility on the FCC and the regulatory restraints it has placed on cable since the mid-1960's. There cannot be much doubt that the regulatory restraints placed on cable television services in order to prevent "disruption" of over-the-air broadcast service have hindered cable's growth. Whatever the wisdom of the FCC policies—about which controversy has never subsided—their clear effect has been to place a substantial burden on the development of cable in most of the nation's largest urban areas. Even without ready access to those markets, cable has continued to grow, but at the present time it still remains an auxiliary service. (At the present time, penetration is less than 20 percent of U.S. households. The annual growth rate in number of subscribers has averaged about 15 percent since 1970. However, the growth of new systems is but a fraction of this rate.)

In fairness to the FCC and in deference to the whole truth, however, it bears notice that federal regulation is not solely accountable for cable's present status. Local regulatory authorities have often been shortsighted in seeing cable as the goose with the gold egg which, if squeezed hard enough, could be made to produce tax revenues and/or services aplenty. Cable entrepreneurs themselves bear a considerable responsibility for that perception. In the go-go spirit of the early years, cable boosters sold the promise of new communications services like the snake-oil salesman of frontier days. A rule of caveat emptor is no doubt good sense in such cases, but it comes with poor grace for the seller of puffed goods to argue that the buyer should not believe what he says. Compounding the problem of government constraints have been the high capital costs faced by the industry throughout its growth period (and cable is exceptionally capital intensive compared to broadcasting).

So it has happened that celebrations of the coming abundance have given way to lamentations of its passing. The pendulum of attitudes may have swung too far. The promise of the nationwide system, carrying diversified entertainment, news, information, education and special services (telebanking, teleshopping, telemedicine and tele-nearly-everything) may be dimmed somewhat but it is not dead. For one thing there are signs that the political and regulatory environment which has been so inhospitable to cable is subtly changing towards more favorable recognition of the advantages of broadband communications. For another, the technological picture is changing also as we shift from the "first generation" technology of coaxial cable to newer generations of broadband transmission systems of the kind described by Walter Baer—most notably optical fiber and direct broadcast satellite systems.

As such new systems emerge, the technical and economic potential is created for both a cheaper and more capacious system (or systems) which, in time, could be as universal as telephone service. Indeed, it may well be that broadband fibers will initially be installed purely for conventional telephone service. Once in place the capacity should be used for more than simply telephony. Some have imagined in fact that such a broadband system will become an "electronic superhighway" bringing an even greater wealth of information, entertainment and related communications services than was projected a decade ago. Assuming this is so, what will be the character of future communications systems? What are their social, economic, and legal implications? What kind of public policies will (or should) shape their development and operation? These questions have been discussed at length in preceding chapters. My main purpose here is to synthesize that discussion and to add some observations of my own. As in the previous discussions, so also in what follows, many of the key questions are left unanswered or—in the fashion of lawyers—are answered only with other questions.

Designing New Communications Systems

Delivery Systems and Market Structure

As Walter Baer points out, advances in communications technology are bringing us more transmission capacity and more choice of transmission modes. There is, however, a tension between greater capacity and choice of different modes—a tension that exposes a basic policy issue that must be confronted. The enormous capacity of modern transmission technologies such as optical fiber suggests to some a future convergence of communications modes that are today quite distinct. Baer indicates that any significant convergence between telephone and cable television is unlikely in the next decade. However, it is not unforeseeable by the end of the century. This is particularly the case given the rapid advances in optical fiber that provide greater and cheaper broadband capacity capable of handling both telephone and television services. A further implication is that one system is quite capable of handling all or substantially all transmission needs and, therefore, future communications delivery should be organized monopolistically.

Although sometimes expressed as a technological imperative, the concept of a single monolithic system is not compelling technologically. One can imagine multiple media providing different or even overlapping services as they do now without any technical difficulty. Basically the argument for a single system rests on economic considerations. The economic argument for a single delivery system is based on the premise that such a system offers economies of scale so large that efficiency dictates a natural monopoly. Switched, public telephone service today is an example.

The economic efficiency argument raises two questions. One, would a single electronic delivery system yield economies of scale so large that it constitutes a "natural monopoly"? Two, even if large scale economies exist, are there offsetting efficiencies for having more than one system even in the face of scale economies (for example, might competitive incentives for technological innovation and competitive pressure on pricing and service offset the economies of scale yielded by a monopoly)? These questions are essentially those now being raised over the respective roles of competition and monopoly in telephone service. However, contemplating future "electronic highways" with capacity for a multitude of different communications services, considerably raises the stake in the outcome of the discussion. If the increased transmission capacity brings economies of scale of the magnitude that many project, the logic of the natural monopoly argument clearly would embrace more than conventional telephone service.

Any consideration of competitive versus monopolistic organization should include different components of the delivery system. For example, it has been suggested that local distribution might be integrated efficiently into a single, monopolistically controlled system, while multiple long-distance systems would be competitively organized. At least in the case of point-to-point telephone type services, present policy appears to recognize a difference between local and national systems, with competition being permitted only outside the "local loop." On closer inspection, however, the apparent present policy distinction between local and distant delivery disappears for most long-distance interstate service is as closed to competitive entry as local service. The only distinction drawn by present policy is between general and specialized services (the latter being open to competition).

Current policy thus suggests that competition should be independent of local-nonlocal service considerations. A different conclusion would be based on the premise that local and nonlocal transmission service involve substantially different economic structures. That premise has not, in my opinion, been solidly established. If economies of scale are, as they frequently are supposed to be, related primarily to capacity, there seems little basis for differentiating between local and nonlocal markets. If any difference in structure is justified, it should not be based on the geographic extent of the market but on distinctions in services. But if service distinction is the basis for defining the respective roles of competition and monopoly, what are the relevant characteristics of service differentiation? Are the present distinctions between switched "public" service and private-line service appropriate for the system(s) of the future?

Another important consideration in evaluating different market structures is the impact of further technological evolution. One of the difficulties of any discussion of economic structure is that it must be pegged to defined technologies in a field of rapid innovation. With some difficulty, we can project future technology from present technological trends. However, these projections could trap us into shaping an economic structure unsuited for unforeseen alternative technologies. While ground-based "electronic superhighways" may be the primary delivery system twenty years from now, the evolution of microwave radio, particularly satellite broadcasting, suggests that the electronic highway could be transmitted from the sky. Instead of a fiber conduit bringing the promised abundance of communications services, microwave transmission might relay from satellites to small roof-top receivers located on the subscriber's roof or at a central location and broadcast to users via terrestrial microwave. How would such an alternative system affect the economic structure question? Would the same scale economies exist for that system as for a purely terrestrial system based on fixed conduit? (Notice that in the case of satellite systems, any economic difference between local and long-distance service virtually disappears.) More important, are the two systems together compatible as competing delivery systems?

Aspects other than economic efficiency must be considered. Prominent among these is social "equity" in the distribution of communications services. Considerations of social equity are a central part of the present debate over competition in private-line telephone service and terminal equipment supply. As Lee Johnson explains in Chapter 3, one of the most vigorously contested issues today is the impact of competition on the structure of telephone rates. The telephone carriers argue that competition will result in an increase in local residential telephone rates by eroding the revenue base from which residential service is subsidized. The FCC and other proponents of competition argue the contrary, that competitive services do not in fact subsidize residential service and, in any event, competition would not affect the telephone carriers' revenue base enough to cause a substantial increase in residential service costs.

The present debate over the factual questions whether a subsidy exists and whether it would be affected by competition is not especially important to the future context of policy as I have portrayed it. What is important in the context of future services is the normative question, which has received rather little attention in the present debate: Should certain services, certain users, be subsidized? If so, how and by whom?

These questions are obviously germane to the issue of industry structure. Whether or not present competition will undermine the basis for subsidized service rates, there cannot be much doubt that full competition is fundamentally at odds with the discriminatory pricing structure needed to maintain internal subsidization. Competition is not, of course, incompatible with subsidization in some form. Certain services or users could be subsidized directly from public funds. Also, a monopoly structure does not necessarily dictate internal subsidization. But if subsidization is desired, and if internal subsidization is selected to implement it, some form of monopoly over the relevant services seems required. This implication gives a sharper focus to the questions just raised. Instead of asking merely whether some subsidization is desirable and, if so, how it might best be accomplished, we must ask whether there are any advantages to internal subsidization which would justify a monopolistic industry structure and pricing system.

We cannot contemplate the possibility of monopoly structure for our communications system without considering government regulation and control. The central issue is the adequacy of the regulatory scheme generally. For many years, doubts have been raised as to adequacy, in terms of size and professional capability of present regulatory agencies to cope with the increasingly complex problems of communications. Some of those doubts stem from the perceived mismatch between regulatory agencies and regulated firms. In turn, these doubts have yielded further skepticism about whether regulation could be relied on to control the "super monopoly" that some project for the future. It has been asked: If the FCC has difficulty regulating AT&T (the largest privately owned firm in the world), how could it regulate an entity with the combined resources of AT&T, the specialized common carriers, and the broadcast and cable industries? The question is misleading, for a single system confined to a pure delivery function does not necessarily imply an entity larger than, or even significantly different from, the present AT&T. (I assume that such a delivery system would not be responsible for production or programs or service, but would serve as a passive conduit, like the telephone. It seems virtually unimaginable that a single, all-purpose carrier would have a major role as a service provider as well.) The possibility of an all-purpose medium capable of performing the same transmission functions now performed by different industries and firms implies nothing whatsoever about the size of the organization(s) controlling it. Hence, it does not follow that the melding of many different communications systems into one system presents problems based on the size of the enterprise alone.

The regulatory problem probably resides less in any increased imbalance in the respective powers and capabilities of regulator and regulatee, than in the expansion of monopoly-controlled enterprise, the contraction of competitive market controls and the increased burden of regulation. However, past experience is ambiguous on this score. One factor behind the FCC's policies of the past decade to promote competition in specialized telephone services was the perception that the FCC was unable to regulate AT&T effectively by itself. Competition was embraced not simply for its general social benefits, but as a pragmatic regulatory tool through which the FCC could escape some of the burdens of regulation through substitution of competitive market control of service quality and rates. Unfortunately, the introduction of partial competition increased the FCC's regulatory burden because it required regulatory surveillance of the terms and conditions on which competition should take place. The FCC has been required to define the respective market boundaries of competition and monopoly and to scrutinize the competition to insure its "fairness." Some of this competition-induced regulatory responsibility may be transitional and may diminish as the "rules of the game" become established more clearly and more securely, but this remains to be seen.

Essentially the same policy issues outlined above are applicable to the field of mobile radio, discussed by Ray Bowers in Chapter 8. However, until recently, policymakers have given rather little thought to the structure of the mobile radio industry. There has been recurrent controversy over the role of radio common carriers and about the absence of competition in paging and mobile telephone service. It was this concern that led the FCC to create a special class for special mobile radio independent of the radio common carriers. The new system provides dispatch service, similar to that provided by radio common carriers, on a competitive, open-entry basis. Perhaps more important than what the Commission did in opening this part of the mobile radio market to competition, was its approval of the cellular system designed by AT&T for providing dispatch and mobile telephone service. It remains to be seen whether AT&T will establish a monopoly position in cellular technology. Looking to the possibility of different systems in different markets and, conceivably, even in a single market, FCC rules allow for competitive development of experimental systems. It is far from clear whether meaningful competition actually will develop. Many think that, at least within a single market, cellular technology offers economies of scale that require it to be treated as a natural monopoly. Questions similar to those discussed earlier are appropriate.

Assuming that there are scale economies which would justify natural monopoly structure on conventional efficiency grounds, are there any offsetting benefits of competition?

If scale economies exist within a local market, do they also exist over the country as a whole, enough to justify extending the monopoly status nationwide?

What should be the relationship between the structure of mobile radio systems, such as the cellular system, and the fixed network of communications services—the "electronic highway"—discussed earlier? If the latter network is monopolistically controlled, what implications does that have for the structure of the former?

Finally, what are the implications of new technology for economic structure? The hazards of designing economic structure on the basis of uncertain projections of new technologies was mentioned earlier in the context of fixed broadband systems. The same caution applies here. Therefore, one must be careful about conclusions of economic structure based on conventional mobile service systems, or even on the new cellular system. What would happen, for example, if instead of the cellular system being designed by AT&T, someone else were to propose establishment of a packet radio system similar to that being developed by the Defense Department? Possibly it would not alter the underlying economic circumstances—a packet radio system might, for example, be as much a natural monopoly as some think the cellular system is. Then, the pertinent question would be merely a matter of whose monopoly it should be. But the possibility of an alternative technological system must at the very least make one careful not to construct an economic and a regulatory structure that might lock in one technology and, correspondingly, lock out others. How can this be done? Is it possible and practical to design a regulatory structure that permits the efficiencies of a natural monopoly from preserving that monopoly beyond its "natural" life?

The U.S. involvement in international communications, which Henry Goldberg examines in Chapter 4, raises issues closely parallel to those concerning purely domestic communications. In both, the market is highly concentrated and has been dominated by AT&T. Also in both communications systems, the FCC has attempted to introduce a modest degree of competition—by providing record services—to the intense dissatisfaction of the industry. Apart from these broad similarities, however, the organization of international communications is beset by peculiarities without counterpart in domestic communications.

For example, the international sphere is artificially divided into voice and record communications. Voice communications are the exclusive province of AT&T, while record communications are dominated by a small group of international record carriers (IRCs) which historically have operated essentially as a cartel. Although the FCC in recent years has taken steps to foster competition, both among the IRCs and from other carriers, this effort has been impeded by several factors—notably statutory restrictions and lack of cooperation by foreign correspondents. Since all international communications require the cooperation of foreign correspondents, the FCC's entry policies are bound by foreign interests. The foreign correspondents are government-owned monopolies (under the control of ministries of post, telephone, telegraph—PTTs) that have little sympathy for U.S. policies of competition and diversified service supply.

A second peculiarity of international communications is the distinction between cable and satellite transmission. The distinction is artificial in that the choice of transmission modes is not determined by technological imperatives or economic efficiency but by a complex set of institutional arrangements, regulatory policies, and political considerations which defy easy rationalization. Central to the somewhat peculiar institutional structure of international communications is the unique role of Comsat which, as the "chosen instrument" of U.S. participation in international satellite communications, has a monopoly with respect to such service. In contrast to the other participants in international communications, however, Comsat is a "carriers' carrier"; as such it cannot offer satellite service directly to users. The choice of satellite or undersea cable rests entirely with the carriers. Because each of the carriers has a capital investment in cable circuits, they have a strong economic bias to favor cable over satellite, a bias that is reinforced by a similar preference of foreign interests. In the absence of any strong competitive pressure to use the most economical facilities, the choice of transmission modes is made on the narrow grounds that are not responsive to economic efficiency or to the communications needs of others.

There are many proposals to reform the domestic structure of international communications. As noted, the FCC has attempted to introduce some competition into international record service. It is premature to gauge the results of the recent, most significant, actions—authorizing AT&T to provide international "dataphone" service and authorizing entry by two value-added carriers (Graphnet and Telenet). Whatever rate or service improvements may be yielded by these changes, they are unlikely to alter what some regard as the touchstone to efficient international communication—more efficient utilization of satellite facilities. The FCC has attempted to effect a balanced utilization of satellite and cable by prescribing that satellite circuits be activated in reasonable proportion to cable circuits. Although it has recognized the artificiality of any mechanical proportional-fill quota, the Commission has been unable to devise a better alternative.

One proposal that has attracted increased attention is to change Comsat's role as a carriers' carrier, permitting it to provide private-line service directly to the user. This would not only increase competition in international communications, for the first time it would allow direct competition among technologies. It is debatable whether the FCC could alter Comsat's status. While Comsat's present role derives from the Commission's own *Authorized User* decision in 1966, that decision has been widely thought to mirror congressional intent in the Communications Satellite Act. However, in assessing long-term policy options it is not necessary to decide where the ultimate responsibility for change resides. The important question is whether the change is likely to yield important benefits. The FCC, which has generally supported increased competition in international communication, has recently expressed a skeptical view of the practical utility of efforts to promote further competition. Its skepticism rests essentially on two circumstances. First, there is the resistance of foreign governments to open competition between satellite and cable. Second, it is difficult to make adequate charging arrangements with AT&T and other telephone carriers so that the costs of voice communications would vary for calls routed by satellite from those routed by cable. (This problem would not arise for record communications; the FCC, however, discounts the economic importance of private record communications, the revenues of which are a fraction of those produced by voice communications.) From this pessimism about realistically developing full competition, the Commission has proposed a fundamental restructuring of international communications by merging all international satellite and cable operations into a single "carriers' carrier." In effect this would extend to all international communications the "chosen instrument" concept that presently underlies satellite communications alone. Corresponding to this change, the FCC has also suggested that it should have the central power and responsibility to negotiate with foreign organizations.

Clearly, the FCC's proposal raises some very fundamental questions concerning the future of international communications—questions that are similar to those raised for domestic communications services. Is competition among U.S. carriers a futile prospect? Apart from foreign policy considerations, what are the advantages and the disadvantages of the chosen instrument approach? What are the consequences of a monopolistic international structure in terms of services, rates, and innovation?

New Communications Terminals and Regulatory Boundaries

Questions of market organization similar to but distinct from those discussed above are raised by the rapid development of new communications and information-processing facilities. Lee Johnson has outlined the directly relevant background of controversy over competition, monopoly, and regulation in the terminal equipment markets. When the FCC decided the far-reaching *Carterfone* decision in 1968, allowing attachment of customer-supplied equipment to the telephone network, it probably had only the dimmest perceptions of what that decision would spawn in the marketplace and in the forums of government policy. But that is often the way with seminal policy decisions and it does not disparage the decision or the subsequent trends that have evolved, to borrow Justice Cardozo's phrase, "in the fullness of time." There is no reason to believe that we have yet experienced the "fullness of time" as it applies to the *Carterfone* principle. However, we have reached a level of evolution in which at least the basic issues of controversy have been identified for the next ten to twenty years.

One issue that distinguishes the controversy over competitive supply of communications terminal equipment from competitive supply of communications services is the question of technical integrity. Despite the strong arguments of the telephone companies that interconnection of customer-supplied equipment with their system may impair the technical integrity of the telephone system, as Johnson notes this is an easy argument to dismiss; not only is there no solid evidence of such impairment from interconnection, but the FCC has a program for registering terminal devices and protective circuitry to ensure that the technology is not harmed—essentially using AT&T's own standards of protection. The telephone industry continues to insist that mere fixed protective standards are not a sufficient defense against systematic pollution of the network. However, in the absence of more tangible evidence of the possibility of harm, it is difficult to impeach the FCC's judgments without being driven to the unrealistic extremes of the flat ban on all foreign electronics attachments.

Economic issues lie near the surface in the debate over technical problems of interconnection. Should the telephone carriers be able to control terminal equipment markets for reasons of economic efficiency or what we earlier called economic "equity"? The argument on both efficiency and equity grounds parallels the argument made for delivery systems but with greater force and quite different implications.

The case for carrier control of the supply of terminal equipment on economic efficiency grounds can be dismissed rather easily. The only efficiency basis for allowing carriers to have a monopoly of terminal equipment supply would be the assumption that the supply of such equipment is characterized by continuous economies of scale over the relevant range of output, and that the supply, therefore, is a natural monopoly. There is, however, no evidence that the manufacture and supply of terminal equipment yields such large economies of scale. Even the carriers cannot argue that they have a natural monopoly of all terminal equipment, such as computer terminals, since they do not even supply much of it. The efficiency argument is focused on specific "communications" equipment; but can such economies lie in the supply of such equipment if they do not exist in other general purpose terminal facilities? The integration of manufacturing, supply, and installation of equipment with telephone service may give carriers such as AT&T an advantage over other suppliers, at least with respect to ordinary telephone equipment. But if so, the advantage would appear to be based on a pricing arrangement which permits cross subsidization rather than on any inherent efficiencies in the design and manufacture of equipment.

There is one distinct argument for carrier control of terminal equipment—subsidization of certain types of, or classes of, service by means of monopoly profits on terminal equipment. This argument requires the opposite of the previous assumption, that terminal equipment is subsidized by service charges. This argument presumes that the supply of certain kinds of terminal equipment, such as specialized business terminals, subsidizes, for example, residential service and equipment. There is a raging dispute over the nature and direction of subsidies in the present pricing structure. It is yet to be established definitively who subsidizes whom; it may prove to be impossible ever to establish this definitively—particularly given conflicting definitions of subsidy and also given the changing nature of the services and their rate structure. Again, what is most notably lacking in the debate so far is an evaluation of values. It is not enough to know whether business users, for example, are subsidizing residential users. One wants to know why as well. Should users of business terminal equipment subsidize residential service? What social purposes are served by such a subsidy? If it is appropriate to subsidize certain users, who should bear the burden of the subsidy? Should the burden of such subsidy be borne internally by other users or externally by taxpayers?

Parallel to the problem of defining the respective roles of competition and monopoly in the supply of communications facilities is that of defining the role of regulation. The definition problem is most sharply presented by the computerization of communications terminal facilities and the consequent confusion between regulated communications services and unregulated general purpose computer uses—data processing. Lee Johnson traces the background of the present controversy by examining attempts to distinguish between communications and data processing. The task has not been easy. While the problem seems to be a technical one, it is not. The policy issue that underlies the controversy is simply whether or not particular services will be regulated or unregulated. For example, AT&T is bound by a 1956 consent decree which effectively prohibits it from competing in unregulated data processing equipment and services. AT&T appears to be content with the decree as long as it can manufacture and sell terminal equipment under the label of "communications" devices. But allowing it to do so may imply that regulatory jurisdiction should be extended to all suppliers of such devices—including virtually the entire computer equipment industry. The FCC has declined to assert jurisdiction over the latter despite the close interconnection with communications. At the same time, the Commission is loath to restrict communications as much as it is data processing. The problem is becoming more acute, as the forces of technology, and economics, are forcing the convergence of communications and data processing in ways that defy separation by other that arbitrary fiat.

Unfortunately, the search for separation criteria has obscured analysis of purpose. At present, boundaries between communications and data processing define the boundaries of regulation. Only recently has it been asked, why the regulatory boundaries should be set by the technical distinction between these services? Is there any economic or social reason for treating the two differently? The answer involves the earlier discussion of competition and monopoly. If it is argued that all terminal equipment could be supplied efficiently in a competitive market, then why regulate any of it? Instead of endless debate over whether some facility performs functions that are primarily communications or primarily data processing, would it not be feasible to leave both to the dynamics of an unregulated market? (This question logically assumes a change in the 1956 AT&T consent decree that would permit AT&T to compete in unregulated markets and to supply facilities without regard to whether they provide "communications" services.) Allowing such competition by AT&T would not end all problems; presumably one would still wish to avoid cross subsidization of competitive services and facilities by AT&T's monopoly services. However, that creates no problem not already present. The solution, if there is one, may involve separate corporate entities—now required of firms providing communications and data processing.

The Nonregulatory Role of Government

We have considered the regulatory role of government in shaping the structure of new communications systems. Yet to be considered is the related but quite distinct question of the nonregulatory role of government in designing, planning and developing communications systems or particular types of facilities. I here finesse the question which governments—local, state or federal—might be involved in planning and promoting the new system(s). On the strength of past experience, it is a safe prediction that most developmental efforts will be made at the federal level. However, even in an integrated nationwide system, it is possible to diffuse public as well as private investment responsibility over many different institutions.

To some degree, of course, government planning may be inherent in the task of regulating communications services. A policy to allow or disallow interconnection among different telecommunications carriers, for example, can influence the structure of the telephone system as much as more direct planning. So also the regulation of rates to permit subsidization of certain classes of users is a form of public promotion that can be remarkably similar in effect to forms of direct government subsidy—though, of course, it differs markedly in how the burden of the subsidy is distributed. However, whether such a "planning" is implemented under the guise of regulation or through more direct means, the question is still the same: How much government involvement is needed or desirable?

In the context of general system design, the case for public investment or aid is most easily imagined as a continuation of past effects to extend basic services to all areas of the country. The most notable example would be extension of service to rural areas where the sparseness of population and distances between homes and businesses create exceptional economic impediments to purely private development of communications services. If we want to have the "wired farm" as well as the "wired city" some form of public investment, perhaps on the model of REA, may be essential. (Congress' Office of Technology Assessment has proposed a series of demonstration systems to test the feasibility of bringing broadband communications into rural areas; as this is written the proposal still awaits executive and congressional endorsement.) Even in this fairly obvious case for public investment there are, however, some troublesome problems to be resolved. There is, for example, a threshold question; how far do we want to extend the same services available in urban areas into the countryside? Universality of basic communications services—telephone, postal service, and at least minimal radio-television—has long been a goal of public policy which has been promoted by a variety of means. But that goal has been essentially achieved, it is a remote area indeed that is beyond the reach of postal service, telephone and at least minimal radio and television service. The public policy question for the future is, to what extent public investment should attempt to expand those services to bring them into equality with urban services?

Assuming a public policy goal of equal service, other questions arise in considering the type of public investment to achieve it. Whatever the facility or mix of facilities chosen, any large public investment in one type of facility will constitute a substantial commitment that will not be easily changed if technological or economic circumstances warrant. Any such public investment should be made only after a careful analysis of the extent to which it allows freedom to experiment and to implement other technological advances as they occur. This caution is not, of course, unique to public investment. Any investment in facilities, public or private, can act as restraint on implementation of new technology. For example, past investment in long-lived facilities have constituted an economic drag on implementation of new or improved telephone technology. Regulatory policies favoring shorter amortization periods for equipment could alleviate the problem. However, changes in amortization rates alone may not be enough without some form of regulatory or market pressure to implement new technology.

In considering the development and design of discrete components or facilities, the question of public involvement is complicated by the fact that the federal government, as a user of services and equipment, plays a large role in research and development. Particularly in its research on military and space technology, the federal government already has played a huge role in the conception and development of new communications facilities and systems. The pioneering of space satellite technology is one general example; the development of packet switching is another, more specific, illustration. However, the successful transfer of government technology to private use has been a fortunate accident. In the case of packet switching, for example, the transfer was wholly the product of individual entrepreneurship by those involved in the research for the government. One may ardently admire the role of the private entrepreneur in economic development and still question whether we should depend on these individuals for the transfer of publicly supported technology from government to civilian use. Unfortunately, institutional mechanisms to obtain the benefits of military or space communications research and development are not now available. In fact, it seems that the idea of transfering technology from government to civilian use is not even being discussed.

Beyond the advantages to the public, is there any case to be made for a major government role in the design and development of the needed communications equipment? Clearly, the market is unable to value all the potential public benefits of new technology—what economists would label a problem of "externalities." There are numerous public services which can be and should be provided by our advanced communications system. The ability to provide them depends, however, not only on the delivery system, but on the receiving equipment as well. If the market is too small, or inadequately organized to create an effective demand for such equipment, the service will not develop—even though there may be adequate public sector support for the service. A current illustration is "captioning for the deaf." Whether the market will support the development of the equipment needed for this "teletext" service is still being debated. At least part of the difficulty may be that the service is too narrowly conceptualized as simply a service to the deaf when, as in Great Britain, it has a much broader potential for specialized services. In this country, however, the broader market may not develop on its own for a combination of reasons. Manufacturers are unlikely to promote development of equipment for general consumer use without an assurance that the service will be provided. Such assurance is not likely to come from the commercial broadcast industry, for a general teletext service would compete with regular programming without providing compensation to offset the loss of advertising revenue. A teletext service could be provided over public television, but the regular audience for public broadcasting may be inadequate inducement for developing the equipment.

The government might intervene here in one of several ways. It could simply direct all broadcasters to provide the service if and when receivers or other special equipment become available. That approach is almost certain to produce industry resistance, uncertainty, and delay. Alternatively, government might simply underwrite the development costs for providing special converter equipment or providing the components as an integral part of receiver design. (Economies of scale from mass production would undoubtedly reduce the cost of manufacture from what it would be for a limited market such as deaf users.) The government commitment would increase public pressure for the service as an adjunct of public, if not commercial, television.

Media Applications and Impacts

Program Diversity: Implications for Regulatory Policy

Any foreseeable convergence of media forces us to think carefully about the relationship between the delivery system and what is delivered. The possibility that the media will not merely converge but will merge into a monopolistically controlled medium heightens the importance of this question. In Chapter 5, Benno Schmidt advocates an enforced separation of responsibility for "content and carriage" giving the carrier no role in the production of programming or information services. Even if we assume separation of responsibility for content and carriage, questions are raised with respect to program services and their regulation. Before looking at the regulatory implications directly, some consideration of program diversity is appropriate.

It is commonplace to observe how broadband communications will enhance diversity and individual choice in television programming. Of course, cynics can be found to argue that the principal yield from new television channels will be more old movies, more super-heros, and hard-headed/soft-hearted police detectives. That may be. The fact is that popular tastes are fairly well served by a limited range of distinctive program types. In this respect, the present broadcast system scores quite well. But this does not diminish the value of increasing the individual choices. Moreover, no matter how uniform and standardized our program tastes may be, most of us harbor at least an occasional desire for something truly distinctive that the present system is ill-designed to provide. With a limited number of channels, special interests are preempted by mass audience preferences because of a programming strategy that induces program providers to seek a share of the mass audience rather than smaller segments of the audience with different program preferences. More programming channels could alter that strategy by reducing the opportunity cost of providing special audience programming.

Perhaps more important than what expanded channel capacity can do for one-way television services is what it can offer by providing two-way communications, enabling the individual user to call up programs on demand. Such programs could be provided by a special subscription service that provides a certain number of selections each week. Or they might be provided for a fee from general libraries or other sources of special interest programming accessible on a dial-up basis.

Those with a special interest in maintaining the status quo ask rhetorically why this diversified programming has not already appeared where cable television has developed the necessary capacity? One answer is that cable does not yet have either the necessary audience or the capacity to support specialized services. Also, few cable systems now have, or can now afford, the facilities necessary for two-way communications. There is nevertheless every reason to expect that this capability will develop within the next twenty years, perhaps by expanding the present cable systems or by developing a new system such as fiber optics. Assuming this happens, what are the implications? The social ramifications of increasing the quantity of information, news and entertainment, and the range of choice available to individuals, will be far reaching. The regulatory issues raised by this promised abundance and choice makes this clear.

In our new world of abundance is there any role for content regulation by the government? As Benno Schmidt notes, the accepted rationale for regulation of communications content is particular to the broadcast medium. The conventional wisdom has been that because the medium uses a scarce resource—the radio spectrum—it is necessary to impose certain regulations on how that resource is used. But can regulation have any application to a system which hypothetically does not use scarce spectrum and is not bounded by its capacity limitations? The answer is contingent on accepting the scarcity rationale as a necessary, as well as sufficient, condition of content regulation. Schmidt calls our attention to rationales of program regulation that are wholly independent of scarcity considerations—the "intrusiveness" of television into our homes (the "privacy argument") and its social, cultural, and political impact (the "power theory").

One need not endorse such rationales to recognize that they play an important role in current regulatory policy. In broadcasting there are a number of program controls that have little apparent relationship to scarcity conditions. Regulation of "offensive" or "violent" programming is one illustration. Moreover, some of the same regulations imposed on broadcasting also are imposed on cable, which does not use the spectrum. In fact, cable regulation arises from a confusion of roles, but this same confusion could carry over to the next generation of broadband communications. Even if there is a clear and sharp distinction drawn between the role of carrier and program producer, it does not follow that because the former is not subject to program regulation the latter will not be.

It cannot therefore be casually assumed that content regulation will disappear with the disappearance of spectrum scarcity. Indeed, in Chapter 6, Bruce Owen warns that the substitution of electronic media for print may lead to a similar substitution of the traditions of regulation for those of laissez-faire. Owen raises other problems that need separate discussion, but in the present context his warning reinforces the importance of several questions about the future role and purposes of program regulation.

If we assume a separation of responsibility for content and carriage and a competitive market for the supply of program services, are the "privacy" and "power" rationales for regulation sufficient to justify continued regulation of the content of electronic programming? Are there any other rationales for such regulation that are distinctive to the electronic media?

Assuming some role for continued program regulation, what should be its scope? What kinds of program regulation could be justified on a privacy or power theory? Consider the basic requirements under present regulation policy:

(1) fairness for programs presenting controversial, public issues;
(2) equal time exposure for political candidates;
(3) obligation to present news and public affairs programming;
(4) ascertainment of community needs;
(5) preservation of "unique" program formats; and
(6) prohibition of, or restrictions on, obscene, indecent, or violent programs.

This list suggests the difficulty in any simplistic extension of present regulatory policy to the kind of broadband medium that has been envisioned. Indeed, if one assumes that the carrier has no responsibility for content, most of the requirements above would be extraordinarily awkward to impose. One might well contemplate enforcing regulations 1), 2) and 6) against each program supplier, but what would be the sense of enforcing regulations 3), 4) and 5)? For example, if an entrepreneur wanted to lease channel space on the system to provide classical drama, soccer matches, and cowboy movies, would the FCC require news and public affairs, or ascertain community needs to be provided also? Would there be a prohibition on changing the programmer's "unique" format?

Social Services

It is scarcely possible to read a discussion of future communications systems in which the word "abundance" does not appear. Granting that whatever is delivered by our new media will be delivered in abundance, what kind of abundance will that be? What services will the media provide, and to whom?

Earlier, mention was made of the promise of entertainment, news, information, education, and special services—from telebanking to televoting. Specialized services—health and education are the two most notable examples—typically involve quite different infrastructures. There is a different supply and demand market for the services, a different set of public policies and, not least, a different set of government rules. Consequently, the issues surrounding the delivery of these services by telecommunications are notably different from those that are confronted by more conventional mass audience programming. There is, therefore, a need to look at the distinctive issues and problems associated with special social service applications of the kind which Bill Lucas explores in Chapter 7. What are those special applications? How may they be provided by our new communications systems? What obstacles will be confronted?

There is no doubt about the technical capability for such services in the future. Indeed, as Lucas illustrates in surveying some current cases of social services applications, the technology for a wide variety of services is available now. Our present telephone, cable, and broadcasting systems have the technical capability for many more services than are in fact provided.

We should not halt our progress towards more sophisticated, more capacious communications delivery systems merely because present service systems are not fully utilized. But the fact that we do not fully utilize present facilities is a sobering reminder that our largest problems in this area are not technological. Unless we can find solutions to the existing problems, our new electronic highway will be a very lonely road as far as special social services are concerned.

What are the problems? One conventional response has been to say that technology has been seeking the need rather than vice versa. If this is correct, it is also irrelevant. Professor Galbraith notwithstanding, there is nothing inherently wrong about a supplier seeking to develop a "need" (demand) for his project. We may rightly question some of the huckstering that goes on in the course of inventing new sources of human satisfaction; we may even doubt the social legitimacy of some of the wants thus created. But neither of these concerns is pertinent here. The problem here is that every attempt to introduce modern telecommunications in support of public services have met with institutional rigidity and resistance from public service professionals and supporting agencies.

Examples of this resistance are plentiful. Lucas cites the experience with instructional television. The field of health care delivery provides similar examples. In part, as Lucas points out, the problem lies in the design of the telecommunications service that often does not fit into the pattern of instruction or health care. Plainly telecommunications service that often does not fit care. Plainly telecommunications specialists cannot be completely in charge of system and service design, but just as plainly, the users themselves must take some affirmative steps to use telecommunications and shape it to their needs.

There are no simple measures for creating a partnership between telecommunications specialists and service providers. If the question were merely one of the telecommunications system design, it could be addressed in fairly straightforward budgetary terms. In some cases organization of the social service profession creates a barrier to effective telecommunications. In health care delivery, professional restrictions on who can provide medical diagnosis or care place severe limitations on the effective utilization of telecommunications. For example, telecommunications facilities can be used to extend primary health care to rural areas where physicians are in short supply. However, the effective use of such facilities depends heavily on the degree to which medical ethics permit para-professionals—or "health-care extenders"—to use those facilities to diagnose, screen, and prescribe care for patients. Even an elaborate telecommunications system that facilitates remote diagnosis and prescription is of very limited use in this regard if a licensed physician is required to prescribe for every headache, bunion, or bruised shin. While such problems may have solutions, the solutions do not lie in the realm of communications policy although the communications connection can be used to dramatize an underlying problem in the professional service structure.

The same somewhat disappointing conclusion seems to emerge from a survey of the bureaucratic problems identified by Lucas. No one who reads a newspaper, or views television, or listens to politicians, can be innocent of the growing concern over "big government" and the problems of the federal bureaucracy. To make matters worse, public services with which we are now concerned are heavily dependent on the biggest bureaucracy in Washington—the Department of Health, Education and Welfare. That makes such services especially vulnerable to the constraints, delay, uncertainty, and bureaucratic lethargy that is the inevitable consequence of the 1,200 pages of regulations and 12,000 pages of guidelines governing HEW programs cited by Lucas. Again, the communications problem is merely secondary and serves primarily to direct attention to the underlying problem.

Aside from specialized public services, what of public broadcasting: What will be its place, its role, in the communications of tomorrow? One might subsume this question under a discussion of educational services generally. This would reflect the origins of public broadcasting and its main focus prior to the Public Broadcasting Act of 1967. However, treating public broadcasting as merely an instructional service in a narrow pedagogical sense would reflect neither present reality nor expectations for the future. While it is still common to refer to the educational purpose of public broadcasting; the term "educational" must be very broadly conceived in this context; public broadcasting embraces a diverse range of programming that does not fit a narrow concept of education. It would in fact be more realistic to call public broadcasting "alternative choice broadcasting" in that the chief underlying rationale for publicly funded broadcasting stresses the importance of providing an alternative choice to the limited range of programming offered by commercial broadcasting.

While the public broadcast community—including the Corporation for Public Broadcasting, the Public Broadcasting System, National Public Radio, station organizations, and public advisory groups—was able to agree on the name for public broadcasting, it has had difficulty agreeing on much else about its purpose and operations. Major disagreements involve programming, the respective roles and prerogatives of the different groups, the sources and internal allocation of funds, and even the basic purposes of public broadcasting. The present troubles have been examined thoroughly and are now being reviewed further by a second Carnegie Commission. (The first Carnegie Commission study led to the Public Broadcasting Act of 1967.) The controversial issues involve organization—particularly the respective roles of CPB and PBS; funding—the allocation of funds between radio and television or between construction and programming; and programming—the amount of sports, news, or light entertainment and the time given to national versus local perspectives. In these debates, little attention is being devoted to the long-term uses of public broadcasting.

As the name public "broadcasting" suggests there is a strong bias of attention towards past and present technology and service delivery. Public broadcasting is wedded to the present broadcast structure, not by technological necessity or legal constraint, but by institutional biases. Public broadcasters as a group are broadcasters first, public service providers second, as their alliance with commercial broadcasters and their resistance to cable television manifests. Even those who are not directly involved in local delivery seem to be fixed on the notion that "broadcasting" must be a specific technology rather than a generic name for mass audience services. Therefore, little attention seems to be given to questions that probe beyond the traditional boundaries of broadcasting to consider the role of public programming in other milieu, including the broadband electronic highway.

The recently approved project that will provide satellite interconnection for public broadcast stations is a noteworthy step towards application of new technology, but it is within the conventional boundaries of the present broadcast system. It does not reflect any innovation in the concept of the public "broadcasting" role—for example, any notion that it has a role in education, or in providing public services. Lucas notes public broadcasting's efforts to develop captioning for the deaf. Important as that service may be, it does not begin to test the full potential of the underlying technology. Captioning is but one application of "teletext" service being pioneered in Great Britain (under the trade names "Ceefax" and "Oracle") for providing news and information services. Should our public broadcasting service be any less innovative in developing the full potential use of new mass communications technologies than the BBC?

These questions clearly take us well beyond the dominant conception of public broadcasting as merely providing an "alternative program choice." Logically pursued, they could lead to an expansion of public broadcasting's role beyond the capacity of any single institution to contain it; at the point where public broadcasting found itself replicating the work of other social service agencies, it would have outlived its usefulness. But if too much ambition and imagination could destroy public broadcasting, too little can have the same effect.

Certainly public broadcasting will have to be examined in light of changing technology and in the context of newly evolving service patterns. If our imagined superhighway of communications services develops, bringing with it the potential for a new diversity of programming, what then will be the purpose of public broadcasting? Would it simply be one of many users of the highway providing its own special programs? In what respect would this programming be special?

Mobile Communications

Popular forecasts of our communications future tend to be oriented toward fixed broadband systems. This probably reflects a bias in our present perspectives that have given far greater prominence to fixed broadcast, cable, and conventional telephone systems, than to mobile communications. This relative neglect of mobile radio is somewhat puzzling when one thinks of the tremendous mobility of American society, a society in which the automobile ranks as one of the most loved possessions. (A 1970 survey showed that the automobile ranks above telephone, television, newspapers, and refrigerators as a necessity, and above television programs, women's fashions, popular music, and movies as an object of most satisfaction.) There are, to be sure, some signs of change in this regard. The growth of CB radio in the past four years—since the nationwide truckers' strike in late 1973—has brought popular attention to the world of mobile radio. However, full appreciation of the present and potential economic and social importance of mobile radio services, which Ray Bowers has explored in Chapter 8, has yet to develop.

Aside from the special case of CB radio, the remarkable increase in demand for mobile radio in the past decade reflects important economic gains that can be expected to continue as a function of the level of general economic activity. Bowers offers some illustrative statistics of the economic benefits. Even allowing suitably wide margins for error, the benefits appear substantial in magnitude, and the economic gains alone do not fully reflect social benefits.

"CB" is at least partly illustrative of the point. Despite the much-heralded popularity of CB, Bowers cautions that it may not receive adequate attention or resources because it is not directly associated with commercial productivity in the same way that other mobile radio use is. This may be a somewhat misleading formulation of the problem, for CB does have economic importance which can be measured directly by the value added from equipment manufacture and related services. Therefore, CB does not lie wholly outside the GNP calculus which Bowers illustrates for other mobile radio services. Yet, in a larger sense, Bowers' caution is undoubtedly justified. First, whatever its GNP contribution, CB may be at a disadvantage relative to other users of the spectrum in competing for the favor of policymakers. Not only is CB regarded as a "mere" recreational facility, but there are many who regard it as a low order of recreational pursuit, one to be tolerated not dignified. This attitude puts CB at the marked disadvantage in confrontations with such "higher order" pursuits as viewing television or listening to the radio. Too, the faddish side of CB and the colorful folkways that it has acquired have tended to obscure the larger social significance that the marketplace has not measured and that public policy has not yet recognized. CB may well be a more forceful demonstration than television of McLuhan's famous statement the "medium is the message." For many users the message itself is unimportant, trivial; what is important is communicating anonymously yet personally to whoever is "out there." In a democracy based on individualism, there is something socially vital in allowing every person to be, in a sense, his own broadcaster.

CB may be the precursor of (if not necessarily the model for) a future system of mobile communications that is more flexible, more individual and more widely used than the system of today. On the basis of present trends, it is reasonable to project continued expansion and growth of present services identified by Bowers—paging dispatch, mobile telephone, CB. But technological developments in the past few years suggest more than mere growth of present services; they suggest a transformation in the services themselves. The development of a "cellular system" discussed by Bowers is an advanced form of multi-channel trunking which will permit expanded and more efficient use of mobile radio. Other technological developments promise even greater expansion of an innovation in mobile services. For example, the technical feasibility of linking satellites with personal, hand-held or even wrist-worn radios has already been demonstrated. Such a satellite-mobile radio system, if fully developed, would enable a person walking on the street in Bangor, Maine, to communicate with a stroller on the beach in Honolulu or an oil rig engineer working off the coast of Texas. A second development that promises dramatic expansion of mobile services is "packet radio"—employing digital transmissions for both voice and data communications, and packet switching techniques similar to those now used by certain specialized, private-line telephone carriers. Such a system offers not only a more efficient utilization of the spectrum, but the advantages of a multiple access broadcast capability (one person can send to and receive from any number of radios).

The expansion of mobile radio will not, however, be possible without a substantial change in radio spectrum allocations.

Historically, mobile radio has been distinctly handicapped in competing for spectrum with other users. In that portion of the spectrum available to non-federal government users, broadcast allocation occupy the dominant share of frequencies currently usable for land mobile use. This is in part a product of the very large spectrum requirements of television broadcast channels. However, a mere comparison of the technical requirements of individual users is far from a complete explanation for the dominance of broadcast allocations as a matter of policy, and it is hard to dispute that broadcast users, as a class, have been favored in spectrum allocations. Historically, television has had the edge because its vast needs materialized first. The present allocations were basically established in 1946, when mobile radio and broadcasting were fledglings; however, the future demands for television were foreseen and easily measurable. Not until television was firmly fixed in the minds and hearts of the public—and the FCC—did the mushroom growth of mobile radio in the 1960s create claims against the large block of frequencies allocated to television. The claims were targeted particularly on ultra high frequencies (UHF), many of which have yet to light up a television screen. The FCC has responded to the claims in several ways—most notably in the controversial "900 MHz" decision, discussed by Bowers, which reallocated a substantial segment of upper UHF frequencies and also made several important policy decisions affecting their use.

The adequacy of these measures to meet the burgeoning spectrum demand of mobile users is difficult to assess. It is partially dependent on the implementation of more efficient utilization systems—such as the cellular system or other spectrum efficient techniques. In large measure it depends on the growth rate of mobile services. If the recent past is prologue, it seems unlikely that the allocations made to date will fully meet the demand two decades from now. If they do not, there will certainly be enormous pressure on the FCC to reallocate more of the UHF band to land-mobile use. As Bowers indicates, a case can be made for such reallocation in economic terms—comparing the benefits of increased mobile radio use with the cost of moving broadcast service (or potential service, since many broadcast channels will be unoccupied) to shielded cable or fiber.

If the "electronic highway" materializes, the spectrum problem may be resolved, as mass communications ("broadcasting") moves onto broadband fiber networks. As noted previously, this might evolve both as a product and a result of efficient utilization of broadband systems—most notably systems employing optical fiber channels—and as the demand for diversified programming and other services exceeds the capability of the present separate broadcast and telephone systems. Now it appears we can add a third important reason for evolving such a system, the need to conserve the spectrum for those uses that cannot employ fixed conduit—mobile radio.

"Electronic Mail"

Among the many uses of new communications systems is a variety of message delivery and information transfer services explored by Henry Geller and Stuart Brotman in Chapter 9. For want of a better term, these services are labeled "electronic mail." The term, though common, is a bit misleading in two respects. First, mail historically is associated with physical delivery. Although electronic message delivery involves facsimile communications, the dimensions of an electronic system are far beyond facsimile communications. Second, the term "electronic mail" tends to suggest some novel special technology for information transfer or message delivery; yet electronic message delivery is as old as the telegraph and as common as the telephone. Recent advances in both terminal equipment and in transmission systems have expanded the capability for delivery that will be expanded even further. But the impetus behind the growth of electronic mail is less a matter of changing technology than of changing economics. Above all it is a change in the relative costs of physical and electronic mail carriage. As the costs of electronic mail delivery have declined steadily because of technological improvements, the costs of physical delivery have increased dramatically, primarily because of rising labor costs which account for some 85 percent of Postal Service operating costs.

The substitution of electronic delivery for physical delivery has numerous advantages, both in terms of economic efficiency and social convenience. It also presents a number of vexing policy problems that parallel the commercial and social problems of electronic funds transfer discussed by Geller and Brotman. Clearly, a fully developed electronic funds transfer system will have a significant impact on banking and commercial transactions. However, there are competing speculations about the impact of electronic funds transfer on the economic structure of banking: by facilitating branch banking, electronic mail could increase competition—facilitating entry into markets from banks in other markets; but opponents insist that liberalized branch banking will simply lead to a takeover of markets by the big banks. There are competing arguments on social utility:

electronic funds transfer could make credit available more cheaply and could reduce the incidence of cash-related crimes, yet it will eliminate the time-honored practice of the "float" and it could increase the incidence of computer crimes.

A most troublesome aspect of electronic funds transfer is one endemic to all forms of electronic mail—the problem of privacy. Assurances by electronic engineers that electronic messages can be encoded to make them secure both in transit and in storage have not eliminated anxiety about invasion of privacy. Part of the problem of making information secure is not a matter of technical capability but human disposition. No doubt some of the fears are exaggerated. Images of *1984* may have influenced our perception of the future too much. And invasion of privacy is not uniquely linked to the wizardry of computers or electronic communications—as the Watergate "burglary" should remind us. Nonetheless, there is a legitimate concern that modern electronics could facilitate privacy intrusions and enhance their effectiveness. This is the central fear of data banks; they provide essentially "one-stop shopping" for the privacy intruder. Of course, computerized data are not specially an initiative of electronics delivery. Nevertheless, electronic delivery and data storage are intimately related; and the security concerns are similar.

Among the policy issues raised by electronic mail none is more serious, yet seemingly intractable, than the impact of electronic delivery on the Postal Service. The problems of the Postal Service are well known, if not yet widely understood. A special presidential commission report on the Postal Service and a recent study of electronic mail by the National Research Council recognized that electronic mail is both a threat and an opportunity to rescue what most people believe is a decaying mail service.

There are several alternatives open to the Postal Service. It can respond to electronic mail by ignoring it. It is generally agreed that this course of action will result in growing diversion of mail from the Postal Service to electronic carriers. Even if the USPS could improve the quality of its service and maintain postal rates—which would require billions of dollars a year in new public subsidies—it cannot avoid substantial impact in the next decade. The significant loss of business will increase USPS losses—and subsidies—and/or decrease service.

A second alternative discussed by Geller and Brotman is for the USPS to seek limited participation in electronic mail by providing, perhaps in joint venture with an electronics carrier, a combination of electronic and physical mail delivery. "Mailgram" service now provided by Western Union in cooperation with the Postal Service is the conventional model. It is questionable whether this form of limited participation is fully responsive to the diversion threat. Much of the "transactions" mail that is likely to be diverted does not require any physical delivery; the message is sent "terminal to terminal" with physical facsimile produced, if at all, only at the receiving end.

This last point suggests a third alternative open to the Postal Service—full entry into specialized electronic message services in competition with communications carriers such as AT&T, Western Union, Telenet, Satellite Business Systems, and others. Needless to say, the prospect of entry by a firm the size of USPS (annual revenues currently exceed $11 billion) would raise more than a few concerns similar to those discussed earlier and reviewed in greater detail by Lee Johnson in Chapter 3. For example, steps would have to be taken to separate USPS monopoly services, or subsidized services, from competitive business to prevent unfair subsidization. As indicated by the age-old controversy over cost allocations in telephone rate determinations, this is not an easy task. And there is a problem of jurisdiction: Postal Service rates are now within the exclusive jurisdiction of the Postal Rate Commission, while electronic carriers are regulated by the FCC and state regulatory agencies. Such a division of responsibility over competing modes is not unprecedented—for example, domestic transportation regulation is the divided responsibility of the CAB and ICC—but it is not without its problems.

Is full competitive message service by the USPS a practical reality in any event and is it worth the effort? It may be that it is too late for the Postal Service to become a major participant in electronic communications. It is certainly open to question whether the USPS could compete effectively with established, experienced electronics firms. Not least of USPS problems would be the economic burden of a labor force which is unnecessarily large for electronics delivery.

There is a body of thought that holds it unnecessary and inappropriate in any event for a quasi-governmental body such as the Postal Service to enter into the private sector in competition with private industry. But this still leaves the question: Why must the USPS be considered for this purpose a public or quasi-public enterprise? Clearly, as the character of our mail delivery changes, so must the Postal Service as an institution. All of the alternatives discussed would require a reduction in the size of the Postal Service labor force. This need not mean loss of employment for current workers for any attrition will be phased in during the next decade by retirement. What would be a sensible workforce level ultimately cannot be guessed now; it is totally dependent on which of the above alternatives is chosen, the level of service maintained, and the amount of subsidies the public is willing to pay.

A final choice could substantially alter the future of USPS; although it is somewhat beyond our subject, it bears mention. Instead of pursuing electronic mail options, where its workforce is largely a liability, USPS might seek out other local service functions where its labor force would be an asset. These might include, for example, "non-mail" pick-up and delivery services for the public generally or for business. Of course, this now raises a new set of problems for the transportation industry and no doubt there would be resistance from that sector. But there are no simple solutions that are free of political controversy. Someone somewhere is going to feel the pinch of changing times. The challenge for policymakers is to minimize the impact on individual "pinchees."

The Future of Print Media

Contemplating the troubles of the Postal Service in coping with the electronic age suggests at once the problems of the print media which many foresee as another victim of electronic communications (and a victim of the Postal Service). The possible adverse impact of electronic media on print has been the subject of considerable anxiety on the part of print journalists, publishers, and many others. In his discussion of the role of print media, Owen concludes the demise of print is remote. Clearly, present indicators show the print media is remarkably healthy despite the unhalted advance of electronic competition in the past half century. However, recurrent suspicion that the end of print is inevitable, if not imminent, forces us to consider the possibility and the implications of such an event.

Even apart from the apprehension of some publishers for the fate of their media as a matter of economic self-interest, the concern for the future of print is partly utilitarian. Electronic media do not permit easy random access; they discourage re-reading, or thoughtful deliberation of content; and electronic facilities are not as portable as print. And some question whether we will have the same cheap access to a rich diversity of content with electronic media as we now enjoy with print. (If one includes all electronic substitutes—such as video discs and tapes—this fear would seem insubstantial; there is every reason that costs of these substitutes will decline until they are as cheap as paperback books.)

Yet the concern over the fate of the print media clearly expresses more than a utilitarian calculus. There is also an anxiety about the cultural implications of reducing Melville to magnetic tapes, or of relegating *The Washington Post* and Doonesberry entirely to television. This may reflect a romantic attachment to the traditions of the printed page; it probably reflects elitism as well since electronic communications, as we now know them, are preeminently popular media. Even if true, these characterizations are not dispositive of the issue: even in a technologically modern and thoroughly democratic society we may wish to allow some room for romanticism, tradition—even elitism. However, assuming some government intervention for a declining print media is appropriate, what can be done without compromising the tradition of independence on which the special value of these media rest?

This last point suggests Owen's concern about the future of print—that its "takeover" by electronics will bring with it the fetters of government control that have attended electronic communications. It could, of course, also happen that the melding of the two would yield the opposite result with the tradition of print freedom ascendant. However, government support of the print media risks introducing the same kind of public proprietary interest that is invoked to justify regulation. Without suggesting that no government support or subsidy of print would be appropriate, advocates of such support might do well to consider Laocoon's warning about the Trojan Horse: "Beware of Greeks bearing gifts."

Conclusion

Despite the temptation to bring this lengthy discussion to a close with a neat summary of issues, implications and options confronting policymakers over the next score years, no such tidy synopsis has been attempted. To those who are used to the clean crispness of the "executive summary," it may strain credibility to claim that the foregoing discussion is itself only a summary of the many issues that are emerging in this field; nevertheless that is the case.

The policy agenda for communications today and for the future is exceedingly rich with diverse, complex and important issues. Now is the time to begin to define and assess them. This statement may provoke some skepticism for we have become accustomed to postponing complex issues of public policy until external events literally force them to our attention, and that has not yet happened in communications. It is sometimes remarked that, in contrast to a rather long list of other social concerns that daily buffet public consciousness—energy, inflation, unemployment, crime, the environment, health care, urban problems—there is no crisis in our communications system or in our communications services. Therefore, some conclude that communications has no prominent claim on the public policy agenda. The premise is correct; it is the conclusion that is dubious. The absence of a crisis, far from being a reason for ignoring the issues, is a circumstance which most commends them to thoughtful study. We should take advantage of the opportunity to assess our long-term needs and problems without being distracted or overwhelmed by excited appeals to short-term solutions to pending imminent crises.

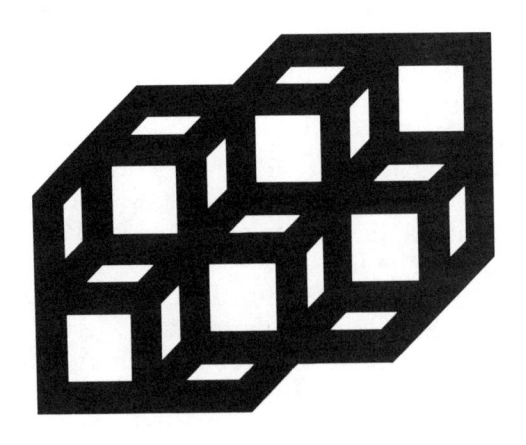

15.
Institutions for Communications Policymaking: A Review

Daniel D. Polsby and Kim Degnan

Communications policy is made in the United States in many forums. However, in this commentary on the Chisman, Gellhorn and Robinson papers in Chapters 10 through 13, we necessarily consider only the major federal institutions—the Congress, the Executive Branch, and the Federal Communications Commission. We should emphasize nevertheless that omission of the communications role of state and local governments does not imply that the state-local role is unimportant; on the contrary, the role of state and local regulation is crucially important in several areas of policy—intrastate telephone service, for example.

Congress

By the first Roosevelt administration, it had become clear that a variety of essentially federal communications concerns would require a more or less permanent and on-going federal presence.

The Communications Act of 1934 formally established the Federal Communications Commission as the agent of Congress, independent of the Executive Branch, for formulating and executing policies relating to electronic communications by wire and radio under a mandate "to serve the public interest, convenience and necessity." In delegating legislative power to the administrative process, Congress has not, in the judicial sense, diminished its basic responsibility to formulate policy. However, Congress rarely has elected to use the legislative process, in the formal constitutional sense, to make or even influence communications policy. As Ernest Gellhorn characterizes it, congressional intervention is "almost entirely reactive rather than creative." Although Congress certainly is capable of formulating policy in this area, it has given no substantive policy direction to the FCC—thereby creating a political policy vacuum.

Gellhorn offers three plausible explanations for the absence of congressional intervention. First, without a visible crisis to unify support, or several strong supporters of specific communications issues to act as representatives in "tradeoff" agreements in session, it is not likely that Congress will legislate long or short-term goals for communications policy. Second, the substantial and clearly articulated support that is essential for the passing of legislation is not forthcoming from the communications industries or the FCC. Finally, many members of Congress not only fail to keep informed about communications developments and issues, most believe that communications policy problems are not significant, relative to more pressing, and publicly visible, concerns such as resource depletion, energy shortages and reelection. Communications issues that excite the public enough to spur Congress to action—such as what television programs are retained or cancelled by stations or networks, or the fairness and objectivity of broadcast news—are singularly unsuited to legislation of any sort. Other issues involve more narrow business interests, such as what a carrier ought to be allowed to count in its rate base, or the extent to which a monopoly position should be protected. These issues are a little too dreary and technical to warrant congressional concern as long as no crisis develops. Hence, Congress has operated "at a secondary level" in communications policymaking, preferring to affect policymaking indirectly, through its powers of investigation and oversight, appropriations, and control of agency appointments.

Oversight control has several potentially troublesome characteristics, as Gellhorn notes. First, these contacts are often less visible than direct interventions and only infrequently noticed except by the regulator and the regulated industry. The devices

Institutions for Communications Policymaking

for congressional control and influence over communications policy are generally ineffective because the mandate is too ambiguous to allow adequate assessment of the degree of FCC adherence to congressional intent. The protean character of the statute's mandate also raises the question of how meaningful any discussion of Congress' intent can be, given the extensive role of the Commission.

Compounding the problem is the diffusion of responsibility for policy oversight. Currently, jurisdiction over various aspects of communications policy is divided among numerous congressional committees. In addition to the communications and appropriation subcommittees, there are a variety of other committees with relevant oversight responsibility. For example, each of the following committees has an important role in communications policy: 1) Agriculture, which handles rural development/electrification; 2) Armed Services, which controls military research and development; 3) Banking, which considers electronic funds transfer systems; 4) Governmental Affairs, which handles government information and the Postal Service; 5) Human Resources, which oversees the National Science Foundation; and 6) Judiciary, which monitors monopolies, including copyrights. The congressional committee system undoubtedly provides valuable diversity in terms of staff, industry contact, and citizen access. It produces new information on communications technology as well as social attitudes and values concerning communications programs and policies. However, the large number of committees concerned with communications almost certainly inhibits the development of an integrated policy perspective because of jurisdictional conflicts and diffusion of responsibility.

Despite the potential for effective leadership in communications policymaking through control of the budgets of the FCC and some executive agencies, Congress historically has paid little attention. The FCC budget, when it is compared in importance with that of, for example, the Department of State, Justice, and Commerce and the Federal Judiciary, necessarily gets a low priority. At least in formal hearings, the former Office of Telecommunications Policy, recently reorganized as the National Telecommunications and Information Administration (NTIA), has traditionally received similarly brief attention, usually a few minutes each year in the Senate subcommittee, sandwiched between the budgets of the Office of the Vice President and the Secret Service. As a consequence of directing its staff and resources at bigger and more conspicuous game, Congress' influence over communication policymaking through its appropriations role has been limited. In the Senate, this difficulty is lessening or easing thanks to the recent practice of having the same Senator chair both the Subcommittees on Appropriations and Oversight. However, as Gellhorn notes, the most important factor in setting the current year's budget is the budget of the previous year, not the merits of the FCC's particular regulatory policies.

Both formally and (especially) informally, Congress bears important responsibilities connected with assuring the caliber of FCC appointments. However, Congress has never hewed to any intelligible, generally applicable criterion of minimal suitability for FCC appointments. Indeed as Gellhorn suggests, "even acknowledged mediocrity is seldom a bar to FCC appointment." As a result, appointments vary widely in distinction, with the average markedly below what might be expected given the importance of the job. Congress' refusal to take a stronger role in demanding merit appointments obviously reflects Congress' view that other issues are more important than identifying technocrats qualified to be commissioners and harnessing the political capital necessary to get such people nominated and confirmed. But while this may explain, it does not justify this abdication of political responsibility.

There is no reason why Congress could not develop specific criteria for selection of commissioners. Congress could also more actively solicit general public views on appointments. The very complexity of the government places certain affirmative obligations on those who are entrusted with running it. For Congress, this might mean reaching out and consulting many segments of the public whose views on selection of commissioners ordinarily would not be heard.

A recent report of the Senate Subcommittee on Oversight and Investigation found that the President and the Congress often underestimate the importance of the regulatory positions and what is required of those who fill such positions. The committee concluded that the law and custom fail to vest the position with status, independence, and immunity to pressure that will appeal to desirable candidates. Further, it found that candidates who were qualified by experience, character, and education might be unavailable for appointment to a high regulatory position; while those who are available, may be less qualified personally and professionally. The committee concluded that most persons in top positions in regulatory agencies were insensitive to consumer interest. This represents, to some degree, hardly more than a conditioned response from members of Congress who owe their office to an electorate which, in recent years, has become consumer oriented. Yet the tangle of administrative law is studiously anticompetitive in its thrust. A 1975 study of FTC and FCC appointments attributed this situation to the caliber of commissioners and suggested that increased congressional involvement in selecting nominees for commissionership might answer for reform. That seems doubtful, since Congress has been involved in the selection of past appointees. Congress has been getting what it wants, and the question is whether its standards have been high enough. If not, reform will occur only by inducing Congress to want something better.

Another approach to the problem of congressional slack in the communications policy area is to strengthen the connection between the President and the Congress. It has been suggested that House subcommittees might initiate the interaction by requiring a comprehensive annual report on the state of communications policy in the Executive Branch and the FCC. Such a report would institutionalize a frequent and thorough review of communications developments affecting a broad range of values. However, it is doubtful that mere lack of information is an important constraint; also, there seems to be no good reason to insist on the centrality of the executive in a process fundamentally legislative in character. Congress clearly is able to make its will felt. The problem, if it is a problem, is subtler and more elusive and basic to the very process of collective decision-making. The real problem is how Congress may know its own will.

One other proposed congressional reform, which has attracted widespread attention outside the communications field, is the "legislative veto." With Gellhorn, we seriously question the premises on which the legislative veto seems to rest—even ignoring the doubtful constitutionality of the veto. Placing Congress on record as disapproving, for example, the Commission's definition of "Grade B protected contour" or "objectionable interference" seems highly unlikely to make a meaningful contribution to democratic theory in action. The nature of Congress as a working legislature requires individual members of Congress to rely on the judgment of specialized colleagues. A legislative veto promises to add an extra layer of inconvenience and expense to the congressional role without greatly changing either the substantive results, or greatly reducing the power of committee chairpersons to influence those results.

The Executive

Examining the executive role in communications policymaking, participants in a July 1976 Aspen Institute conference concluded that the mandate of each executive entity should determine the nature and extent of its contribution to communications policymaking. However, there are at least 18 separate offices, councils, advisory committees, quasi-independent agencies and executive departments that perform a variety of communications-related functions. This diffusion of policymaking and policy-related responsibilities has caused considerable difficulty in fixing institutional responsibility in such a way as to give it an effective influence.

The FCC, the Federal Reserve System, and OTP all were involved in an ad hoc Electronic Functions Transfer (EFTS) Commission for long-range policy development, but the interaction among these bodies was at best sporadic. Spectrum development policy is seldom separated from the spectrum allocation and management responsibilities of NTIA. Also, it is rarely integrated with other policy issues, such as the tradeoffs between use of the federal government's half of the radio spectrum for maritime, citizens band, or land mobile communications. National security concerns involving communications are primarily within the Defense Department, which has shown no desire to cooperate with other executive agencies in spectrum or hardware demands. The Executive Branch already possesses the necessary "clout" to make its policies effective because it controls both the allocation of, and a large part of the usable spectrum, and the technology resource and development budget. With such resources at its disposal, it is puzzling that the executive maintains such a low profile in nurturing communications policy.

Apart from direct policy responsibilities there are several major operational functions that pertain to communications policy. Chisman lists the following:

(1) operation of telecommunications system—the General Services Administration's intergovernmental telephone network;
(2) design of procurement systems—competitive bidding;
(3) "technology tracking"—information exchange about new developments;
(4) spectrum management—conducted by Interdepartment Radio Advisory Council;
(5) development of scientific hardware and social-applications research;
(6) social service activities—support of rural telephone systems; and,
(7) negotiations on international issues—State Department.

Institutions for Communications Policymaking

Sufficient justification exists for continued separation of these functions in most instances. For example, centralized procurement and operational planning would be of dubious value to an application with highly specialized requirements. However, coordination between functions is critically needed. Most applications are not so specialized as to be functionally unique. Systems designs and procurement and telecommunications operations are plagued by duplications and overlap. The close relationship between technology tracking, research and development, and policy assessment seems obvious, yet there is little coordination between these functions.

One deficiency in executive policymaking is evident in the fragmentation between policies generated by NTIA (and its predecessor, OTP) and operational functions. Shorn of operational capability, a communications policy office depends heavily on obtaining the concurrence of the FCC and of congressional committees to implement its policies. While depriving this office of operational power provides both a check against overzealousness and an opportunity for interests to be heard, the asymmetry between the short-range issues addressed by the FCC and the Congress and the long-range NTIA perspective encourages muffled or deferred responses. A coordinating executive office would have an armory of potential remedies and could select those best calculated to fill a specific social need.

A case can be made for maintaining the current arrangement of executive communications activities based on the premise that some degree of specialization of functions is inevitable. However, specialization without coherent policy direction results in a high degree of fragmentation. Many observers believe that this factor currently bars the Executive Branch from any significant contribution to communications policy. Fragmentation can impede effective information flow within the agencies and departments, causing unnecessary duplications of effort and unproductive competition for scarce resources. Clearly, some duplication is tolerable and even desirable, because it promotes a healthy degree of pluralism in policy analysis. Just as competition in the private sector is thought to spur beneficial innovation, so in the public sector, competition can play a similar role. Yet somewhere there must be drawn a line between "many hands make light work" and "too many cooks spoil the sauce." Competition and diversity in policy analysis is not incompatible with coordination of policy and central leadership. The question is what kind of institutional structure can provide effective coordination and leadership?

The FCC

At the most general level, the FCC's problems begin to bump into one another. Indeed, the greatest difficulties with the agency may be more with the political context in which it exists than with any of the fundamental premises upon which the Communications Act rests. A key problem identified by Robinson is resource misallocation. Enormous quantities of administrative energy are spent on activities of dubious merit. At the same time, matters of genuinely great importance barely are treated by the agency. Not all of the blame for this state of affairs belongs to the FCC, however; much of it belongs to Congress. The agency is compelled to engage in a great deal of unproductive supervision of the broadcast industry. It is, however, true that the agency has not been among the vanguard asking for the elimination of these activities, but why should it? Broadcasting, after all, is show biz, while rulemaking is chiefly the biz of accountants and economists, whose dreary conversations take many years to comprehend. No one would willingly leave show biz for that sort of fate.

Robinson's suggestions for reform fundamentally are directed to changing the misapplication of energies that has characterized the past. More economists and statisticians should be added to the Commission staff to leaven the present overwhelming dominance of lawyers and engineers. Amendment of the statute should be sought to permit the Commission the discretion to phase out unproductive regulatory programs and policies. Also, much more attention should be paid to planning. Yet, it should not be overlooked that the Commission has always had at least a limited discretion to implement many of these reforms itself. Why has it not done so?

In a real sense, the deficiency of communications policy planning within the FCC may be explained partly by a mature industry structure that aggressively maintains its own equilibrium and partly by a failure of strong congressional or presidential leadership in one direction or another. The FCC has not proven tough enough, institutionally, to make certain difficult decisions entirely on its own. The political process has not, clearly, enlisted on the side of reform. Another difficulty, also of a political kind, concerns the way in which members for the FCC are selected, which we commented on earlier. Persons with detailed familiarity with any of the industries supervised by the Commission are apt to be ineligible for Commission membership because of conflict of interest. The commissioners, including the chairman, are all presidential appointments. As a result, commissioners are not often selected on the basis of background or experience, although occasionally some commissioners are in fact qualified. Rather, the appointments are based on political criteria. Considering the importance of the job, it does not seem unreasonable to require that merit be the first criterion of selection for membership on the FCC. It seems unlikely that this will occur unless there is greater public and congressional demand for it.

Given the importance of communications regulation, it seems pertinent to ask why these positions have been based on patronage rather than merit. To characterize past practice charitably, it might be said to betray a democratic rather than meritocratic model for FCC decisionmaking. But the FCC has never functioned as a representative institution. As recently as 1976, three of its seven members were from Chicago and two others from Minneapolis. Furthermore, where the day-to-day grist of the job consists of obscure economic and legal technicalities, it ignores reality to make too much of populist virtues in commissioners. In order to function as something other than mere rubberstamps for the Commission staff, they have to be reasonably learned in public policymaking.

Closely related both to the agency's failure to devote its resources to the right things, and to its inability to determine what those right things are, is the problem of "industry capture." A significant deficiency in the Commission's information gathering and processing method is its almost unavoidable dependence on the regulated industries for data and analysis. Much of the detailed information necessary to make policy decisions originates in the regulated industries. The FCC would be in a position to regulate long distance telephone rates much more effectively, if, for example, it could rely to a greater extent on its own information rather than having to rely on what it is told by the Bell System. It must be noted that this symbiotic relationship is not intrinsically bad. It is only when information and analysis are tainted by advocacy that this situation poses serious questions of reliability and accuracy.

The pervasive question—how to identify the public interest in communications policy—remains, as do several related questions. For example, what should be the relationship between the agency and the industries it is supposed to regulate? Should increased public participation in FCC decisionmaking processes be encouraged? If so, in what directions? The latter issue, indeed, is potentially a most instructive one. If the need for more public participation in Commission processes is perceived, what is its source? Is it that the agency does not speak for the public interest? If not, is the appropriate remedy introducing liberalized (and more complicated) public participation, rather than rethinking the strengths and weaknesses of the agency itself and the missions with which it has been entrusted.

The Judiciary

Any consideration of the judicial role in the making of communications policy ought to begin with the observation that, when they review administrative action, courts are performing a complex of different functions, some traditional and others not. The traditional functions are not difficult to identify. As an aspect of fundamental fairness, administrative agencies are required to treat like cases in like manner; and they are required to safeguard their decisional processes from contamination by off-the-record contacts with interested parties. These norms of fundamental fairness are embedded in the Constitution, as well as in the Administrative Procedure Act, and the substantive statutes that establish independent regulatory agencies and define their authority. Elaborating on the enforcement of these norms is a part of the traditional judicial function.

In addition to this role, however, it is apparent that the judiciary has staked out a substantially larger role for itself than merely acting as guarantor of the regularity of the policymaking process. The language of the Communications Act creates at least an ostensible, admittedly amorphous, legal standard against which the substantive policy decisions of the FCC are to be measured. The Commission is empowered to exercise the plenary grant of Congress of its legislative powers, "in the public interest." The term "public interest" communicates very little information about what policies should be incorporated into it. However, courts are not strangers to navigating through trackless wildernesses of language. Ideas such as "probable cause," "the ordinarily reasonable and prudent person," or "cruel and unusual punishments" always have been the prerogative of the judiciary to define. It could be that Congress, by structuring the Communications Act as it did—including the provisions for judicial review by the United States Courts of Appeals—contemplated some vague substantive role for the courts. Whether that was Congress' intention or not, courts have claimed and exercised this right to define.

In relation to the performance of its nontraditional role, it is worth making some distinctions, even if they are only of emphasis, in the way that this function may be perceived and exercised. The judicial role, clearly, is something that ought to be different and distinct from the substantive policymaking role of the agency whose product is to be reviewed, for it makes no sense whatever to build a duplication of functions into the process. One view of the judicial role was concisely articulated at the 1977 Aspen conference that initiated this volume. Judge Shirley M. Hufstedler, of the United States Court of Appeals for the Ninth Circuit, pointed out that the entire justification for a judicial review of administrative decisions rested on the courts' separateness from the agency. In her view, the reviewing judges sit as reasonably intelligent "people-in-the-street;" the judges decide if the action was reasonable or unreasonable. If, after full briefing or oral agrument, an agency cannot persuade at least two generally disinterested lawyers (a majority of the court in ordinary cases where a panel of three judges review agency action) that what they have done makes sense, then the action is overturned.

While this conception gives the reviewing court a distinctive role to play in the process—a role that points toward rational decisionmaking without duplicating the agency's work—it is not a model that has been embraced universally by the courts of appeals. In particular, the District of Columbia Circuit, into which the vast majority of administrative review proceedings are by statute channeled, has articulated a subtly different review function for itself. While the D.C. Circuit view is not totally separate and distinct from that proposed by Judge Hufstedler, there is an unmistakable difference of emphasis in its approach. While Judge Hufstedler sees the court as aloof and apart from the agency's doings, the D.C. Circuit's conception has been increasingly that it is an integral part of the administrative process. The classic exposition of this view comes in the 1969 case of *Greater Boston Television Inc. v. F.C.C.*, where it is proposed that the relationship between court and agency is a sort of partnership. From this partnership concept, a fertile list of implied agency obligations has developed. They are not specified by either Constitution or statute, but nevertheless assertedly belong to the very relationship between agency and reviewing court that is established by the statute. Judge Harold Leventhal, author of the opinion in *Greater Boston*, consistently has taken the view that the most important obligation that this relationship places on the agency is a reasonably clear articulation of its purposes— some verifiable evidence that the agency has come to grips with the hard questions embraced in its decision. This requires that the agency regularly observe the distinction between the "tolerably terse" and the "intolerably mute," in justifying their results.

The court, on this theory, is entitled to reverse and remand dockets for further agency discussion and consideration when they cannot meet this standard.

The virtues of this sophisticated view of administrative review should be obvious. It is seldom harmful and generally constructive to get an agency to examine what it is doing. Even when the agency declines to change its result, the appearance of fairness, and possibly even the actual fairness of its decision may be enhanced by the judicial command to "take a hard look" and think again.

If the partnership idea of administrative review has certain unmistakable strengths, it also contains some unmistakable hazards. It might seem to invite the court to play a substantive role in policymaking that exceeds the traditionally conceived judicial competence. These elements of partnership have not firmly settled into law but are still in various stages of appellate litigation. The most outstanding example in the area of communications is found in *National Citizens Committee for Broadcasting v. F.C.C.*, which reviewed the Commission's media cross-ownership rulemaking. The Commission had concluded that, broadly speaking, it was not a good thing for newspapers to own broadcast stations in the same market. Rather than requiring that these combinations be changed, however, it merely forbade them from coming into existence in the future while "grandfathering" all existing combinations. Divestiture was required only in the very few cases in which a single owner had a literal monopoly over mass-communications media in a given market. The Court of Appeals decided that if combinations were a bad thing, given the sturdy premise that diversity of voices is a good thing, then the grandfathering provisions were irrational and had to be changed. The presumption ought to favor divestiture, and allowing existing combinations to remain in place required detailed justification. The court even sketched the sort of rule it had in mind. If the court's corollary is correct, and its assertion of authority is not overturned by the Supreme Court, it is fair to ask what the implications are for any form of multiple ownership of broadcast facilities. If diversity of voices were to be the all-ruling norm in this area—and neither Congress nor the Commission has ever suggested this—it would seem to rule out AM-FM combinations in a single market, and, by the same logic, combined ownership in more than one market. It would go too far to write these proposed policies off as foolishness, of course, for there are few scale economies connected with the operation of broadcast stations (although there may in some circumstances be some). Yet to suggest that it is appropriate for a court to make this determination after a Commission finding to the contrary would appear to give the judiciary some rather more substantive policy discretion than has been typical of courts.

Despite the *National Citizens Committee for Broadcasting* case, Robinson's investigation does not discern any great increase, overall, in the D.C. Circuit's behavior with respect to administrative review over the past generation. The rate of reversals of the FCC by the court was not much different in an earlier period from the rate between 1970 and 1976. Whether one accepts "counting" as an adequate means of accounting for judicial behavior depends on how much one can rely on what the judges say while they are deciding. Numerous social scientists have long held the view that the opinion aspect of the judicial decision is simply too fuzzy to manage scientifically, an epistemological view which is, perhaps, colored with the nonlawyer's bias against trusting anything that lawyers say. Hence, the results of the decisions count as data, while the opinions are discouraged as background music. The strengths of this procedure are undeniable. Yet one thing that is left unaccountable is inquiry into the question of whether agencies modify their behavior in order to avoid being reversed by courts. Robinson points in the direction of this consideration by observing that the vagueness of a substantive grant of statutory authority to an agency, if it simultaneously invests the agency with a broad discretion, and for the same reason, invests the court with a discretion equally as broad. If "the public interest" leaves the FCC in a trackless normative wilderness in which it is free to make up the rules of the game, the courts' discretion to pass on those rules for reasonableness or substantial correspondence with record evidence is not less broad.

If agencies have trimmed their sails to what they believe to be the policy preferences of reviewing courts, the reversal rate could well remain constant for a generation, while the substantive powers of the courts, in relation to those of the agencies, may have burgeoned. Robinson, however, does not suggest that this is so, nor do the present authors have any ground for asserting that it is. Therefore, such imperial lurches as that in the *National Citizens Committee for Broadcasting* case should, at least until there is more evidence, be treated as sport rather than as sound new growth in the law of judicial review.

Conclusion

In discussing the adequacy of the various federal institutions in which communications policy is made, fundamental questions must be asked. Do the institutions, whether separately or collectively considered, have the legal (or constitutional) authority to perform their missions? If they do, have they the practical capacity to perform? To the first question we can give a short answer without much hesitation: The institutions largely have all the legal authority they need in order to proceed. When this is not true—in connection, for example, with the FCC's uncertain jurisdiction over cable television—the defects are easily remedied by Congress. As to the second question, the reply must be more complicated and more tentative. By and large the institutions have proved at least adequate to their missions. However, in certain areas, most glaringly in the FCC's regulation of interstate and international common carriage, very significant investment of additional expert staff is sorely needed and would almost certainly produce a better regulatory work product, which would ultimately benefit the entire society. While other institutional constraints to adequate policy formulation have been noted, such as the passivity of the Congress and the essentially political provenance of FCC commissioners, these problems are rooted, not in the framework of existing institutions, but in the political foundations on which these institutions rest. That the actors who run these institutions might frequently find political motives dominating their substantive decisionmaking ought to shock no one. It is in the nature of political institutions to be political, and the benefits of pluralism are generally thought worth such costs. That these institutions have very often failed to bring forth the particular policies of which we would most approve is, in terms of any assessment of the adequacy of the institutions themselves, beside the point. The point is that, given the right people and the right values, the Congress, the FCC, and the executive could have any kind of communications system they wanted. Changing these wants is an essentially political agenda, a matter of building a consensus among the politicians, bureaucrats, and business people who run, regulate, and legislate for the communications industry. If no ringingly clear consensus exists now about the substantive policy changes that should be adopted, the reason is not the inadequacy of the institutions so much as the complexity and diversity of interests in the world in which they operate.

Index

A

ABA. *See* American Bar Association.
ABA Committee on the Judiciary, 338
Above 890 decision (Allocation of Frequencies Above 890 Mz decision), 131
Access, 242, 434
   to cable TV, 211–213, 223, 248–252, 254
   to courts, 417
   to electronic media, 127, 192, 198–202, 214–216, 440, 447–448
   to undersea cable, 165
Administrative Procedure Act, 183, 371, 374, 418, 433, 510
AEROSAT. *See* Satellites.
Aetna Life and Casualty, 95, 133, 317
Agency for Consumer Advocacy, 376
Agriculture, Department of, 56, 268, 405
All America Cables and Radio, Inc., 162, 171
All-channel receiver law, 449
American Bar Association (ABA), 388, 455
American Satellite Corporation, 132
American Telephone and Telegraph Company (AT&T), 24–25, 30–31, 55, 56, 62, 78, 81, 82, 83, 86, 95, 98, 100, 102, 105, 106, 111, 132, 133, 134, 162, 164–168, 170–171, 175, 177, 185, 187, 281, 316, 319, 331, 332–333, 348, 362, 379, 382, 474–478, 479–482, 497
   See also Telephone industry.
American Telephone and Telegraph v. *F.C.C.*, 442
Amnesty International, 57
Ash Council, 389, 390, 391, 399–400
AT&T. *See* American Telephone and Telegraph Company.
AT&T Consent Decree of 1956, 481
Authorized User decision (Authorized Entities and Authorized User decision), 151, 167–168, 478
ATS-6. *See* Satellites.

B

Bailar, Benjamin Franklin, 313, 333, 335
Bartley, Robert, 394
BBC. *See* British Broadcasting Corporation.
"Bell bill," 404
Bell, Daniel, 4, 55, 468

Bell System, 23, 73, 81, 100, 102, 103, 105, 106, 111, 117, 128–129, 132, 133–153, 154, 348, 509
Bilingual-Bicultural Coalition on Mass Media v. *F.C.C.*, 438
Black Coalition of Richmond v. *F.C.C.*, 438
Board of War Communications, 159
Books and Periodicals, 26, 27
   television's effect on, 233–234
   trends in, 230–231
   See also Newspapers, Print media.
Boston City Hospital, 261
Brademas, John, 461
Brandeis, John, 35
British Broadcasting Corporation (BBC), 85, 257, 346
British Independent Broadcasting Authority, 85
British Post Office, 85, 105, 176, 316, 332, 338, 339
Budget and Impoundment Act of 1974, 452
Burch, Dean, 394

C

Cable Landing License Act, 174
Cable television, 62, 82, 179, 194, 245–254, 267, 271, 299, 308, 317, 318–319, 358–359, 460, 471, 486–487
   Congress and, 447, 449
   consumer expenditures for, 237
   FCC and, 209–214, 378–379, 469–470
   judicial review of FCC decisions on, 423–424, 428, 435, 441
   print media and, 234, 240
   systems, 95–97, 107–109, 247–249, 253–254, 318–319
   two-way communications, 87, 89–90, 108–109, 113, 221–223, 247–249, 251–252, 259
   See also Pay/subscription television, Television.
Canada, 53, 317, 331, 338, 402
Canadian Postal Act, 331
Carnegie Commission, 272, 490
Carter, Jimmy, 55, 376, 454–455
Carterfone decision (Use of Carterfone decision), 133, 154, 479
Cater, Douglass, 91
CB. *See* Citizens Band radio.

517

CEEFAX. *See* Teletext.
Censorship, 46, 197, 200, 202, 204, 244, 440
Children, 193, 204-205, 208-209, 217-218, 250, 255-256, 263-264, 457
Citibank, 79, 323
Citizens Band (CB) radio, 89, 110, 112, 276, 280, 281, 283, 285, 290-291, 294, 296-297, 303, 304, 469, 492-493
  See also Land mobile communications.
Civil Rights Act of 1964, 433
Coaxial cable. *See* Cable television, Undersea cable.
Cogar Foundation, 305
Commerce, Department of, 51, 55, 174, 183, 401, 405, 411, 445-446, 452, 460, 461
  See also National Telecommunications and Information Administration, Office of Telecommunications.
Commercial Cable Company, 162
Commission on Postal Service, 343
Common Cause, 455
Communications, Department of, 401, 412
Communications Act of 1934, 127, 143, 146, 150, 152, 172-173, 174, 204, 212-213, 330, 334, 342, 348, 355, 358-359, 403, 430-431, 449, 456, 458, 469, 502, 508, 510
  Section 214, 169, 177-178, 184
  Section 222, 163-164, 185
  Section 314, 158, 160
  Section 315, 198-203, 447
  Section 601(b), 328
Communications Satellite Act of 1962, 166-168, 171-172, 173, 174, 178-179, 185, 478
Communications Satellite Corporation (COMSAT), 132, 151, 157, 166-169, 171-172, 173-174, 175, 179, 185, 187, 317, 333, 442, 477-478
Communications Satellite Corporation v. *F.C.C.*, 442
Community Service Broadcasting of Mid-America v. *F.C.C.*, 440-441
Comparative hearing process. *See* Licensing.
Computer-based PABXs, 75, 77-78, 106, 107
Computer technology, 28, 49, 62, 63-91, 129-130, 141-142, 250, 252, 266, 293, 345, 359, 469, 481
  FCC regulation of, 237, 244
  print media and, 237-238
  U.S. export of, 51-54
  USPS and, 308, 314-318, 321-322, 341, 348

COMSAT. *See* Communications Satellite Corporation.
COMSAT General Corporation, 95, 133
COMSAT rate case *(COMSAT* v. *F.C.C.)*, 362
Congress, 22, 146, 179, 192, 197, 198-199, 214-215, 219, 223, 256, 260, 261, 355-359, 366, 372, 402-403, 408, 412-413, 445-462, 501-505, 508, 510
  appropriations to FCC, 449-452, 460, 461, 503-504
  cable TV and, 212-213, 224, 447
  communications policy and, 445-450, 457, 460
  confirmation of FCC appointments, 385-386, 389-390, 449, 452-455, 460, 504-505
  informal controls by, 449-450, 452
  land mobile communications and, 296, 452
  legislative oversight of FCC, 180, 455-460, 502-503
  OTP and, 173-174
  regulation of common carriers, 163, 447
  USPS and, 308-313, 327-328, 330, 331, 333-334, 336, 340-341, 342
Constitution of the United States, 358, 446, 448, 459, 510, 511
Consumer Product Safety Commission, 432
Copyrights, 27, 192, 240, 250, 460, 503
Cornell University, 305
Corporation for Public Broadcasting (CPB), 256, 258, 490
Council of Economic Advisors, 404-405
Council on Environmental Quality, 404-405
Council on Wage and Price Stability, 341
Courts, 215, 336, 355-357, 372, 375, 402, 415-444, 459, 510-513
  controversial role of, 424-429
  measuring judicial performance, 419-424
  "partnership" with agencies, 419-420, 422, 429, 432-434, 509-510
  reforms of, 429-432
  technical issues and, 435
  See also U.S. Court decisions.
CPB. *See* Corporation for Public Broadcasting.
CRIMENET, 407
"Cultural imperialism," 45

D

Dataphone Digital Service (DDS), 102, 171

518

Index

Data Speed 40/4, 142, 316
Data Transmission Corporation (Datran), 132, 133
  See also Specialized Common Carriers.
Defense, Department of, 134, 164, 181, 404, 405, 406, 409, 412, 452, 460, 476, 506
"Deintermixture" controversy, 447
Democratic National Committee, 202–203
Digital Recording Corporation, 67
Dispatch service, 110, 280, 286, 288, 293, 298, 300, 477
  See also Land mobile communications.
District of Columbia Court of Appeals. See U.S. Court decisions.

E

Educational Television Facilities, 269
EFTS. See Electronic funds transfer systems.
Electromagnetic spectrum, 193–197, 203, 305, 414, 493–494
  allocation of, 112, 153, 234, 236, 258, 286, 294, 295–303, 361, 394, 406–407, 409–410
  international negotiations for, 47
  land mobile communciations and, 279, 280–282, 286, 291, 295–303
  management of, 404, 405, 506
  policy development of, 403, 404, 407, 410, 506
  scarcity of, 209–217, 221, 225, 486–487
Electronic funds transfer systems (EFTS), 27, 41, 83, 102, 116, 139–140, 142–143, 151, 155, 314, 322–326, 332–334, 342, 346–347, 403, 407, 469, 495–496, 503, 506
Ellington Jr., J.T., 333
Employment,
  equal opportunities for, 433, 438–439
  "information society" and, 4–8, 40–43, 55
Energy, Department of, 56
Environmental Protection Agency (EPA), 401, 411, 432, 435
Equal time requirements, 198–199, 434, 447, 487
EROS. See Satellites.
ERTS. See Satellites.
Ethyl Corporation v. *E.P.A.*, 435
ETV. See Television.
Execunet, 141
Executive branch, 22, 146, 342–343, 401–414, 505–507
  integration among agencies and functions of, 406–413

operational telecommunications functions, 405–407
policy functions of, 403–405, 448
public information and, 37
  See also President of the United States
Executive Order 12046, 414
Ex parte contacts, 357, 370–373, 435
Exxon, 25

F

Facsimile transmission, 213–214, 320–322, 330, 331–332, 335, 346, 369
Fairness Doctrine, 199–203, 209, 213, 216, 440, 447–449
"Family Viewing policy," 208–209, 447–448
Faulkner Radio v. *F.C.C.*, 440
Federal Aviation Administration, 452
Federal Communications Commission (FCC), 132–133, 214, 216, 219, 328, 330–331, 333–334, 342, 343, 348, 353–400, 402, 403, 408, 413, 460, 474–475, 508–510
  appropriations from Congress, 376, 449–452, 460, 461, 503–504
  cable TV and, 96, 209–214, 224, 240, 469–470
  commissioners, 382–388, 398, 399, 452–455, 460, 504–505, 508–509
  Common Carrier Bureau, 142
  congressional investigation and supervision of, 446–450, 455–460, 502–503
  control of programming, 198–209, 224, 257–258, 428–429, 434, 485–488
  creation of, 355, 502
  Docket 18128, 139, 144
  Docket 18261, 296
  Docket 18262, 296, 297–300, 303
  Docket 19528, 134
  Docket 20003, 134–135, 143–144, 154
  functions of Chairman, 360, 391–448
  international communications and, 157–169, 172–173, 175–186, 187, 476–477
  judicial review of decisions by, 415–444
  land mobile communications and, 279, 282, 295–304, 434, 452
  organization of, 377–382, 388–395, 508
  processes of, 368–370, 370–373, 412
  reforms of, 354–355
  regulation of computers, 239, 244, 434
  regulation of telephone prices and services, 128, 133–153, 154–155, 236, 394, 434, 479
  relationship to electronic print media, 234, 315

519

satellites and, 317, 477-478
Federal Reserve System, 403, 407, 413, 506
Federal Trade Commission (FTC), 37-38, 56, 461
Federal Trade Commission Act, 196
FEDNET, 15, 407
Fiber optics. *See* Optical Fiber Technology.
First Amendment, 17, 46, 191-197, 200-203, 205-206, 208-209, 212, 215-218, 219-220, 221, 225, 364, 429, 439, 441, 448
FOIA. *See* Freedom of Information Act.
Ford-Carter debates, 199
Freedom of Information Act (FOIA), 15, 37, 39, 421
Freeman, Gaylord, 343
Friendly, Fred W., 202
FTC. *See* Federal Trade Commission.
FTC Communications, 171
Functional illiteracy, 40-42, 57

G

General Electric (GE), 62, 281
General Services Administration (GSA), 303, 405, 406, 407, 412, 506
General Telephone and Electronics, 78, 100, 281
    See also Telephone industry.
Georgia Power Project v. *F.C.C.*, 440
Graham, J., and V. Kramer, 452-453, 462
Graphnet, 183, 331, 334-335, 477
    See also Specialized common carriers.
Greater Boston Television Incorporated v. *F.C.C.*, 511
Greater New York Savings Bank, 84
GSA. *See* General Services Administration.
GT&E. *See* General Telephone and Electronics.

H

Hawaiian Telephone Company, 171
Health, Education and Welfare, Department of (HEW), 260, 262, 269, 302, 405, 490
Hector, Louis, 393
Hewlett-Packard Corporation, 77, 79
Higashi-Ikoma, 346
Holographics, 71
Home Box Office, 95-96, 257
Home Box Office v. *F.C.C.*, 371-372, 441
Home communications terminals, 83-91, 316-321, 336-341, 345
House and Senate Committees on Government Operations, 402

House Interstate and Foreign Commerce Committee, 398
House Subcommittee on Communications, 342, 359
House Subcommittee on Postal Facilities, Mail and Labor Management, 314
Housing and Urban Development, Department of (HUD), 268, 405
Hufstedler, Shirley M., 511
Human rights, 44, 46, 49, 51
Human Services Management Information System, 265
Hush-a-Phone decision (*Hush-a-Phone* v. *U.S.*), 133, 154

I

IBM. *See* International Business Machines Corporation.
Illiteracy. *See* Functional illiteracy.
Incremental cost pricing, 137-139, 143-144
"Industry capture," 373, 374, 379, 386-388, 395-396, 509
Information occupations. *See* Employment.
"Information society," 3-57, 402, 468-469
    GNP and, 4, 8-10, 54, 468
    ideology in , 14-18, 19-20
    illiteracy in, 40-42
    "information elite" and, 11-12
    shifting to, 10-12, 24-29
INMARSAT. *See* Satellites.
Input devices, 68-69
Intelligent copiers, 76-77
Intelligent data terminals, 76
Intelligent telephone, 84-86, 103-104, 117
INTELSAT. *See* International Telecommunications Satellites Consortium.
Interdepartment Radio Advisory Committee (IRAC), 405, 407, 414, 506
International Business Machines Corporation (IBM), 25, 62, 77, 78, 95, 104, 133, 238, 317, 333
International Communications Agency, 403
International record carriers (IRCs), 162-167, 170-171, 176-177, 185, 477
International relations, communications issues, 43-54, 477
    satellite services, 46, 94-95
    U.S. role in international relations, 43-54, 96, 157-187, 403, 406-407, 409-410, 476, 477
International Telecommunications Satellite Consortium (INTELSAT),

Index

50, 94-95, 170-172, 174, 178-180, 183, 185, 187, 317
International Telephone and Telegraph Company (ITT), 78, 100, 162-163, 168, 333
Interstate Commerce Act, 196
Interstate Commerce Commission, 302, 328, 393
IRAC. *See* Interdepartment Radio Advisory Committee.
IRCs. *See* International record carriers.
ITT. *See* International Telephone and Telegraph Company.
ITT Worldcom, 171
ITT World Communications Inc. v. *F.C.C.*, 442

J

Jackson, Charles, 327
Japanese communications technology, 53-54, 97, 118, 317, 339, 346
Javits, Jacob, 455
Johnson, E.F., 281
"Judicial activism," 415, 417, 420, 422, 425-427, 429-430, 432, 433-434, 436
Justice, Department of, 31, 141, 162, 302, 336, 341, 377, 382, 393, 402, 405, 413, 461

K

Kappel Commission on Postal Organization, 347
Kennedy Administration, 203, 327

L

Land mobile communications, 106, 109-112, 275-305, 407, 434, 452, 492-495
  commercial purposes of, 275, 282, 286, 295
  costs of, 284-285, 286-287, 288, 291-292
  dangers of, 293-294
  military uses of, 276, 303
  mobile radio, 265
  police uses of, 275, 276, 282, 286
  safety and emergency uses of, 275, 282, 286, 288-290
  two-way vs. one-way, 276, 277-279, 281, 294
  See also Citizens Band radio, Dispatch service, Mobile telephone service, Paging service.
LANDSAT. *See* Satellites.
Large-scale integrated (LSI) circuits, 64-68, 72, 73, 85, 89, 109, 114
Law Enforcement Assistance Administration (LEAA), 405
LEDs. *See* Light-emitting diodes.
Lee, Alfred M., 305
"Legislative veto," 458-459, 505
Leontief, Wassily, 20
Licenses, 192-193, 197-198, 200, 206-209, 360, 395, 416-417, 420, 422, 428, 430-431, 434, 451, 469
  cable "certificates of compliance," 367-368
  cable landing, 161, 173
  Citizens Band radio, 283, 296-297, 366-368
  common carrier certificates, 364
  comparative hearing process, 363-366, 439-440
Light-emitting diodes (LEDs), 69, 70, 99
Lister Hill Medical Library, 269
Arthur D. Little, Inc., 325
LSI. *See* Large-scale integrated circuits.

M

Magnuson, Warren, 450
"Mailgram," 321-322, 327, 329, 332, 339, 340, 496
  See also U.S. Postal Service, Western Union
Management information system (MIS), 38
Mann-Elkins Act of 1910, 358
Marconi Company, 161-162
Massachusetts Medicaid Program, 261
MCI. *See* Microwave Communications, Inc.
MCI Communications Corporation, 132
  See also Specialized common carriers.
MCI Telecommunications Corporation v. *F.C.C.*, 442
McKay Radio and Telegraph, 162
McLaughlin, John M., 313, 343
MCTS. *See* Multichannel trunked systems.
Medicare, 261-262
Memory systems and devices, 66-68, 89, 114
Mexican Telegraph Company, 162
900 MHz decision, 494
Microwave technology, 62, 101, 130, 131, 246, 254, 262, 301, 320, 469, 473
Microwave Communications, Inc. (MCI), 132, 139-141
Minow, Newton, 393
MIS. *See* Management Information system.
Mitre Corporation, 319
Mobile communications. *See* Land mobile communications.
Mobile telephone service (MTS), 110,

521

153, 280, 281, 282-283, 284, 291-292, 294, 298, 300-301, 475, 493
  See also Land mobile communications, Telephone industry.
Monopoly, 134, 136-138, 159-161, 175, 194, 215, 223, 233, 243, 396, 471-472, 474-477
  media cross-ownership, 437-438, 512
  satellite, 166-169, 178-180, 181
  telephone, 128, 140-147, 149-150, 298, 300, 348, 404, 442, 480
  USPS and, 311, 335-336, 341, 497
  See also Communications Satellite Corporation, Subsidization.
MOTHER (Multiple Output Telecommunications Home End Resources), 91
Motion pictures, 197, 231-232
Motorola, 281
MTS. *See* Mobile telephone service.
Multichannel trunked systems (MCTS), 110, 277-279

N

National Aeronautics and Space Administration (NASA), 56, 94, 134, 405
National Assessment of Educational Progress, 57
National Association of Broadcasters, 208
National Broadcasting Company (NBC), 255
National Broadcasting Company v. U.S., 355
National Center for Health Care Research and Development, 269
National Citizens Committee for Broadcasting v. F.C.C., 437-438, 440, 512-513
National Commission on Electronic Fund Transfers, 157
National Commission on Paperwork Reduction, 30
National Communications System, 404, 406
National Endowment for the Humanities (NEH), 224
National Institute of Education, 269
National Organization for Women v. F.C.C., 438
National Postal Forum, 335
National Public Radio (NPR), 490
National Research Council (NRC), 97, 329-330, 337, 496
National Science Foundation (NSF), 305, 503
National security, 44, 46, 50-51, 52, 403, 506

National Security Council, 414
National Telecommunication and Information Administration (NTIA), 23, 55, 173-174, 182, 343, 377, 401, 403, 407, 414, 445, 503, 506-507
  See also Office of Telecommunications Policy.
Navy, Department of, 158, 197
NBC. *See* National Broadcasting Company.
A.C. Nielsen Company, 232, 243
Newspapers, 194-195, 215, 231
  new technology and, 26, 86-87, 213-214, 237-238
  trends in number and circulation, 232-233
"New world information order," 44, 45-46
Nixon, Richard, 201, 389
North Carolina Utilities Commission v. F.C.C., 442
NRC. *See* National Research Council.
NTIA. *See* National Telecommunication and Information Administration.

O

Obscenity and indecency, 203-205, 217-218
Office of Communication of the United Church of Christ v. F.C.C., 438-439
Office of Economic Opportunity, 262
Office of Education, 269
Office of Legal Counsel, 402
Office of Management and Budget (OMB), 407, 413, 414
Office of Plans and Policy (OPP), 380-381
Office of Presidential Nominations, 455
Office of Science and Technology, 414
Office of Technology Assessment (OTA), 253, 343, 483
Office of Telecommunications (OT), 55, 269
  See also HEW.
Office of Telecommunications Policy (OTP), 55, 173-174, 181-182, 401, 404, 405, 407, 410, 412-413, 460, 503, 506
Office communications terminals, 74-83, 115, 316-321, 332-336
"Open skies policy," 132
OPP. *See* Office of Plans and Policy.
Optical fiber technology, 92, 99-100, 106, 107, 112-113, 117-118, 148-149, 209-210, 212-213, 245, 308, 317, 319, 469, 471, 486
ORACLE. *See* Teletext.
OT. *See* Office of Telecommunications.

522

Index

OTA. *See* Office of Technology Assessment.
OTP. *See* Office of Telecommunications Policy.

P

PABX. *See* Computer-based PABXs.
Pacifica Foundation v. *F.C.C.*, 439
Packet communications, 73, 79, 102, 111-112, 133, 149, 315, 345, 476, 483-484, 493
Packet radio. *See* Packet communications.
Packet-switching. *See* Packet communications.
Paging service, 110, 279-280, 281, 291, 364, 475, 493
  See also Land mobile communications.
Panko, Raymond, 80
Pasadena Broadcasting v. *F.C.C.*, 439
Pastore, John, 450, 457
Pay/subscription television, 68, 87, 96, 107-109, 117-118, 215, 223-224, 240, 249-251, 257, 403, 428, 441, 449
  See also Cable television.
PBS. *See* Public Broadcasting System.
Percy, Charles, 455
PES. *See* Private Express Statutes.
Postal Rate Commission, 343, 348, 497
Postal Reform Act (PRA) of 1970, 328-331, 348
  Section 401(2), 335-336
Postal Reorganization Act of 1970, 309, 313
Postal service. *See* U.S. Postal Service.
Postal Telegraph Company (Postal), 161-163
Post Roads Act, 327
President of the United States, 412-413, 414, 505
  FCC appointments and, 384-386, 391, 453-455
  international communications and, 173-174, 181, 409-410
  responsibility for communications policy, 23, 193, 343, 390
  use of communications by, 181, 219
  See also Executive Branch.
President's Commission on Postal Organization, 308
President's Communications Policy Board, 158
Print media, 229-244
  content of, 203, 205
  current role of, 230-232
  effect of cable on, 234-235
  electrification of, 86-87, 229, 236-239, 241-242
  expenditures for, 87
  freedom of expression and, 191-196, 242
  scholarly publications and electrification of, 243-244
  subsidization of "electronic press," 236-237, 244
  See also Books and periodicals, Newspapers.
Privacy Act of 1974, 36
Privacy Protection Study Commission, 342
Privacy rights and issues, 15, 192, 204, 217-218, 221-222, 403, 407, 414
  abroad, 52
  electronic mail and, 324, 342, 496
  impact of future technology on, 35-37, 73, 103-104, 240, 273, 486-487
  land mobile communications and, 278-279, 292
Private Express Statutes (PES), 307-309, 311, 334-336, 339, 341, 346-347
Private Line Rate proceedings, 362
Property rights
  to information, 34-39
  to licenses, 366
PTTs (government ministries of posts, telegraph and telephone/telecommunications, 157, 165, 182-183, 184, 477
Public affairs programming, 192, 487
Public Broadcasting Act of 1967, 490
Public Broadcasting System (PBS), 96, 256, 257-258, 490
Public information, 37-38, 256
"Public interest" standards and considerations, 192, 197-198, 205-209, 341, 356, 363, 365, 396, 417, 419, 428, 429, 433, 442, 447, 456, 502, 509, 510, 513
Public participation and public interest groups, 14, 21, 222, 224, 225, 247, 251-252, 294, 296, 299, 343, 355, 356-357, 361, 373-377, 388, 424, 425, 431, 455, 459, 509-510
Public Service Satellite Consortium, 97
Public television. *See* Television.
Pye, Roger, and Ederyn Williams, 80, 82

R

Radio, 158-160, 161-162, 196-198, 205-208, 276, 310
Radio Act of 1927, 158, 192, 196, 197, 198, 358
Radio Corporation of America (RCA), 67, 82, 88, 95, 166-163, 167, 281

Rate-base regulation, 175-176, 187
Rate-Setting Commission, 261
RCA Global Communications, Inc. v. *F.C.C.*, 442
RCA Globcom, 132, 171
REA. *See* Rural Electrification Administration.
Red Lion Broadcasting Co. v. *F.C.C.*, 201-203, 212
1910 Regulatory Enactment, 328
1973 Report of the Board of Governors on Restrictions on the Private Carriage of Mail, 335
1977 Report of the Commission on Postal Service, 347
Resource satellites. *See* Satellites.
RFD. *See* Rural free delivery statutes.
Ribicoff, Abraham, 455
Richardson, Elliot, 260
Roosevelt Administration, 501
Rural Electrification Administration (REA), 105, 153, 254, 342, 405, 482-483
Rural free delivery (RFD) statutes, 236
See also U.S. Postal Service.
Rural Health Association, 262

S

Safety and Special Services Bureau, 367
See also Federal Communications Commission.
Satellites, 50, 51, 57, 61-62, 95-96, 102, 117, 130-131, 132-133, 149, 151, 166-169, 171-172, 177-180, 183-185, 187, 230, 239, 245, 255, 257-258, 272, 279, 308, 317-318, 332-333, 341, 342, 364, 405, 469, 473, 477-478, 483-484, 491
   AEROSAT, 180
   ATS-6 satellite, 245, 269
   direct broadcast satellites (DBS), 19, 46, 47, 92, 96-98, 254
   EROS, 39
   ERTS, 39, 47
   geosynchronous vs. orbiting, 179
   INMARSAT, 180
   LANDSAT, 47, 56
   MARISAT, 94
   resource satellites, 39, 56, 98
   See also Communications Satellite Corporation, International Telecommunications Satellite Consortium.
Satellite Business Corporation, 147
Satellite Business Systems (SBS), 95, 97-98, 149, 317, 497
Securities and Exchange Commission, 461

Senate Appropriations Committee, 450
Senate Commerce Committee, 330, 450, 454
Senate Subcommittee on Oversight and Investigation, 504
"Sesame Street," 256
Southern Pacific Communications Company (SPCC), 132, 141
See also Specialized common carriers.
Southern Satellite System, 96
Speech-recognition equipment, 69
Specialized common carriers, 132, 140-141, 147, 171, 298, 315-316, 332-333
See also Above 890 decision, MCI Communications Corporation, Graphnet, Data Transmission Corporation, Southern Pacific Communications Company.
"Speed mail," 327-328, 330
State, Department of, 47, 173-174, 181, 183, 185, 406, 409-410, 413, 414, 452, 460, 461, 506
Subscription television (STV). *See* Pay/subscription television.
Subsidization, 214, 236-237, 244, 248, 252-254, 348, 473-474, 480-482
   cross-subsidization, 135-140, 143-144, 150, 151, 300, 334
   separate subsidies for competitive services, 148-150
   USPS and, 309-310, 313, 333-334, 340, 341, 496-497
"Sunset law," 458
Switching technologies,
   computer-controlled, 104-105, 116
   digital, 105-106
   See also Value-added networks.

T

Tama New Town, 339, 346
TAT-1, TAT-7. *See* Undersea cables.
TAT-4 decision, 165, 187
Technological determinism, 12-18, 19, 26-28, 472-473
Technology tracking, 405, 406, 407, 506-507
Technology transfer, 47-48, 49-50
Teleconferencing, 80-81, 82, 97
Telemedicine, 261-262, 269, 469, 488-490
Telenet, 181, 497
   See also Value-added networks.
Telephone industry, 23, 92, 128-129, 133-134, 138, 141, 153, 170, 252-253, 261, 271, 281, 291-292, 298, 300, 308, 310, 316, 339, 342, 345, 404, 469, 478, 479-483

Index

Congress and, 447, 449
court decisions regarding, 434, 442
FCC and, 128, 236, 364, 369-370, 475
new transmission and switching technology, 62, 73, 84-86, 89-90, 100, 102-106, 113, 117, 130-131, 240-242, 246, 248, 254, 265, 320, 470, 471
potential effects of competition, 143-148, 473
POTS (Plain Old Telephone Service), 106-107
See also American Telephone and Telegraph Company, Bell System, General Telephone and Electronics, Hawaiian Telephone Company, Satellites, Specialized common carriers, Value-added networks, Western Union International.
Teleprompter Cable Communications Corporation v. *F.C.C.*, 441
Telerent decision *(Telerent Leasing Corporation. Affirmed. North Carolina Utilities Commission* v. *F.C.C.),* 146, 148, 150
Telesat Canada Corporation, 95
Telex, 79-80, 114, 170-171, 314, 318, 322
Teletext, 346, 469, 484
  CEEFAX, 85, 346, 491
  ORACLE, 85, 346, 491
  ViewData, 17, 85, 105, 316-317, 339
Television, 218-219, 254-259, 298-299, 301, 310, 316
  cable TV and, 210-213
  effect on children, 204-205, 208-209, 217-218
  games, 89-90
  illiteracy and, 42
  instructional, 97, 248-250, 256, 263-264
  international service, 62, 170, 187
  print media and, 233-235
  public TV, 224, 254-256, 272, 440, 490-492
  satellite transmission, 92, 94-95, 130-131, 255, 257-258, 272
  sex and violence on, 42, 203-205, 217-218, 447-448, 457, 486-487
  two-way communications and data processing, 82, 85-86, 221-223
  See also Cable television, Pay/subscription television.
Third World, 44, 45-50
Toffler, Alvin, 368, 402
Transaction Network Service (TNS), 102, 140, 142-143
Transportation, Department of, 302, 394, 401, 411
Tropical Radio Telegraph, 162
Tymshare. *See* Value-added networks.

U

Undersea cables, 72, 100, 161-166, 169, 179, 477-478
  TAT-1, 164
  TAT-7, 169
  See also TAT-4 decision.
United Nations Universal Declaration of Human Rights, 57
United States v. *Miller,* 36
United States Transmission Systems, 132
UNIVAC, 129
U.S. Court decisions,
  Court of Appeals, 160, 201-202, 204, 207, 212, 297-298, 416-417, 420, 423, 425-444, 512-513
  Supreme Court, 36, 159, 164, 197, 201-203, 204, 213, 355, 359, 420, 422, 425-426, 429, 434-446, 448
U.S. Post Office Department, 308-309, 311, 327-328
U.S. Postal Service (USPS), 26, 307-349, 403, 412, 503
  books and periodicals, 236
  Congress and, 308-313, 327-328, 330-331, 333-334, 336, 340-341, 342
  costs of, 307, 309-312, 331, 344, 347, 348
  "electronic mail" and, 27, 75, 86-87, 116, 242, 307-308, 317-326, 328-344, 412, 495-498
  first class mail (FCM) and, 309-310, 311, 312-313, 326
  history of, 308, 327-331
  new electronic transmission technology and, 314-327
  problems of, 307-313
U.S. Supreme Court. *See* U.S. Court decisions.

V

Value-added networks, 92-106, 132-133, 315-316, 330-331, 332-334, 337, 339-340
  Telenet, 79, 132, 477
  Tymnet, 79
  Tymshare, 132, 318
Videophone, 81-82
Video recorders and players, 87-89
ViewData. *See* Teletext.

W

Wall Street Journal, 98, 230
Wanamaker, John, 327

525

WARC. *See* World Administrative Radio Conference.
Ware, Harold, 305
Warner Cable Corporation, 68, 250, 318
Warren, Samuel, 35
Western Electric, 150, 152
Western Union International (WUI), 163, 171
Western Union Telegraph Company (Western Union), 95, 96, 128-129, 132, 161-164, 170-171, 314, 321, 327, 330, 332-333, 339, 497
WGBH, 257
White, Lee, 399
Wilson, James Q., 369
Word processors, 75-76, 115
World Administrative Radio Conference (WARC), 47
WNET, 255
WUI. *See* Western Union International.

X

Xerox, 77, 238, 332

About the Editor

Glen O. Robinson is a Professor of Law at the University of Virginia and Special Adviser to the Aspen Institute for Humanistic Studies. From 1974 to 1976, Mr. Robinson served as a commissioner on the Federal Communications Commission. In 1977, he was appointed head of the U.S. delegation to the World Administrative Radio Conference. He is the author of *The Forest Service: A Study of Public Land Management* (John Hopkins—1974) and co-author, with Ernest Gellhorn, of *The Administrative Process* (West—1974) and numerous articles in the field of communications and administrative law.

RELATED TITLES
Published by
Praeger Special Studies

Published in cooperation with the Aspen Institute Program on Communications and Society.

TELEVISION AS A CULTURAL FORCE
edited by
Richard Adler and
Douglass Cater

*THE FUTURE OF PUBLIC BROADCASTING
edited by
Douglass Cater and
Michael J. Nyhan

ASPEN HANDBOOK ON THE MEDIA, 1977-79 Edition: A Selective Guide to Research, Organizations, and Publications in Communications
edited by
William L. Rivers
Wallace Thompson
Michael J. Nyhan

*THE MASS MEDIA: Aspen Guide to Communication Industry Trends
Christopher H. Sterling
and Timothy Haight

Also Published by Praeger Special Studies:

COMMUNICATIONS TECHNOLOGY AND DEMOCRATIC PARTICIPATION
Kenneth C. Laudon

*THE MEDIA AND THE LAW
edited and with an introduction by
Howard Simons and
Joseph A. Califano, Jr.

*Also available in paperback as a PSS Student Edition.